Neuroanatomical Research Techniques

(A). Green fluorescent cell bodies of the substantia nigra, pars compacta, showing the presence of dopamine in the neurons and proximal dendrites. Among these cell bodies are small, green fluorescent structures which represent a catecholamine innervation of this component of the substantia nigra. Vibratome-formaldehyde method.

(B). Motor nucleus of the trigeminal nerve in the pons showing autofluorescence (orange-yellow) cell bodies of the nucleus innervated by a dense plexus of varicose, green fluorescent axons originating in the noradrenaline cells of the nucleus locus coeruleus. Vibratome-formaldehyde method.

(C). Lateral septal nucleus showing numerous, fine, non-varicose dopamine axons in the neuropil with very large numbers of axons investing lateral septal nucleus neurons and their proximal dendrites. In some cases, as on the right, this is very intense, but the axons can be seen to have varicosities when approximated to the cell. This is particularly evident on the round cell at the left-hand side of the picture. Just above that cell is a typical locus coeruleus axon with numerous, round, evenly spaced varicosities and fine, intervaricose segments crossing in a transverse direction. Vibratome glyoxylic acid method.

(D). Single, completely stained cerebellar granule cell from brain of young cat, showing relation of dendrite claws to adjacent partially stained granule cell somata. Golgi modification. Original magnification × 600.

(E). Single completely stained cerebellar granule cell from hemispheric region of young cat. Note how clearly the adjacent, partially stained granule cell soma is enfolded by the claw-like processes of several granule dendrites. Golgi modification × 600.

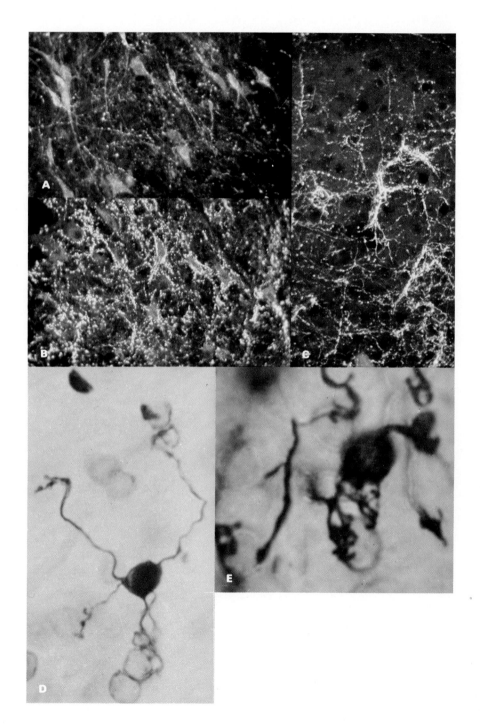

METHODS IN PHYSIOLOGICAL PSYCHOLOGY

EDITOR: Richard F. Thompson
DEPARTMENT OF PSYCHOBIOLOGY
UNIVERSITY OF CALIFORNIA, IRVINE

Neuroanatomical Research Techniques

Edited by

RICHARD T. ROBERTSON

Department of Anatomy
College of Medicine
University of California
Irvine, California

ACADEMIC PRESS   New York   San Francisco   London   1978
A Subsidiary of Harcourt Brace Jovanovich, Publishers

ACADEMIC PRESS, INC.
111 Fifth Avenue, New York, New York 10003

United Kingdom Edition published by
ACADEMIC PRESS, INC. (LONDON) LTD.
24/28 Oval Road, London NW1 7DX

Library of Congress Cataloging in Publication Data

Main entry under title:

Neuroanatomical research techniques.

    (Methods in physiological psychological psychology ;
v. 2)
    Includes bibliographies.
    1. Neuroanatomy――Technique. 2. Neuroanatomy――
Research. I. Robertson, Richard T. [DNLM: 1. Ner-
vous system――Anatomy and histology――Laboratory manuals.
2. Histological technics. 3. Cytological technics.
W1 ME961 / WL25 N494]
QL927.N48   591.4'8'028   78―17320
ISBN 0―12―590350―2

PRINTED IN THE UNITED STATES OF AMERICA

Contents

PREPARATION AND STUDY OF BRAIN TISSUES

Chapter 1   Microscopy and Photomicrography: An Introduction
Friedrich K. Möllring

Chapter 2   Tissue Preparation and Basic Staining Techniques
George Clark

Chapter 3   Computer-Assisted Recording of Neuroanatomical Data
Thomas A. Woolsey and Michael L. Dierker

vii

TECHNIQUES FOR THE STUDY OF NORMAL TISSUE

Chapter 4   The Methods of Golgi

Madge E. Scheibel and Arnold B. Scheibel

Chapter 5   Fluorescence Histochemistry

Robert Y. Moore and Rebekah Loy

Chapter 6   Single-Cell Staining Techniques

Charles D. Tweedle

Chapter 7   Electron Microscopy and the Study of the Ultrastructure of the Central Nervous System

Jerald J. Bernstein and Mary E. Bernstein

STUDY OF CONNECTIONS IN THE NERVOUS SYSTEM
A. Techniques Based on Orthograde Processes

B. Techniques Based on Retrograde Processes

C. Electrophysiological Techniques

PERSPECTIVE

List of Contributors

Numbers in parentheses indicate the pages on which the authors' contributions begin.

JERALD J. BERNSTEIN (175), Departments of Neuroscience and Ophthalmology, University of Florida College of Medicine, Gainesville, Florida 32610

MARY E. BERNSTEIN (175), Departments of Neuroscience and Ophthalmology, University of Florida College of Medicine, Gainesville, Florida 32610

JACQUELINE C. BRESNAHAN (407), Department of Anatomy, College of Medicine, Ohio State University, Columbus, Ohio 43210

GEORGE CLARK (25), Veterans Administration Hospital, and Department of Anatomy, Medical University of South Carolina, Charleston, South Carolina 29401

MICHAEL L. DIERKER (47), Department of Anatomy and Neurobiology, Washington University School of Medicine, S . Louis, Missouri 63110

STEPHEN B. EDWARDS (241), Department of Anatomy, University of Virginia, Charlottesville, Virginia

ROLAND A. GIOLLI (211), Department of Anatomy, California College of Medicine, and Department of Psychobiology, University of California, Irvine, California 92717

ANITA HENDRICKSON (241), Department of Ophthalmology, University of Washington, Seattle, Washington 98195

R. KEVIN JONES (291), Department of Psychobiology, University of California, Irvine, California 92717

AZARIAS N. KARAMANLIDIS (211), Laboratory of Anatomy and Histology, School of Veterinary Medicine, University of Thessaloniki, Thessaloniki, Greece

RONALD M. KOBAYASHI (317), Neurology Service, Veterans Administration Hospital, San Diego, California 92161, and Department of Neurosciences, University of California, La Jolla, California 92093

JENNIFER H. LAVAIL (355), Department of Anatomy, School of Medicine, University of California, San Francisco, California

REBEKAH LOY (115), Department of Neurosciences, University of California, La Jolla, California 92093

GARY LYNCH (291), Department of Psychobiology, University of California, Irvine, California 92717

PATRICIA MORGAN MEYER (407), Department of Psychology, The Ohio State University, Columbus, Ohio 43212

FRIEDRICH K. MÖLLRING (3), The Carl Zeiss Company, Oberkochen/Württ, Federal Republic of Germany

ROBERT Y. MOORE (115), Department of Neurosciences, University of California, La Jolla, California 92093

RICHARD J. RAVIZZA (337), Department of Psychology, The Pennsylvania State University, University Park, Pennsylvania 16802

ARNOLD B. SCHEIBEL (89), Departments of Anatomy and Psychiatry, and Brain Research Institute, University of California, Los Angeles, California, 90024

MADGE E. SCHEIBEL (89),* Departments of Anatomy and Psychiatry, and Brain Research Institute, University of California, Los Angeles, California 90024

JOHN SCHLAG (385), Department of Anatomy, and Brain Research Institute, University of California, Los Angeles, California 90024

HUGH J. SPENCER (291), Department of Psychobiology, University of California, Irvine, California 92717

CHARLES D. TWEEDLE (141), Departments of Biomechanics and Zoology, Michigan State University, East Lansing, Michigan 48824

THOMAS A. WOOLSEY (47), Department of Anatomy and Neurobiology, Washington University School of Medicine, St. Louis, Missouri 63110

*Deceased.

Foreword

The major approaches that characterize the organization and functions of the brain can be grouped into four categories of techniques—electrophysiology, anatomy, chemistry, and behavior. All these approaches to the study of the brain and its functions will be treated in this series on *Methods in Physiological Psychology*. The series began with a three-volume treatment of *Bioelectric Recording Techniques* (Thompson & Patterson, 1973). It appropriately continues with this volume on *Neuroanatomical Research Techniques*. The classic discipline for the study of the nervous system has been anatomy. It has the longest history and a most exciting present and future.

The past few years have witnessed a literal revolution in neuroanatomy. This revolution, whose roots are multiple, corresponds to the current unprecedented expansion in the neurosciences. In the twentieth century the field has been fortunate to claim a succession of extraordinary scientists—beginning with Ramon y Cajal and Lorente de No to such present figures as Nauta and the Scheibels. Further, in the past few years a number of new, powerful, and *rapid* methods have been developed, particularly for the tracing of connections. Finally, the advent of the electron microscope has added a vast new set of dimensions to our understanding of the structure of the most important and complex of all physical structures—the brain.

This is an extraordinary volume. Dr. Robertson is to be complimented in the strongest terms for bringing together a great many of the leading contemporary scientists in neuroanatomy, each to discuss his or her favorite techniques. The chapters are much more than descriptions of technique, although they are, of course, outstanding in this regard. The chapters are written with devotion and communicate considerable knowledge and wisdom about the study of neural structure. They convey the excitement and ferment, as well as the accomplishments, of this very basic field. In the context of physiological psychology, this volume presents a most valuable contemporary treatment of the study of the structure basis of behavior.

RICHARD F. THOMPSON

Preface

Much of our current knowledge of physiological psychology, and indeed most of the neuroscience field, has a strong base in results of neuroanatomical research. An understanding of the basic morphological substrates is essential for an understanding of neural systems and their behavioral significance. Our knowledge of these neural substrates, in turn, is dependent on the proper use and interpretation of neuroanatomical research techniques.

Neuroanatomy is a particularly exciting field. It is part science, part art. Today's laboratories often blend the older classic methods, often steeped in tradition and mystique, with newly introduced techniques, which may lack the tradition of their older siblings but in many cases not the mystique. Our present understanding of the morphology of the nervous system, particularly the fascinating intricacies of its connections, rests on experimental techniques. It is essential that we understand the power, the proper application, and the limitations of these techniques.

The purpose of this volume is to provide an up-to-date presentation of the major neuroanatomical research techniques available today. The authors are leaders in their fields, and describe the experimental techniques and approaches with the interest, flavor, and detail that can only come with direct personal experience.

The first section deals generally with the preparation and study of brain tissue. Möllring introduces us to the microscope, discussing optical magnification, limitations of microscopy, and optical contrasting methods. Clark summarizes basic techniques for tissue preparation and sectioning, and presents guidelines for a number of standard, but essential, staining procedures. Woolsey and Dierker present sophisticated and very contemporary computer techniques that are proving so helpful as neuroanatomy evolves from a qualitative to a quantitative discipline.

The second section deals with techniques that are often used for the study of normal tissue. These techniques have in recent years, however, received wider application in the study of experimental tissue, and can be an integral part of any approach to the nervous system. The Scheibels lucidly and lovingly present the glamour, the frustration, and the insights into the nervous system that are all part of the Golgi method. Fluorescence histochemistry, which has been a major part of a revolution in the neurosciences, is described in pragmatic detail by Moore and Loy. The recently introduced techniques for staining single neurons are presented in the up-to-date and detailed review by Tweedle. Finally, the Bernsteins take us to the next step of anatomical resolution, describing their elegant use of the electron microscope in studying the fine structure of the nervous system.

Much of the work in neuroanatomy has dealt with the intrinsic connections of the nervous system. It is here that the introduction of new and powerful techniques has perhaps had the greatest influence. These techniques are the subject of the third section. Giolli and Karamanlidis present in detail the basis, the uses, and many of the modifications of techniques for silver impregnation of degenerating fibers. Hendrickson and Edwards bring us up to date with a detailed review and analysis of axonal transport and the powerful autoradiographic technique for studying axonal projections. Somatopetal movement of horseradish peroxidase as a tool for studying connections and neuron morphology is presented by Spencer, Lynch, and Jones. Kobayashi explores the effects of lesions on regional brain chemistry and discusses its applications to the study of neural connections.

The classical method of analysis of retrograde degeneration following placement of lesions, and its place in contemporary laboratories, is critically reviewed by Ravizza. The advantages and limitations of the retrograde transport techniques, including the use of horseradish peroxidase as well as other macromolecules, are clearly presented by LaVail. Completing this section is a chapter by Schlag, in which he reviews electrophysiological approaches to the study of connections in the nervous system and points out the advantages and hazards of these techniques for studying the anatomy and function of neural systems.

As each chapter has pointed out, the techniques are powerful, potentially enlightening, and potentially deceiving. In the final section, Meyer and Bresnahan discuss the interpretation of results from neuroanatomical research techniques and present examples of the applications of neuroanatomical methods to major problems in physiological psychology.

We have organized this volume with the hope that it will be useful both for the novice and for the experienced investigator. The contributors deserve thanks for sharing with us their expertise and experience; especially helpful are the personal hints that often are not included in journal articles. I also thank Cynthia Payne for her invaluable organizational and secretarial assistance, and the Fels Research Institute and the University of California for their support.

RICHARD T. ROBERTSON

Preparation and Study of Brain Tissues

Chapter 1

Microscopy and Photomicrography: An Introduction

Friedrich K. Möllring

The Carl Zeiss Company
Oberkochen/Württ
Republic of Germany

I. The Use of the Microscope

The microscope is one of the most important tools of the anatomist and every reader of this book will be more or less conversant with it. Those who decide they belong to the rather "less" than "more" familiar class will find in this introductory chapter a summary of the facts most vital for practical work. In order to be as concise as possible, there will be no extended general remarks but we shall adhere to practical rules.

A. Practice—The Best Introduction

1. THE MICROSCOPE

For the beginner in microscopy, familiarization with the basic components of the microscope is recommended as a starting point. Figure 1 illustrates the basic parts of a standard laboratory binocular microscope. The user should be familiar with the components identified on the microscope.

2. FOCUSING THE SPECIMEN

With relatively few exceptions, the material of interest to anatomists consists of thin tissue sections mounted on glass slides and covered by a thin cover glass. The slide is placed on the microscope stage with the cover glass on top. The condenser should be raised to its maximum height and the condenser front lens swung into place. The specimen and objective should be cautiously moved closer together using the coarse adjustment until they are only a few millimeters apart. (Observe from the side!) In modern microscopes, a stop and the spring mount of the objectives prevent ordinary specimens (1-mm slide, 0.17-mm cover glass) from being damaged by the objective. Only a few special purpose objectives are exceptions.

At the beginning of an examination a low-power objective of 2.5 to 10× should be used because it permits coverage of a wider field and is thus best suited for scanning. In addition, its depth of field is greater than that offered by objectives of higher power, and thus it is easier to find the correct plane of focus. The power of an objective is its initial magnification indicated by the first value engraved on each objective. The total magnification of the microscope is determined by multiplying the magnification of the objective by that of the eyepiece. A 10× objective used in conjunction with a 12.5× eyepiece yields a total magnification of 125×.

The condenser iris should be closed approximately halfway. Now the microscopist can look into the eyepiece and increase the distance between the objective and the specimen by the coarse adjustment until the details of the specimen can be recognized even if they are still blurred. Exact focusing is achieved with the aid of the fine adjustment. When the area to be examined is in the center of the field of view and focused, the objective of the next higher magnification may be moved into position, and focused with the fine control.

The adjustment of the condenser is a frequent source of error. The condenser is correctly adjusted when it is just below its highest position, and when its iris has been closed only far enough to obtain sufficient contrast. The beginner usually closes (stops down) the condenser far too much. The condenser iris should never be used to dim the image! Other means, e.g., neutral density filters, must be used for this purpose. Lowering the condenser "to increase contrast," popular with many microscopists, has the same effect as further closing the condenser iris and

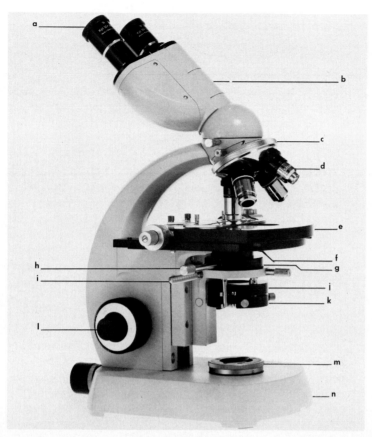

Fɪɢ. 1.   A standard binocular research microscope. (a) Eyepiece; (b) binocular tube (inclined tube for monocular observation also can be used); (c) nosepiece for exchange of objectives; (d) objective designation: 40/0.65, 160/0.17 means: objective magnification 40×, numerical aperture 0.65, optics computed for mechanical tube length of 160 mm and 0.17 mm cover-glass thickness; (3) specimen stage; (f) iris lever (condenser or aperture diaphragm); (g) condenser; (h) lever for swinging out the condenser front lens, for illumination of larger object fields; (i) knobs for centering the condenser (for Köhler illumination); (j) swing-out holder for filters; (k) swing-out holder for centerable auxiliary lens; (l) fine and coarse adjustment knobs; (m) diaphragm insert with field iris diaphragm; (n) base with illuminator.

is inappropriate. If the illuminated field should be too small with objectives of very low power, the front lens of the condenser should be swung out and its diaphragm fully opened.

The depth of field in the specimen, corresponding to a certain setting of the fine adjustment, is extraordinarily shallow. This shallow depth of field is more pronounced with high-power objectives than with lower powers, and more with

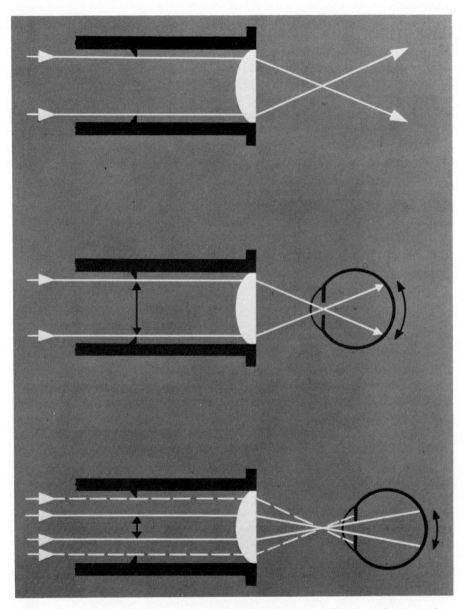

FIG. 2.   Illustration of the pattern of emerging light from the eyepiece. The observer's eye must be at the proper distance from the eyepiece for a full image view.

high-quality optics than with poor-quality optics. In microscopic work, the fine adjustment is continuously moved up and down in order to cover the entire depth of the specimen.

Beginners often make the mistake of focusing the microscope as if the magnified image they want to view was located very close to the eye—as one would view an object closely without optical aids. This near setting of the eye (accommodation) is extremely tiring if maintained for a long time and, therefore, it is best to work with perfectly relaxed eyes. At the beginning, this is done more easily if it is imagined that the image being viewed is at infinity. This encourages looking "through the microscope" instead of "into the microscope." Another important point is to maintain the eye at the correct distance from the eyepiece. The best procedure is to move in from a few inches distance until the field of view in the eyepiece is widest and sharply defined. At this distance all the light rays emerging from the microscope find their way into the eye. As is evident from Fig. 2, this is possible only if the pupil of the eye is brought very exactly to the point of constringence of the microscope's light rays, which has been named, therefore, the "exit pupil of the microscope." When you have acquired a certain amount of practice in the use of a microscope, you will automatically find and keep the right distance from the eyepiece.

It is usually advantageous for those who wear prescription eyeglasses to keep their glasses on when using the microscope. Special rubber guards on the eyepieces prevent scratching.

B. Optical Components of the Microscope

1. OBJECTIVES

The objective provides the initial, and usually the major, magnification of the specimen image. Of the three optical components in a microscope—condenser, objective and eyepiece—the objective is the part whose quality has the greatest effect on image quality. By using a cheap magnifying lens as an example, it becomes immediately apparent that a simple lens exhibits quite a number of image errors: black figures show colored fringes along their edges, sharpness rapidly decreases toward the edges of the field, and intolerable distortions are produced if the object is viewed at only a slight angle through the lens. If such a simple magnifier exhibits aberrations that are so obvious, how much more evident must these errors be at the far higher magnification of a microscope objective when, to make things worse, the magnified image is viewed through an additional magnifier—the eyepiece. The ingenious selection and combination of lenses, lens curvatures, different glass types, etc., allow the microscope manufacturer to eliminate many of the image errors to varying degrees. It is obvious that the elimination of more errors—that is, higher correction—requires greater sophistication, which results in higher cost. Thus, the differences in micro-scopist's requirements and available funds have prompted microscope manufac-turers to offer several "quality categories" of objectives.

The greatest number of microscopes are supplied with so-called achromats. Up-to-date versions of this simplest type of objective have a remarkably high performance and are entirely satisfactory for many routine purposes in teaching and research. The fact that certain color fringes will still be noticed in critical work is hardly an obstacle to observation, especially since this can easily be remedied by using a filter, normally a green one. However, even the beginner will notice that while using a 40× achromat, for example, the center of the field and the outer zones cannot be seen in sharp focus at the same time. The surface of best definition is "dished." This aberration, called "curvature of field," is not a great handicap in microscopic examination as movement of the stage or ad-justment of the fine focus can allow the viewer to clearly see all areas. Field curvature may be a nuisance in photomicrography, however, where a sharp image of the entire field of view is required. The extent to which curvature of field becomes visible also depends considerably on the type of specimen ob-served. It will be less noticeable in a thick section than an extremely thin speci-men.

On the other hand, there are some very positive characteristics that, in general, can only be found in achromats. Because their numerical aperature is low, achromats offer greater depth of field and have a longer working distance than more highly corrected objectives. This allows greater freedom in the selection of cover glasses, which, in this case, will not degrade image quality as quickly should they deviate from the optimum thickness of 0.17 mm (see Section D3). These factors favor the use of such objectives by the beginner and routine worker.

Specifying the next higher "quality category" after the achromat is somewhat difficult because opinions vary as to which of the aberrations inherent in ach-romats should be eliminated first. While flat-field achromats, so-called planach-romats, have been corrected for curvature of field, fluorite objectives and apochromats will produce images distinguished from those of achromats by greater color fidelity, better contrast, and higher resolving power. The difference between fluorite objectives and apochromats is less of a basic than of a tran-sitional nature; the fluorite objectives lie roughly between achromats and apo-chromats. Investigators frequently using photomicrography will give preference to planachromats. Workers demanding optimum definition, for example, those in critical research work, will choose fluorite objectives or apochromats. The un-compromising microscopist who demands both a completely flat field and op-timum color fidelity combined with extremely high definition, may choose from the range of planapochromats—microscope objectives of absolutely top quality.

Although the layman generally believes that results will improve with increasing magnification, the experienced microscopist will use high powers only if absolutely necessary. For technical reasons, resolution does not increase in proportion with magnification. High-power objectives, therefore, give relatively weak images, have a very low depth of field, and offer comparatively less brilliance and sharpness than do low-power objectives. The magnifications to be chosen for a set of objectives will always depend on the type of specimen to be examined.

2. EYEPIECES

Although a few years ago it was necessary to use different types of eyepieces (oculars) for low- and high-power objectives, microscope objectives now are used generally with compensating oculars. Consequently, eyepieces no longer need to be exchanged when changing objectives. In spite of this, two classes of eyepiece are still being manufactured. Although simple eyepieces are used primarily in conjunction with achromatic objectives, compensating flat-field eyepieces are intended for the other objectives of higher correction but may, of course, also be combined with achromats. The most widely used eyepiece magnifications are $10\times$ and $12.5\times$.

Particular mention should be made of the eyepieces specially designed for spectacle wearers. The exit pupil of the microscope and the eye pupil must coincide if the entire field of view is to be covered. With normal eyepieces, the distance between the edge of the eyepiece and the exit pupil (eye point) is so short that eyeglasses cannot be kept on if the two points are to coincide. The high eye point of the special eyepieces for eyeglass wearers allows the eyeglasses to be worn; the rubber guards supplied with these eyepieces prevent scratching of the spectacle lenses. If these special types of eyepieces are used by observers not wearing glasses, care must be taken to keep the eyes at an adequate distance from the eyepiece. If necessary, finding the correct distance can be facilitated by slipping eyecups onto the eyepieces.

Apart from its proper magnification, an eyepiece is characterized by its field-of-view number. With the aid of this number it is easy to calculate the diameter of the field covered in the specimen plane. Field diameter is equal (in millimeters) to the field-of-view number of the eyepiece divided by the initial magnification of the objective. For example, a $10\times$ objective in combination with an eyepiece of the field-of-view number 12.5 permits a field of 1.25 mm diameter to be covered in the specimen plane.

3. CONDENSERS

The primary purpose of the condenser is to provide appropriate illumination of the object specimen. To meet the high standards of most microscopists, the condenser performs two functions: It provides homogeneous illumination of the object plane and, more importantly, it delivers light rays parallel to the optical axis and with a particular angle of incidence. This latter function results in complete filling of the rear lens of the objective (the back focal plane or the exit pupil of the objective) with light, which is a prerequisite for obtaining an image of high resolution.

Low-power objectives take a large object field with a small angle of incidence (low aperture), whereas high-power objectives require a small object field and great angle of incidence. No single condenser can meet these contradictory demands, so most condenser units are combinations of a regular condensing lens and an auxiliary removable front lens. The front lens can easily be removed from the optical path for use with low-power objectives.

The demands generally made on condensers are partly inadequate and partly exaggerated. A standard condenser has a numerical aperture of 0.9 and a front lens that can be removed to illuminate large object fields. Achromatic–aplanatic condensers are recommended for color photomicrography and microphotometry. Condenser apertures above 0.9 are useful only for fluorescent and dark-field work. In addition, they are effective only if the condenser is immersed in oil (see Section D2).

If phase-contrast work should be contemplated—perhaps at a later date—it is advisable to buy a phase-contrast condenser right at the beginning because this type of condenser is useful for conventional bright-field microscopy as well.

C. Care and Cleaning of the Microscope

An instrument that has to satisfy the most exacting requirements regarding mechanical and optical precision naturally demands a certain amount of care. Dust on optical elements will degrade the image quality. Although all surfaces exposed to dust are also easily accessible for cleaning, no glass surface is improved by "cleaning." The best advice is always to avoid exposure of the microscope to dust by covering it with a hood when it is not in use or by keeping it in a cabinet. Special care should be taken to ensure that the tubes of the microscope are always closed either by an eyepiece or a dust plug.

While dust particles on the eyepiece will only give rise to patches in the image, a dirty objective front lens may hopelessly reduce the sharpness of the image, or at least its contrast. Because it is close to the specimen and possibly to the immersion oil on the cover glass, as well as being particularly close to the hand operating the nosepiece, the front lens of the objective is especially prone to getting soiled. Even the lightest fingerprint may have grave consequences (Fig. 3). Before starting important work it is advisable to unscrew every objective and check it carefully with the aid of a magnifier.

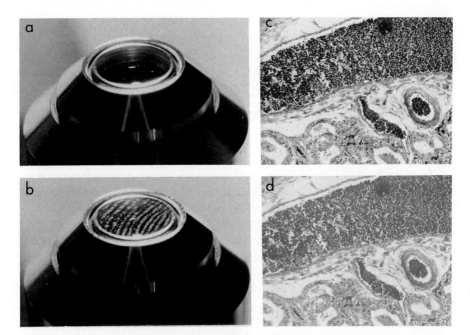

FIG. 3. Example of image deterioration by a dirty lens. The finger print on the objective in (A) markedly reduces the contrast and resolution of the image (B).

Should structures be found in the image which are suspected of being extraneous to the specimen, the fault may be traced as follows: If the trouble can be eliminated by slight adjustment of the condenser, the cause must be in the bulb of the lamp, the lamp condenser, or the filter in front of it. However, if adjustment of the condenser does not produce any result, the next step is to turn the focusing adjustment, which should identify all faults due to soiling of the condenser front lens or the specimen. If this does not identify the source of the problem, slightly turn first the objective and then the eyepiece, and you will immediately notice in which case the foreign body follows the rotation. Dust particles are most clearly seen when the aperture diaphragm has been fully closed, because in this case the depth of focus is at its greatest.

In almost all cases it will be sufficient to clean the outer lens faces with the aid of a grease-free brush (if necessary, wash in ether first) or with a frequently washed, absolutely dust-free linen rag and distilled water, produced easily by breathing upon the surface to be cleaned. If an organic solvent must be used, it is advisable to use *very little* ether or benzene, rather than water. Alcohol might destroy the cement between the lens elements if older optical systems are being used. Ether is usually preferred to alcohol, for cleaning, because it evaporates more quickly, and any harmful effect is thus less likely. Finally, residues are always removed with water as described above. Should compressed air be available for cleaning, be sure to use a filter of cotton wool.

D. Topics for the Advanced Microscopist

1. USEFUL MAGNIFICATION, LIMIT OF RESOLUTION, AND NUMERICAL APERTURE

To understand this topic better, it is helpful to think of a projector and a projected slide. At first the landscape is observed on the screen from far behind the projector. In this case, magnification is still low because the angle of view is quite small. Magnification improves upon moving closer to the screen, and previously unrecognized details become evident. Finally, however, a point is reached where higher magnification (moving closer) does not give better results. Instead of single leaves on the tree, only the structure of the color film is seen—an accumulation of dye particles (the grain). Further reducing the viewing distance could, of course, further increase magnification, but this would be "empty magnification," i.e., magnification that does not reveal any new detail.

In the microscope the situation is similar. The projected slide image corresponds to the aerial image formed in the tube by the microscope objective. As with the projected slide, there is also a limit for the projected aerial image where "useful magnification" ends and "empty magnification" begins. It is true that the microscopic image will not exhibit any actual grain structure, but one may speak of "grain" in this case because there is no optical image in which an object point is really reproduced as a point (i.e., with no measurable diameter). Owing to the phenomenon of diffraction, an object point is always transformed into a small disk in the image. It is these so-called "airy disks" that form the "image grain." This type of grain can be found in any optical image if the magnification is high enough. It is obvious that image details will no longer be clearly recognized as soon as they reach the size of the airy disk, just as it is impossible to recognize the leaves in the slide when they are the same size as the dye grain.

If a photographic slide is viewed with higher magnification than is useful, no harm is actually done because it is very well known that the above-mentioned grain structure does not belong to the object. However, in microscopy the situation is different. Empty magnification may show structures in the image that do not exist in the specimen, and with unfamiliar specimens there is a danger of erroneous conclusions.

It follows from the above remarks that the performance limit of the microscope objective in use should be known, i.e., the point at which its useful magnification ends. This is indicated by a value engraved on every objective and called "numerical aperture." The numerical aperture is the factor that determines the size of the airy disks in the microscopic image, i.e., the size of the "screen."

The following rule is helpful at this point: The total magnification of the microscope (i.e., the magnification of the objective multiplied by that of the eyepiece multiplied by any tube factor) should not exceed one-thousand times the numerical aperture. Microscopists who have particularly good visual acuity should even use five-hundred times the numerical aperture as a limit. Exceptions to this rule may be allowed in order to facilitate measuring and counting work.

An aid to understanding numerical aperture is given in Fig. 4, which gives a diagrammatic view of an objective front lens and specimen. The angle u' subtended by the optical axis and the outermost rays still covered by the objective is a measure of the aperture of the objective; it is half the aperture angle. However, the magnitude of this angle is not indicated in degrees but in the form of a sine value, that is, a numerical value. This explains the origin of the term "numerical aperture." It is the sine of one-half the aperture angle multiplied by the refractive index n of the medium filling the space between the cover glass and the front lens.

$$\text{Numerical aperture} = n \sin u'$$

Since air has a refractive index of 1, n may be neglected when dry objectives are used. Immersion oil, which in the case of an oil immersion objective fills the

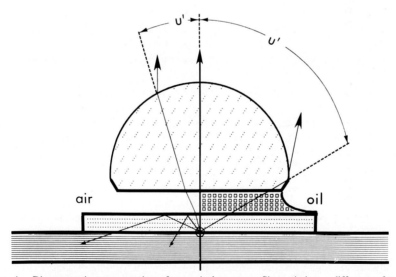

FIG. 4. Diagrammatic representation of numerical aperture. Since air has a different refractive index than glass, dry objectives are limited by the amount of light they capture. The use of immersion oil with a refractive index similar to glass renders considerable improvement.

space between the cover glass and the front lens, typically has a refractive index of about 1.5. It is obvious that such an objective makes it possible to achieve a gain of 50% in numerical aperture.

With the aid of the numerical aperture we can compute the limit of resolution or maximum resolving power. This is the smallest distance, designated d, by which two structural elements may be separated in order to be seen as two distinct elements instead of one.

If NA_{obj} and NA_{cond} are the numerical apertures of objective and condenser, respectively, and λ is the wavelength of the light used for observation (in 1/1000 mm or microns), then the smallest resolvable separation between two object points (in micrometers) is

$$d = \frac{\lambda}{NA_{obj} + NA_{cond}}$$

Example: Using a green filter, the wavelength is 0.55 μm; assuming the objective used has an aperture of 1.25 and the condenser one of 0.9, then $d = 0.25$ μm. However, structural elements which are just resolved are *not* reproduced with full fidelity.

2. OIL IMMERSION

Oil immersion objectives are used with immersion oil (instead of air) between the objective front lens and cover glass. The refractive index of the oil is similar to that of glass, resulting in a gain in numerical aperture (see Fig. 4) and less diffraction. In regard to correction, and for the same scale, an oil immersion objective is always superior to a comparable dry objective. The user should, if possible, employ only the type of immersion oil supplied by the manufacturer of the objective because the refractive index and the dispersion of the oil must have certain values.

If the condenser in use has a numerical aperture higher than 0.9, it is advisable also to immerse the condenser front lens for use in combination with an oil immersion objective. For this purpose, a drop of oil is applied to the underside of the specimen slide and another one to the condenser front lens. This will effectively reduce the danger of bubbles forming when the condenser is immersed. Even if the immersion oil is of the nonresining type, it should be removed at the end of the day by means of a clean rag of the kind used for cleaning optical elements. In case the use of oil seems troublesome, the use of water (refractive index 1.333) is preferrable to leaving a high aperture condenser unimmersed.

3. COVER GLASSES

Knowing the high precision required to produce microscope objectives that are to give first-rate images, it should not be surprising that the cover glass, which is located between the specimen and the objective front lens, has optical effects and must have certain specifications. Simple glass plates, that is plane plates, have an optical power. The optical power of the cover glass must be taken into account in the computation and design of the objective. With normal microscope objectives, the cover glass is part of the image-forming system. It is a "lens element" (though with infinite radii of curvature) which is located outside the objective mount but still forms a part of the optical system. The optimum thickness of a cover glass is 0.17 mm.

Strictly speaking, this applies only to objectives whose numerical aperture exceeds a certain value. With objectives having an aperture not exceeding 0.3, specimens may be examined both uncovered or with cover glasses as thick as a millimeter, without the spherical aberration being noticeable. With apertures between 0.3 and 0.7, the cover-glass thickness should not deviate from the nominal value by more than 0.03 mm. With higher apertures, even a deviation of 0.01 mm will considerably impair the quality of the image.

In order not to restrict the cover glass to specified thicknesses when using dry objectives of particularly high apertures, some objectives have been designed in which the optimum cover-glass thickness is not a fixed 0.17 mm, but can be varied within a range of 0.12 to 0.22 mm. These are objectives with correction collars. If the thickness of the cover glass on the specimen to be examined is unknown, first set the correction collar at 0.17 mm and find a high-contrast area in the specimen. Then test whether the contrast is improved if the correction collar is set for a greater or lesser cover-glass thickness. This also requires refocusing. It is obvious that in the hands of inexperienced users, who usually leave the correction collar at one end of its range of adjustment, these objectives may give very disappointing results (Fig. 5).

The situation is somewhat different with oil immersion objectives because a medium of high refractive index, namely oil, is in front of the first lens. In this case, the thickness of the cover glass is not very critical. Still, it should be remembered that immersions are not "homogeneous" enough (i.e., glass and oil do not have identical refractive indexes) to allow the issue of cover glass to be dispensed with as a matter of course.

For practical work with specimens that are to give optimum images, it is necessary always to use cover glasses that are exactly 0.17 mm thick. These may be purchased from most scientific supply houses.

It should be mentioned that in those cases in which the cover glass is not in direct contact with the specimen, the intermediate layer of mounting medium has the same effect as additional thickness of cover. It is, therefore, advisable to weigh down the covers of specimens during drying.

4. KÖHLER ILLUMINATION

In illumination by microscope, two things are important. First, the field of view should be evenly illuminated, and when the condenser iris is open and the condenser is in its uppermost position, the back lens of the objective should be

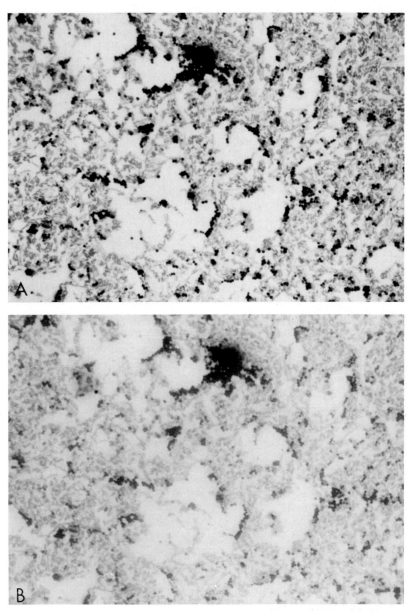

FIG. 5. Image deterioration that occurs when the objective is not matched with the thickness of the cover glass. A planapochromat 40/0.95 objective with a correction collar and 0.17-mm cover glass are used. In (A) the collar is adjusted to 0.17 mm. In (B) the collar is adjusted to 0.19 mm.

filled with light as completely as possible. For microscopic research work, photomicrography, microprojection, etc., precise control of the light path should start before the light reaches the specimen, i.e., at the light source. Professor August Köhler first used this exact control of the light path in the illuminating beam of the microscope, and the method is today called "Köhler illumination."

Use of the Köhler method depends on the availability of a microscope illuminator with a lamp condenser and an iris diaphragm in front of it. The lamps used in such illuminators usually have a small, concentrated filament and operate on 6 or 12 V; they are, therefore, called low-voltage lamps. It is not the wattage of the bulbs that is important but the luminance. This is in contrast to normal bulbs used for electric lighting which are important for the intensity of their light.

The procedure for applying this method of illumination is as follows (see Fig. 6).

1. Raise the condenser to its maximum and take the eyepiece out of the tube. The back lens of the objective will be filled with light. Dimming the light by a neutral density filter would be advisable. Insert the eyepiece and focus the specimen with a 10× or 16× objective.

2. Close the iris diaphragm of the field stop (Fig. 1,m) almost completely (Fig. 6a) and focus the edge of this diaphragm image by slightly lowering the condenser (Fig. 6b).

3. Now move the sharp image of the field stop into the center of the field (Fig. 6c) by operating the centering knobs (Fig. 1,i) of the condenser. Both the specimen and the lamp field stop are now sharply defined. Then open the field stop until the entire field of view is just clear (Fig. 6d).

4. Slightly readjust the lamp socket or the lamp condenser, until the field of view is evenly illuminated.

5. At first open the condenser irirs (aperture stop) completely (image very bright) and close it only far enough to eliminate glare in the most important image elements and to make them appear with satisfactory contrast. With stained specimens this is the case even at a relatively wide aperture, while fresh specimens require slightly more stopping down. Stopping down further than is absolutely necessary (for example, to reduce the brightness of the image) is one of the biggest mistakes a microscopist can make since it entails a loss of resolving power. If the intensity of the light is too high, reduce the lamp voltage or insert gray or green filters into the filter holder.

6. If with low-power objectives only part of the field is illuminated, swing out the condenser front lens, and open the condenser iris fully. Contrast is then controlled with the aid of the lamp diaphragm. [Note: Some microscopes have two possibilities of centering the image of the lamp field stop. It is done first with the usual centering screws on the condenser (coarse adjustment) and second with a centerable auxiliary lens below the condenser. The latter is an easily accessible fine adjustment that may be recentered after changing objectives.]

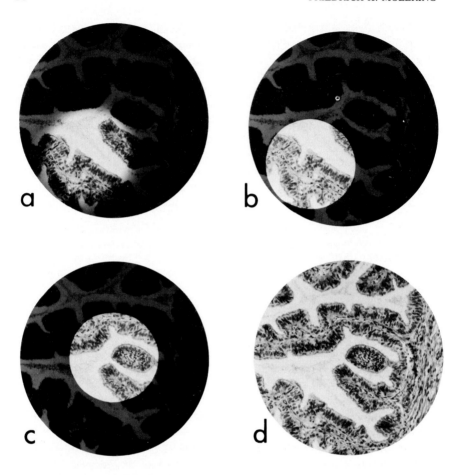

FIG. 6.   Steps for adjustments for Köhler illumination. (a) Close the field iris diaphragm. (b) Raise the condenser to sharpen the image of the iris border. (c) Center the light beam. (d) Open the iris until the field is just totally illuminated.

When adjusting the equipment for Köhler illumination, it is possible that a certain granularity may be seen superimposed on the specimen over the entire field of view. This has the texture of a ground-glass screen or an etched collector lens, which is used to reduce the irregularity of the light source. If this occurs, it can be eliminated by slightly altering the height of the condenser.

5.   OPTICAL CONTRASTING METHODS

Normally in microscopy, work is done with bright-field illumination. With this method, either the brightness or the color of the light penetrating different areas of the specimen is differentially affected resulting in bright–dark contrast or, with colored specimens, in color contrast. Most microscope specimens, however, are not colored and since staining is not always possible or preferred, methods offering contrast without staining are sometimes desirable. The simplest way to increase contrast in such cases is to reduce the aperture (condenser) even though a resulting loss of sharpness of detail cannot be avoided. Similar contrast enhancement can be achieved, again at a cost of sharpness, by fine-focusing slightly off the plane of maximum sharpness (so-called extrafocal adjustment). Several optical means are available that are preferred.

a. PHASE-CONTRAST ILLUMINATION. The best solution to the contrast problem with transparent objects was discovered by the Dutch physicist Frits Zernike, an achievement for which he was awarded the Nobel Prize for Physics in 1953.

The solution is based on the observation that when light passes through transparent material, its amplitude and wavelength are not appreciably altered, but its phase is altered as a function of the refraction characteristics of the tissue. For example, as light passes through a group of unstained cells, the cell membranes will accelerate or retard light to a different extent than will the cytoplasm or the nuclei. This results in a phase difference. With appropriate optics, these differences in phase of light rays are transformed into differences in amplitude of light rays and thus are perceived by the viewer as differences in brightness of different parts of the specimen (see Fig. 7).

To facilitate the separation of the unchanged light and the phase-shifted light, a phase condenser has an annular diaphragm rather than an iris so that the specimen is illuminated with a hollow cone of light. Phase objectives have the means both to intensify the phase differences of light transmitted through the specimen and to dim the unchanged light. When adjusting a phase setup, the bright annulus of the condenser is covered by a dark ring in the objective; this is possible only if the annulus and the ring have the same aperture (angular size). Since various powers of phase objectives have various apertures, a phase condenser contains a rotating disc with annular diaphragms of different sizes.

Figures 7C and D present examples using normal and phase contrast illumination (short: phase or Ph).

b. DIFFERENTIAL INTERFERENCE CONTRAST (AFTER NOMARSKI). The problem of increasing contrast in low-contrast specimens can also be solved with the differential interference contrast (DIC) method. With the DIC method, as in phase contrast, the light path is interferred with, but in a different manner. Polarized light is produced and is split into two separate beams by a special prism in such a way that the planes of polarization of the two beams are perpendicular to each other. These two closely adjacent beams (distance of separation is below

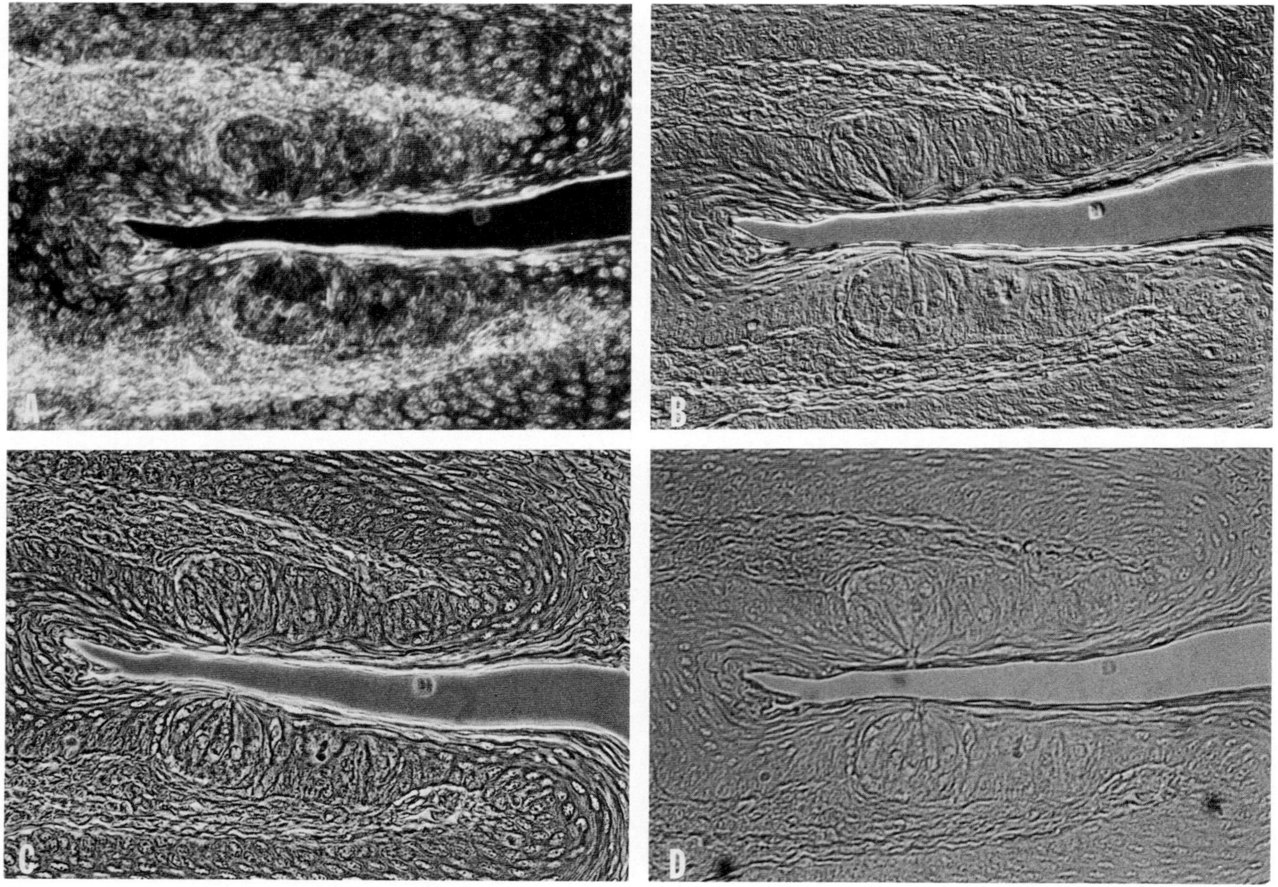

Fig. 7. Examples of different illumination techniques. Cat tongue, unstained section. (A) Dark field. (B) Differential interference contrast. (C) Phase contrast. (D) Bright field.

TABLE I

ADVANTAGES OF PHASE-CONTRAST ILLUMINATION AND DIC

	Phase-contrast	DIC
Special objectives required	Yes	No, accessories only
Special condenser	Yes	Yes
Contrast type	Black and white	Black and white or color contrast, relief or pseudo-relief effect
Contrast variable	No (possibly by change of embedding medium)	Yes
Illuminating aperture variable	No	Yes
"Halo effect"	Yes	No
Specially suited for	Very thin objects (microbiology, cytology)	Slightly thicker and structured objects (cytology, histology)

the resolving power of the optics) are subject to different phase shifts because of the optically different properties of neighboring object areas. To produce contrast by interference, the two beams are first reunited by another prism, and then brought into the same plane of polarization by an "analyzer." In addition, the second prism is adjustable so that the user can add or subtract phase differences to achieve any desired contrast up to a dark field image or to color contrast. DIC produces the final image in relief contrast with shadow effects (Fig. 7B).

Table I lists some of the relative advantages of Ph and DIC methods.

c. DARK FIELD ILLUMINATION. Dark field illumination is an excellent means of increasing the contrast of small light-reflecting particles in tissue. A special purpose condenser contains a plate that blocks out the central portion of the illuminating light. The specimen is thus illuminated by a hollow cone of light striking it at an angle. Only the light diffracted by the specimen enters the objective to form an image. The tiniest particles inside a specimen diffract light into the objective making themselves visible. On the other hand, light rays transmitted by the transparent portions of the specimen go off at an angle and do not enter the objective with the result that the areas in the field of view that contain no object structures remain dark (Fig. 7A). Dark field illumination requires a powerful light source since only a small portion of the light reaching the specimen actually enters the microscope's lens system.

II. Photographic Techniques

A. *Photomicrography*

Photomicrography is the topic of comprehensive text books, therefore, within the present space limitations, only some of the more important items will be treated. The assumption is made that the reader is conversant with the basic principles of photography so we may concentrate on items specific to the production of photomicrographs.

Before starting photomicrographic work two prerequisites should be met: (1) The investigator should posses a basic competence in microscopy, and (2) the specimens to be photographed should be of the finest quality. The first point seems rather obvious, but too often the second point is disregarded. In day-to-day microscopy, one becomes accustomed to "filtering out" imperfections in the material. In producing photomicrographs, however, those artifacts or slight problems of focus that are normally ignored or suppressed become glaringly obvious, and may make the photomicrographs unacceptable.

Photomicrographic equipment range from the simplest amateur camera set to infinity and mounted as close as possible to an eyepiece of a microscope, to a sophisticated research stand with fully automatic camera that offers the utmost convenience through mere pushbutton work. Let us restrict our considerations to an average type of photomicrographic device—the widely used attachment camera for 35-mm film on a trinocular tube.

1. Focusing with a Graticule

The perimeter of the area being photographed is indicated by a graticule, either in one of the eyepieces of the tube or in a special focusing eyepiece near the camera. If the instrument is correctly adjusted, the plane of this graticule, which serves as a focusing aid, coincides optically with the film plane. To ensure a properly focused photograph, the focusing graticule and the specimen must both be sharply focused, regardless of whether or not the microscopist wears eyeglasses. To enhance the accuracy of focusing, a small $2\times$ telescope can be added to the eyepiece. This is especially helpful when using low-power objectives.

2. Means to Enhance Contrast

Because microscopic specimens generally lack sufficient contrast for good photographs, the contrast should be enhanced. The procedure is easiest with stained objects to be photographed with black-and-white film. Colored filters are a means to give almost any color the range in grey tone desired. Light of the same color as an object detail will decrease contrast, whereas light of complementary color will increase contrast. Table II presents suggested filter colors for enhancing contrast in material typically encountered in the neuroanatomical laboratory.

TABLE II

Suggested Filter Color for Enhancing Contrast in
Neuroanatomical Material

Staining procedure	Specimen color	Contrast filter color
Thionine	Blue	Red
Cresyl violet	Violet	Yellow
Neutral red	Red	Green
Nauta, Fink–Heimer	Yellow-brown	Blue
Golgi	Yellow-brown	Blue

It should be noted, however, that too much contrast can result in loss of detail in dark areas. Some experimenting is always necessary when photographing unfamiliar materials. In case the contrast problem is particularly delicate, one can use an interference wedge filter which provides every color of light depending on what part covers the light path of the microscope. With this device one will quickly find which special filter will be optimally suited for the given specimen.

The proper choice of film will improve the quality of the photographs. For most applications a panchromatic film (sensitive to all wavelengths) with fine grain and low to medium speed is best, although there are certain instances in which an insensitivity to red is helpful. It is, of course, not feasible to use color filters to improve the contrast of color photomicrographs. Any good quality, fine-grain color film specified for tungsten bulb light can be used.

In order to properly interpret the finished product, it is essential that the magnification scale be known. The simplest, and perhaps most accurate, means of doing this is to photograph a scaled graticule (object micrometer) with the same objective and eyepiece used to photograph the specimen. Graticules with a specified scale (e.g., 10 mm in 100 divisions) are available from scientific or optical supply houses. After both prints are made, a calibrating scale can easily be determined from the graticule print and the magnification determined. Preferably, the finished print of the specimen should contain a scale bar indicating a specified distance as a reference.

B. Photomacrography

In neuroanatomical research it is sometimes necessary to take photographs of an object field larger than can be obtained in a microscope. In this case a special set-up, combining a reflex camera with bellows and special macro lens, is needed. The magnification depends on the focal length of the lens and the extension of the bellows

$$\text{Magnification} = \frac{\text{image size}}{\text{object size}} = \frac{\text{lens to film distance} - \text{focal length of lens}}{\text{focal length of lens}}$$

To eliminate the need for computing, many instruments contain measuring scales enabling the user to read the magnification directly. A set of lenses (16, 25, and 40 mm) provide magnifications from 2× to 16×.

The simplest and surest way to illuminate large specimens homogeneously is by using a diffuse light box. All areas outside the frame can be covered with some pieces of dark paper so that maximum contrast will result.

Because a survey photograph is densely packed with small details, and because low magnifications have a relatively high resolving power, it is worthwhile to use a film type of higher resolution than normal. A very fine grain film of low speed, e.g., 25 ASA, is preferred. There is a danger of getting blurred photographs (double contours seen upon inspection with a magnifier) caused by vibration of shutter and mirror movement of a reflex camera. If the available device does not meet the demands, exposure times should be longer than 2 seconds; in this case the proportion of time the camera is shaking to the total exposure time is such that minimal impairment of sharpness will occur.

Many smaller laboratories may not have special photomacrographic equipment available. It is, however, possible to produce acceptable quality photomacrographs by using an ordinary photographic enlarger. Stained sections can be projected onto 4 in. × 5 in. sheet film, the film developed, and then contact printed to produce the positive print. Experience will dictate the optimal exposure and development times.

Chapter 2

Tissue Preparation and Basic Staining Techniques

George Clark

Veterans Administration Hospital, and
Department of Anatomy
Medical University of South Carolina
Charleston, South Carolina

25

I.   Introduction

In any experimental manipulation involving the placement of lesions, electrodes, or cannulae in the central nervous system, exact localization of the
damage or placement is essential. This exact localization requires removal of the
tissue, fixation, sectioning, and staining for microscopic study. These steps are
the subjects considered in this chapter.

II.   Tissue Fixation

The preservation of tissue structure is of utmost importance. Autolysis begins
at death, the bony coverings of the central nervous system make the removal of
tissue slow, and large blocks are needed for neuroanatomical, neurophysiological, and neuropsychological studies. For these reasons preliminary tissue fixation
by perfusion is always desirable. If perfusion is not possible (i.e., when the
animal dies unexpectedly), the tissue should be removed as rapidly as possible
and immersed in a 10% formalin (Clark, 1973) solution using at least 15–20
times the volume of tissue removed.

The following equipment is required for perfusion: funnel support, 4-
in.-diameter funnel, about 4 ft of plastic or rubber tubing, a stainless-steel cannula similar to that shown in Fig. 1, scalpel, two scissors (the author uses
kitchen shears and a small surgical scissor), two curved hemostats, tissue forceps, self-retaining retractor, an aqueous solution of 8.5% sucrose (the super-
market variety is satisfactory; this solution must be refrigerated), and formalin–

FIG. 1.   Diagram of perfusion cannula.

sucrose solution. A word of explanation for the beginner is appropriate here. Formalin is a saturated solution (approximately 37%) of formaldehyde in water with a small amount of methanol as a preservative. Formalin–sucrose solution is 10% formalin and 7½% sucrose in water. Although 10% formalin is typically used for routine work, several "additives" have been suggested. Many investigators use 10% formalin in 0.9% saline, although Koenig, Groat, and Windle (1945) advocate the addition of gum acacia. Baker (1965) suggested sucrose as an additive, and after a series of trial substitutions, we have selected the procedure given above.

In some instances it is useful to prepare fresh formalin from the polymerized paraformaldehyde. To obtain 100 ml of 4% paraformaldehyde (approximately 10% formalin), add 4 gm of the paraformaldehyde powder to 50 ml of distilled water. Heat to 60°C under a fume hood. While stirring, add a few drops of 1 N NaOH until the solution clears. Let solution cool and add 50 ml 0.1 M PO$_4$ buffer (or other buffer) at pH 7.4.

The procedure for perfusion is relatively simple. First, clamp the tubing and pour the 8.5% sucrose solution (50 ml for rat, 500 ml for cat) into the funnel. Holding the cannula above the funnel, remove the clamp and lower the cannula until all air is expelled. Then reclamp the tubing. The funnel should be about 16 in. above the animal board. Next, deeply anesthetize the animal; there should be no reflex response to a painful stimulus to the paw. Make an incision the length of the animal's sternum about 1 cm to the right of midline. In the cat or rat the ribs can be cut with kitchen shears, but in the dog or a larger animal rib cutters will be needed. After exposing the contents of the thorax with the self-retaining retractor, grasp the pericardium with forceps and free the heart of this covering with fine scissors. Grasp the right atrium with tissue forceps, apply a curved hemostat as near the heart as possible and cut off the right atrium just peripheral to the hemostat. Then plunge the cannula into the lumen of the left ventricle. After removal of the clamp on the tubing, the increase in pressure in the left ventricle forces the bevel of the cannula against the inner wall of the ventricle sealing it. Removing the hemostat from the right atrium will allow the escape of fluid so the blood will be quickly flushed from the vascular system. Before the funnel is emptied, begin to add the sucrose–formalin solution, being sure that no air enters the system. The total perfusion time may vary somewhat depending on conditions, but at least 100 ml of formalin for a rat or 500 ml for a cat (with corresponding amounts for other species) should be used. To avoid tissue artifacts, Cammermeyer (1961) has insisted it is best to wait a few hours before removing the desired portions of the central nervous system. Overlying bone should be carefully cut away with rongeurs, neural tissue should be carefully removed and then postfixed in sucrose–formalin (Section II) for several days before further processing.

III. Preparation of Tissues for Sectioning

In order for thin sections to be cut, the fixed tissue must be either hardened by freezing or embedded in a sufficiently hard medium such as paraffin wax, celloidin, or plastic. Embedding in plastic is presently used only in preparation for electron microscopy and will not be considered further. Which of the other three is chosen depends on the information desired and the time available.

Paraffin embedding is simple, sectioning is easy, many stains can be used, cellular structure is well preserved, and resolution of detail is limited solely by the resolving power of the light microscope. On the other hand, differential shrinkage in processing, compression during sectioning, and differential expansion during mounting all occur and seriously limit the use of paraffin material for localization of electrode placements, electrode tracks, or lesions. Further, the dehydration and clearing agents used prior to paraffin embedding render the tissue unusable for some histochemical analyses and unsuitable for some silver impregnation methods.

Freezing the tissue provides a rapid means of obtaining sections that are suitable for a variety of histological procedures, and are obligatory for localization of some enzymes, as well as other procedures. Large blocks, such as a complete cat brain, are readily cut. However, considerable practice is needed to flatten these sections and affix them to a slide.

When time allows, celloidin embedding is the method of choice for localization of electrode placements and lesions. The sections are cut readily at a variety of thicknesses, a wide variety of stains may be used, shrinkage is virtually uniform, distortion is minimal, sections are easily handled, section losses rarely occur, and many stains are unusually brilliant. For these reasons, despite the time-consuming processing, celloidin embedding is optimal. However, as with paraffin embedding, dehydration and clearing agents render the tissue unsuitable for some histochemical techniques.

A. Paraffin Method

Although not particularly useful for the localization of lesions, electrode tracks, etc., the paraffin method is excellent for cytological detail at the light microscope level. Thus, it may be used for the study of normal or degenerating neuron cell bodies. The process of embedding in paraffin is, of course, the progressive replacement of the water in tissue first by the alcohols, then by a clearing agent (the author routinely uses xylene), and finally by paraffin. Each reagent must be miscible with both the preceding and the following reagents. There are innumerable methods for embedding tissue in paraffin, and most of them work. The following schedule is used at present in the laboratory at Medical

University of South Carolina for well-fixed blocks about $1 \times 1 \times 1$ cm or smaller.

1. Wash briefly in water.
2. 70% Alcohol, 1 hr. (The term alcohol indicates ethyl alcohol. Where other alcohols are needed, their full names are used.)
3. 95% Alcohol, 1 hr. This preferably contains sufficient eosin (about 0.1%) to render the outside of the tissue slightly pink. The color helps in orienting the tissue in embedding and preparing the blocks for sectioning.
4. n-Butyl alcohol, 1–2 hr. The n-butyl alcohol currently available contains 0.1–0.2% H_2O. This water can be easily removed by adding about 200-gm desiccated molecular sieve (Type 4a) to each gallon of the alcohol. This will remove most of the water and render the alcohol completely miscible with xylene. If the sieves are placed uncovered in the paraffin oven for a day or so after use they can be used again.
5. n-Butyl alcohol, overnight.
6. Xylene, 1 hr.
7. Paraffin 2 changes of 1 hr each at 56°–60°C.
8. Embed.

This procedure works well for blocks of tissue $1 \times 1 \times 1$ cm or smaller. For larger blocks all times in reagents must be increased, and some experimenting may be required. Fortunately, n-butyl alcohol does not harden the tissue, so complete dehydration is not a problem. Clearing must be continued until the block is translucent, but as prolonged exposure to xylene will harden the tissue, clearing times must be kept as short as possible. Infiltration time, including an additional change or two in paraffin, should be increased to 12 hours or even longer.

Because of the wide variety and various shapes of tissue blocks, it is usually not feasible to use the plastic boxes favored in pathology. Instead, use paper portion cups (usually ½ or ¾ oz.) obtainable from restaurant supply houses or make your own in the manner explained below for celloidin blocks. In embedding, one of these is filled with more than enough melted paraffin to cover the block and is placed in a metal tray (for example, a baking tin about $1 \times 7 \times 10$ in.) with a few ice cubes. Soon a layer of congealed paraffin appears on the bottom of the portion cup. Then the portion cup is removed, and the tissue is inserted into the paraffin. The block is handled with forceps that have been mildly heated in a burner and wiped off with a towel. With these forceps the tissue is so oriented that the area where sectioning is to begin is at the bottom of the cup. After the tissue is placed, return the cup to the metal tray with the ice cubes where it is to remain until hard. After the paraffin is completely congealed, the paper cup can be stripped from the block and the excess cut away. A cutter is available from many supply houses that enables the sides to be cut at right angle to the surface with opposite sides parallel. When completed, the paraffin should extend 2–3 mm on each side of the tissue and about twice that on the top and bottom. The tissue is attached to either a metal object disc or to a wooden block (about 18–20-mm square and perhaps 25-mm long) impregnated with paraffin. A spatula that is warm enough to melt paraffin is placed on the surface on which the block is to be attached, and the paraffin block is placed on the spatula. As the spatula is pulled out, the block is pressed against the surface. When the paraffin congeals, the block will be firmly attached. Since there are several different microtomes for paraffin sectioning and since each of these has an instruction manual with detailed directions, it is suggested that these booklets be carefully studied before attempting to section paraffin blocks. For those desiring further information, the volume by Steedman (1960) will be useful.

B. Frozen Section Method

For some stains, frozen sections are required, and most stains can be applied to frozen sections. A sliding microtome with an object disc cooled by a refrigerant or by carbon dioxide is preferred, but blocks of solid carbon dioxide (dry ice) on a large platform stage can be used. Beginners should carefully study the instruction book for the particular instrument to be used. Blocks no larger than 1.5 × 1.5 × 1.5 cm are optimal, but blocks as large as an entire cat brain can be used. The block to be sectioned is routinely placed in 30% sucrose–10% formalin solution at 3°–5°C until the block sinks. This limits the size of ice crystals that can seriously damage tissue. Before freezing, it is good to mark one side of the block for later orientation. A convenient method is to push (and then remove) a fine glass rod through an area not in the area of interest.

Put a few drops of water on the object disc, place the properly oriented block on the water, and proceed with freezing the tissue. If the block is large or irregular, more water must be placed around the block to hold it firmly in position. The temperature of the frozen tissue block is critical. If too cold and hard, the tissue will shatter rather than cut; if the tissue is not adequately frozen, uneven sections will result. Some firsthand experience is necessary for the beginner to learn the intricacies of temperature control.

After a section is cut, it can be removed from the knife by picking it up with a wet soft camel's hair brush such as those used in water color art work. The section can be placed in a compartmented plastic box or a small wide-mouthed jar (a 1- or 2-oz ointment jar) containing 10% formalin. Quite commonly, ten such jars are used; the first section is placed in jar #1, the second section in jar #2, etc. After 10 sections are jarred, the next section is placed in jar #1, etc. This results in ten series ready for mounting or staining.

A cryostat (simply a microtome housed in a temperature-controlled chamber) is helpful for taking frozen sections of poorly fixed or unfixed tissue. The cut sections remain frozen until mounted on slides so tissue damage from handling is, to a large degree, avoided.

C. Celloidin Method

Unless the stain to be used requires frozen sections or time is essential, the celloidin method is the method of choice. Thick (50 μm sections are routine) sections are optimal for the study of retrograde or transneuronal degeneration, gliosis, delineation of nuclear groups, and localization of lesions, electrode tracks, and electrode placements. Also, cell and fiber stains are brilliant. While other varieties are available, Parlodion is listed by most suppliers and has virtually unlimited shelf-life. In preparing solutions of celloidin (usually 5 and 10% are prepared), the strips of celloidin are weighed and placed in a bottle with a small amount of absolute ethyl alcohol, the bottle is shaken, and then the alcohol is poured off. This removes any dirt or water adhering to the strips. Next the proper amount of ether:alcohol is added and the bottle capped. The bottle should be upended several times each day until solution is complete. If this is not done, the celloidin will coalesce into a viscous mass that will dissolve very slowly.

For a whole rat brain, the schedule shown below is adequate, but for a cat brain the lengths of time should be doubled.

1. 50% alcohol, 3 days	6. Absolute alcohol, 3 days	
2. 70% alcohol, 3 days	7. Ether:alcohol, 3 days	
3. 95% alcohol, 3 days	8. Ether:alcohol, 3 days	
4. 95% alcohol, 3 days	9. 5% celloidin, 1 week	
5. Absolute alcohol, 3 days	10. 10% celloidin, 1 week	

Tissue is routinely embedded in celloidin in paper boats. The paper for these must be heavy and glazed so that some (but not too much) ether alcohol can diffuse through the paper to ensure uniform hardening. The steps in folding the boats are shown in Fig. 2; the final step is to staple the ends. Several sizes of wooden blocks are needed, since each boat should be about 1 cm wider and longer than the block of tissue. The boat is partially filled with 10% celloidin, and the block placed in the boat. The block should be rotated slowly to allow escape of any trapped air and be oriented so that the part to be cut first is at the bottom of the boat. Add additional celloidin until the block is covered by about 5 mm of celloidin. The boat is then placed in a jar, and a vial half-filled with concentrated H_2SO_4 is placed alongside. These should almost fill the jar, for if too much air space exists the celloidin will crust over and not concentrate evenly. The jar should be kept tightly capped except when testing. As the ether alcohol

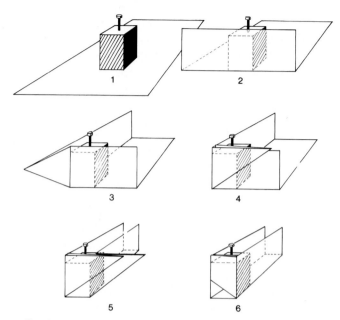

FIG. 2.   Diagram of steps in folding boats for celloidin embedding.

vaporizes from the surface and through the walls of the boat, it is taken up by the acid which then becomes darker with some carbon deposits. For a few days additional celloidin is added to keep the tissue covered, and the acid is replaced when the volume is about doubled. When, in the absence of a crust, the celloidin no longer or only very slightly sticks to the finger- the acid is replaced with chloroform. After a day or so the block will become quite firm to the touch, and the paper boat can be stripped off. If the block is still readily compressible, add more chloroform and recap the jar. When the block is quite firm, it can be stored in 70% alcohol. The block can be mounted directly after hardening, but if it has been stored in 70% alcohol it must be dehydrated in n-butyl alcohol overnight.

   The fiber block on which the block is to be mounted should be a few millimeters wider and longer than the block. A small amount of ether:alcohol is poured into half a petri dish and some 10% celloidin placed into the other half. The top surface of the fiber block is placed down into the ether:alcohol and allowed to remain there for a few minutes. The fiber block is then placed topside down into the 10% celloidin and the celloidin block with the bottom side down is placed in the ether:alcohol. When the surface of the celloidin block is quite slippery change it to the 10% celloidin and place the fiber block surface up on the table. When a thin scum appears on the fiber block, press the celloidin block snugly against the fiber block. Then add a little more celloidin along the sides of the block and place in a jar with a little chloroform. The next day the block will be firmly attached and ready for cutting.

The novice in the celloidin method should carefully read the instruction book for the particular microtome to be used. With all varieties, the knife is set at an angle to the direction of travel of knife (sliding microtome) or block (sled microtome). Thus, the knife enters one corner of the block and leaves from the opposite corner. Brush the top of the block and the knife with 70% alcohol and begin cutting; each stroke should be at a constant speed. There will be some inital irregularities of the surface so begin collecting when a full section is cut. It is necessary to dampen the surface of the block and the knife each time a section is cut and, if cutting is interrupted, to cover the block with gauze dampened with 70% alcohol. Routinely, one stops just short of the final corner of the block and unrolls the section on the knife with a dampened brush. Then the cut is completed. It is routine procedure to put sections 1, 11, 21, etc., into one jar and sections 2, 12, 22, etc., into another. The intervening sections, 3–10, 13–20, 23–30, etc., are put in separate jars in the same sequence. The first two provide materials for a cell-stained series and a myelin-sheath-stained series and the others provide material to check a particular region. While complicated numbering methods can be used, it is usually possible to place the completed slides on a viewing box and arrange them in order without using a microscope. Celloidin sections are typically mounted on slides after they have been stained.

IV. Mounting of Sections

A. Slide Preparation

Sections cut by each of the above-described techniques are mounted on glass slides. For routine work, commercially available "precleaned" slides are adequate, but in special cases it is advisable to put them through a final cleaning process. For example, in work with autoradiography, it is essential that slides be as clean as possible, as remaining dirt or oils will tend to produce artifacts. For cleaning slides, this method is recommended.

1. Soak overnight in dichromate cleaning solution.
2. Rinse in tap water.
3. Rinse each slide in deionized water.
4. Dip in 80% alcohol (twice).
5. Let air dry.
6. Dip in "subbing" solution (see below) and let dry in a *dust-free* place overnight.

Two types of adhesives are commonly used to keep tissue sections firmly fixed to the slides during the staining process. For routine work, a small drop of

Mayer's albumin is spread evenly and very thinly over the slide, until the slide appears to be dry. For critical work "subbed"slides are used. A solution of 1 liter deionized H_2O, 5 gm gelatin, and 0.5 gm chrome alum [$CrK(SO_4)_2 \cdot 12H_2O$] is made by bringing the water to a boil and adding the gelatin. When gelatin has dissolved, add the chrome alum and filter. Slides should be dipped while the solution is warm.

B. Paraffin Sections

Paraffin sections (if not to be used for autoradiography) are mounted on albumin-treated slides. Sections are floated in a water bath at 37°–40°C. A couple of drops of wetting agent (e.g., Brij) are added to alleviate problems with air bubbles trapped under the sections. Sections are simply floated onto the slides, and the slides left on a slide warmer overnight at 40°C. See Clark and Clark (1971) for a more detailed account.

C. Frozen Sections

Frozen sections are mounted on either albumin-treated or "subbed" slides. Sections are floated in a freshly prepared solution of 40% alcohol and 0.3% gelatin. Some care and experience are necessary to spread the sections and avoid folding or tearing of the tissue. A soft camel hair brush or sable brush is helpful. Sections are allowed to dry on the slides overnight in an oven or slide warmer at 37°–40°C. See Clark and Clark (1971) for a more detailed account.

D. Celloidin Sections

Celloidin sections typically are stained before they have been mounted on slides. Sections taken directly from xylene solution are placed on a drop of mounting medium on a slide. Additional mounting medium is applied over the top, and the cover slip is applied.

In all cases, the slide, the tissue section, and the coverslip must be sandwiched together as closely as possible for good optical quality (see discussion in Chapter 1). Small brass or lead weights placed over the cover slips while the sections and mounting medium are drying will be helpful.

V.   Staining Procedures

Stains are used to increase contrast for microscopic study of sectioned material. Since various tissue elements have different chemical compositions, it is possible, with different dyestuffs, to more or less selectively stain one structure or another. Thus, basic dyes stain Nissl bodies and nuclei and hematoxylin stains axon myelin sheaths. For localization of lesions, electrode tracks and placements, the sections should be stained with either a basic dye, such as cresyl violet acetate or thionine, or with one of the myelin sheath stains. Preferably, if interrupted serial sections are available, one set of sections should be stained for cells and another for myelin sheaths.

A. Weil Myelin Sheath Stain

Localization in the brainstem and spinal cord is easiest with a myelin sheath stain; furthermore, many atlases are made from myelin-sheath-stained sections. There are a number of these stains, some are complicated to use and some ruin the tissue for other stains, but the Weil stain (Berube, Powers, & Clark, 1965; Weil, 1928) has none of these drawbacks. It can be used on paraffin, frozen and celloidin sections. Three solutions are needed.

1. 10% Hematoxylin in 95% or absolute ethyl alcohol
2. 4% Aqueous ferric alum [$FeNH_4(SO_4)_2 \cdot 12H_2O$]
3. Potassium ferricyanide [$K_3Fe(CN)_6$], 12.5 gm
   Borax ($Na_2B_4O_7 \cdot 10H_2O$), 2.5 gm
   H_2O sufficient to make 1000 ml

Paraffin sections or dry frozen sections mounted on albuminized slides (see Section IV,A) are treated similarly. One to 25 slides in a slide rack can be processed at once.

Staining schedule

1. Xylene, 5 min
2. Xylene, 1 min; xylenes serve to remove paraffin from the sections
3. *n*-Butyl or absolute ethyl alcohol, 1 min
4. 95% Alcohol, 1 min
5. 70% Alcohol, 1 min
6. H_2O 2–3 changes, 0.5 min each; alcohol and water series hydrate the tissue in preparation for the water-based stain
7. Stain for 20 min in
   10% Hematoxylin, 1 part
   4% Ferric alum, 10 parts
   H_2O (distilled), 30 parts
   This staining solution begins to deteriorate as soon as it is mixed. It is best to add the hematoxylin to the water, then add the ferric alum solution, and immediately pour the staining solution over the slides. It can be used only once

8. Wash in water several times
9. Begin differentiation in
   4% Ferric alum, 10 parts
   H_2O (distilled), 30 parts
   This is continued until a difference between the white and gray matter is just discernible
10. Wash in several (at least three) changes of tap water and at least three in distilled water. If washing is not ample a blue precipitate will promptly appear and ruin the section
11. Differentiate in the modified Weigert's differentiator (solution 3) until gray matter is colorless or yellow-orange and the white matter remains black. If differentiation is carried too far, wash the sections thoroughly and restain
12. Wash in several changes of distilled water
13. Dehydrate, clear in xylene, and coverslip using a commercial mounting medium

Celloidin sections are best handled unmounted in a carrying sieve. This must be homemade. Saw the buttom off a disposable beaker, and, using a plastic cement, cover the bottom with a small piece of women's nylon hose. A 100-ml beaker will easily handle 15–20 celloidin sections up to 2×2 cm. When using these sieves, it is convenient to have the various solutions (alcohols, xylene, staining solutions, etc.) in small nesting bowls so the sieves can easily be moved from one solution to another. It is necessary to keep the sieves moving so that each section is in full contact with the solutions. Since the celloidin sections are routinely stored in 70% alcohol, all that is needed before staining is to wash the sections in water. The procedure for the actual staining and subsequent steps are the same as for paraffin or frozen sections.

Combinations of a myelin sheath stain and a cell stain are quite spectacular on relatively thin sections but on thicker sections, for example, the routine 50-μm celloidin sections, so much tissue is stained that little is visible. Use of Darrow red on Weil-stained sections is perhaps the best. Because there is a limited need for such a staining procedure, the reader is directed to the discussion by Powers, Clark, Darrow, and Emmel, 1960.

B. Cell Stains

Cell stains are the staining procedures second only to myelin sheath stains for solving most localization problems, and can be used on frozen, paraffin, or celloidin sections. The two most widely used dyes are cresyl violet acetate and thionine. Toluidine blue O, neutral red, and Darrow red are also good but are used infrequently (Clark, 1973). Again, there are a number of procedures with various staining times and dehydration procedures. The following procedures work well in my laboratory.

Solutions needed for Cresyl Violet Acetate or Thionine
   Buffer: 0.2 *M* sodium acetate, 9 parts; 0.2 *M* acetic acid, 11 parts. This will give a pH of about 4.5.
   Stock staining solution: Cresyl violet acetate or thionine, 400 mg; 100 mg Buffer.

Staining schedule
   1. Dewax and hydrate as in the Weil procedure
   2. Stain at room temperature for 20 min in a solution of
      Stock staining solution, 1 part
      Buffer, 1 part
   3. Rinse quickly in *n*-butyl alcohol
   4. Dehydrate in *tert*-butyl alcohol
   5. Clear in xylene

When the sections are cleared, they should be checked microscopically. The nuclei of neurons should be relatively clear with only the prominent nucleolus showing, and the Nissl bodies should be discrete. If sections are too light, rehydrate and restain doubling the concentration of the dye. If too dark, return to the *n*-butyl alcohol for a short time, then return to the *tert*-butyl alcohol, and then to xylene. In either case, recheck the sections microscopically. For frozen sections, the dry slides should be handled exactly as the paraffin sections. Celloidin sections are washed in distilled water and then placed in the staining solutions. The carrying sieves (see Section V,A) are convenient to use with celloidin sections.

C. Silver Stains for Axons

Silver stains for axons and end bulbs are not useful for the usual localization problems but are useful for special needs as they stain both myelinated and unmyelinated axons. For paraffin sections, Bodian's protargol method is best, for frozen sections it is the Glees' method (1946), and for celloidin sections the Davenport method (1929).

1. BODIAN'S PROTARGOL METHOD (BODIAN, 1937; CLARK & CLARK, 1971)

Solutions Needed

   1. 1% Protargol (protargol S certified by the Biological Stain Commission is satisfactory)

This must be made up fresh each time. Dust the protargol on the surface of the water and do not shake or stir but allow the protargol to dissolve from the surface downward. If shaken or stirred, the protargol coalesces into viscous masses surrounding dry dye. Just before the slides are put into the stain, add approximately 5 gm copper to each 50 ml of the protargol solution. The copper should be cleaned with dilute HNO_3 before adding it to the protargol solution

2. Reducing solution
   Hydroquinone, 1 gm
   Sodim sulfite, anhydrous, 5 gm
   H_2O, 100 ml
   This solution is not stable and must be made up just before use

3. Gold chloride toning solution
   Gold chloride, 1 gm
   H_2O, 100 ml
   Glacial acetic acid, 3 drops
   This may be used repeatedly

4. Gold reducing solution
   Oxalic acid, 2 gm
   H_2O, 100 ml
   This may be used repeatedly

5. Fixing solution
   Sodium thiosulfate, 5 gm
   H_2O, 100 ml
   This may be used repeatedly

Staining schedule

1. Cut paraffin sections at 4–5 μm
2. Dewax and hydrate
3. After adding the copper to the protargol solution, insert slides and place in paraffin oven (55°–60°C). The staining time may vary from 4–16 hr. Impregnation is complete when the sections become golden brown in color. With under-impregnation, only the larger fibers will be faintly stained; with over-impregnation, the finer fibers will be lost in the excessive staining of the background
4. Wash very briefly in distilled water (in and out)
5. Place in reducing solution for 3–10 min
6. Wash thoroughly, preferably in running water followed by two or three changes of distilled water
7. Tone in gold chloride for 7–10 min. The sections become white and opaque
8. Rinse quickly in distilled water (in and out).

9. Place in gold reducer (oxalic acid) for 2–5 min. The sections assume a faint purple color
10. Fix in thiosulfate solution
11. Wash thoroughly in water
12. Dehydrate, clear, and coverslip

Results: Axons will be stained black to purplish-black; neurons will be reddish-purple.

2. MODIFIED GLEES' METHOD (CLARK, 1973; NAUTA & GYGAX, 1951)

With good silver stains, there is so much tissue stained that very thin sections are needed to permit any possible resolution. Nevertheless, at times it is necessary to have a silver stain on part of a series to confirm and expand the results that have been obtained. With the procedure suggested, it is possible to obtain a good silver stain on frozen sections as thick as 35 μm.

Solutions needed

1. Ammonia solution
   Concentrated NH₄OH, 1 ml
   50% Alcohol, 100 ml
2. Initial silver solution
   Silver nitrate (AgNO₃), 1.5 gm
   Pyridine, 5 ml
   H₂O, 95 ml
3. Diamine silver
   30% Alcohol, 30 ml
   Silver nitrate (AgNO₃), 450 mg
   Concentrated NH₄OH, 2 ml
   2.5% Aqueous NaOH, 2 ml
4. Nauta–Gygax reducer
   10% Formalin, 10 ml
   1% Citric acid, 20 ml
   95% Alcohol, 75 ml
   H₂O, 100 ml
5. Fixing solution
   Na₂S₂O₃·5H₂O, 2.5 gm
   H₂O, 100 ml

Staining schedule

1. Collect sections (perhaps 10–15) in carrying sieves (see Section V,A)
2. Place in alcoholic ammonia (solution 1) for 20 hr at room temperature

3. Wash briefly in water, first in tap, and then in distilled water
4. Place in initial silver solution (solution 2) for 20–24 hours
5. Without washing, place in diamine silver (solution 3) for 3 min
6. Without washing, transfer sections to Nauta reducer (solution 4). As soon as the solution begins to cloud, place in fresh reducer. 2 min is ample
7. Wash in water
8. Fix in sulfite (solution 5) for a few seconds
9. Wash thoroughly in water
10. Mount on albuminized slides from 40% alcohol (see Section IV,C). Blot slide with bibulous paper, dehydrate, clear, and coverslip

Results: Axons and terminals will be stained black; neurons will be light yellow-brown.

3. DAVENPORT'S SILVER METHOD FOR CELLOIDIN SECTIONS (CLARK, 1973)

Although this stain is reasonably reliable for larger fibers in formalin fixed celloidin sections, fine fibers are usually not stained.

Solutions needed

1. Impregnating solution
   $AgNO_3$, 10 gm dissolved in 10 ml H_2O
   95% alcohol, 90 ml
   1 N HNO_3, 0.5 ml
2. Reducing solution
   Pyrogallol (pyrogallic acid), 5 gm
   95% alcohol, 100 ml
   Full strength formalin, 3 ml

Staining schedule

1. Place in 95% alcohol, 5 min
2. Impregnate sections singly or well spread out in a shallow dish at 35°–40°C in dark. Impregnation is complete when sections are light brown in color
3. Rinse in 95% alcohol, two changes of 3–5 sec each.
4. Reduce for several minutes in solution 2. Keep section in motion and check depth of stain microscopically
5. Wash in several changes of 95% alcohol
6. Place in tertiary butyl alcohol. 5–10 min
7. Clear and mount

D. Hematoxylin and Eosin

The innumerable hematoxylin and eosin procedures are usually designed primarily for routine use and can be used on paraffin, celloidin, or frozen sections. The method given here uses solutions so simple to prepare that for occasional use a fresh hematoxylin solution should be prepared each time.

Solutions needed

1. 10% Hematoxylin in absolute ethyl alcohol
2. 0.1% Aqueous sodium iodate ($NaIO_3$)
3. 3% Aqueous potassium alum [$KA1(SO_4)_2 \cdot 12H_2O$]. This is an excess and either Na or NH_4 alum can be used
4. Saturated aqueous lithium carbonate (Li_2CO_3)
5. 0.5% Aqueous eosin
6. Hematoxylin staining solution (should be prepared fresh)
   Solution 3 (potassium alum) 40 ml
   Solution 2 (sodium iodate) 4.5 ml
   Solution 1 (hematoxylin) 1 ml
   Mix and place in paraffin over overnight and then dilute to 100 ml. The solution can be used immediately and remains good for several weeks.

Staining schedule

1. Dewax and hydrate
2. Stain in solution 6 for 5 min. This hematoxylin is a relatively good progressive stain but after a time it may be necessary to decrease the staining duration
3. Rinse several times in tap water and then "blue" in half-strength Li_2CO_3 (solution 4 diluted 1:1)
4. Rinse several times in tap water and then in distilled water. It is essential that no Li_2CO_3 be carried over to the eosin solution.
5. Stain in eosin (solution 5) for 1 min. The depth of the eosin stain is a matter of personal preference and the staining time can be increased or decreased. Increasing the time in the dehydrating alcohols will also lighten the eosin
6. Dehydrate, clear, and mount

Results: Nuclei will be stained blue; other structures will be pink.

E. Glial Stains

Glial nuclei are, of course, stained with basic dyes and an increase in number of glial nuclei (gliosis) is easily recognized. Specific glial stains are primarily tools of the neuropathologists and are of little value to the experimentalist interested in problems of localization. These various stains are more or less specific for different pathological glia and the staining of normal glia is routinely poor and unspecific. Many of these procedures may be found in Clark (1973), Ráliŝ, Beesley, and Ráliŝ (1973), and in Scharenberg (1968). They are all capricious and erratic.

1. CAJAL GOLD SUBLIMATE STAIN

Cajal's gold sublimate stain for activated astrocytes is almost dependable. Since with this stain normal astrocytes are occasionally differentiated and since no difficult endpoints are involved, it is included here. Unmounted frozen sections (10–35 μm) are used.

Solutions needed

1. Gold sublimate
   1% aqueous gold chloride 5 ml
   5% aqueous $HgCl_2$ 5 ml
   H_2O (distilled) 30 ml
   Use dichromate cleaning solution on all glassware used in preparing this solution
2. Fixing solution
   5% aqueous sodium thiosulfate
3. 5% aqueous urea
4. Reducing solution
   5% oxalic acid solution 100 ml. Just before use add 0.5 g of hydroquinone

Staining schedule

1. Wash sections in distilled water (use carrying sieve)
2. Impregnate in solution 1 for 2–6 hr
3. Wash in distilled water
4. Fix in solution 2 for 5 min
5. Wash in several changes of distilled water
6. Mount on albuminized slide (see Sec. V,A) from 40% alcohol
7. Blot with bibulous paper
8. Complete dehydration with 2 changes of alcohol (absolute ethyl or *n*-butyl) clear in xylene and mount

Results: Activated (and occasionally normal) astrocytes are sharply stained purple against a rosy red background.

Optional steps

When results are not satisfactory, wash fresh sections several times in water and immerse in solution 3 (urea) for 30 min. Then wash in several changes of distilled water and immerse in freshly prepared reducer (solution 4) for 1 min. After thorough washing, follow staining schedule, beginning with impregnation step.

2. PHOSPHOTUNGESTIC ACID HEMATOXYLIN

Astrocytes are a type of glial cell that often undergo changes (react) in areas of neural degeneration. With the proper procedure, this stain is specific for reactive astrocytes and fibers. Normal astrocytes and their fibers are not differentiated. Any type of section may be used.

Solutions needed

1. Oxidizing solution
   Potassium permanganate ($KMnO_4$), 300 mg
   Sulfuric acid (H_2SO_4), conc., 0.1 ml
   H_2O 100 ml
   This solution must be prepared, preferably using a mechanical stirrer, immediately before use.
2. 5% aqueous oxalic acid $(COOH)_2$
3. Ferric alum [$KFe(SO_4)_2 \cdot 12H_2O$], 4 gm
   H_2O 100 ml
   This solution can be used repeatedly
4. Staining solution (PTAH)
   Hematoxylin 0.1 gm
   H_2O 60 ml
   0.1% aqueous $NaIO_3$ 4.5 ml
   Leave solution 4 overnight at room temperature. After 24 hr add 1 gm phosphotungstic acid and dilute to 100 ml. It is ready for use the following day.

Staining schedule

1. Dewax and hydrate
2. Oxidize in $KMnO_4$ (solution 1) for 15 min
3. Wash in tap water until water remains colorless. Then wash in distilled water
4. Reduce in oxalic acid (solution 2)
5. Wash several times in tap water and then in distilled water
6. Soak in ferric alum (solution 3) for 2 hr
7. Rinse quickly in distilled water
8. Stain in PTAH (solution 4) for 2 hr
9. Rinse quickly in 95% alcohol
10. *n*-Butyl alcohol 2 changes of 2 min each
11. Xylene 2 changes of 1 min each, then mount

Results: Reactive astrocytes and connective tissue-red; red blood cells, myofibrils and glial fibers-blue to black.

VI. Conclusions

The importance of histological processing as part of a research program in neurobiology or physiological psychology can not be overemphasized. Precise localization of damage to the nervous system or of placement of electrodes or cannulae is essential to the interpretation of physiological or behavioral results of these experimental manipulations. Proper fixation of the experimental tissue and use of appropriate staining procedures are the first steps in this localization process.

Further, these staining procedures, appropriately applied, can reveal in vivid detail much information on the normal structure of the nervous system. Indeed, much of our current knowledge of the nervous system has come from studies of normal and abnormal material, using the cellular, fiber, and glial staining procedures described in this chapter. A more detailed account of some stains, embedding and cutting procedures can be found in Clark and Clark (1971). Answers to many specific questions can be found in Gray (1973).

References

Baker, J. R. The fine structure produced in cells by fixation. *Journal of the Royal Microscopical Society,* 1965, **84,** 115–131.

Berube, G. R., Powers, M. M., & Clark, G. Iron hematoxylin chelates. 1. The Weil staining bath. *Stain Technology,* 1965 **40,** 53–62.

Bodian, D. The staining of nervous tissue with activated protorzol. The role of fixatives. *Anat. Res.* **69,** 153–162, 1937.

Cammermeyer, J. The importance of avoiding "dark" neurons in experimental neuropathology. *Acta Neuropathologica,* 1961, **1,** 245–270.

Clark, G. (Ed.) *Staining procedures used by the biological stain commission.* (3rd ed.) Baltimore, Maryland: Williams & Wilkins, 1973.

Clark, G., & Clark, M. P. *A primer in neurological staining procedures.* Springfield, Illinois: Thomas, 1971.

Davenport, H. A. Silver impregnation of nerve fibers in celloidin sections. *Anatomical Record,* 1929, **44,** 79–83.

Glees, P. Terminal degeneration within the central nervous system as studied by a new silver method. *Journal of Neuropathology and Experimental Neurology,* 1946, **5,** 54–59.

Gray, P. (Ed.) *The encyclopedia of microscopy and microtechnique.* Princeton, New Jersey: Van Nostrand-Reinhold, 1973.

Koenig, H., Groat, R. A., and Windle, W. F. A physiological approach to perfusion-fixation of tissues with formalin. *Stain Technology,* 1945, **20,** 13–22.

Lillie, R. D., Pizzolato, P., Welsh, R. A., Holmquist, N. D., Donaldson, P. T., & Berger, C. A consideration of substitutes for alum hematoxylin in routine histologic and cytologic diagnostic procedures. *American Journal of Clinical Pathology,* 1973, **60,** 817–819.

Nauta, W. J. H., & Gygax, P. A. Silver impregnation of degenerating axon terminals in the central nervous system: (1) Technic (2) Chemical notes. *Stain Technology,* 1951, **26,** 5–11.

Powers, M. M., Clark, G., Darrow, M. A., & Emmel, V. M. Darrow red, a new basic dye. *Stain Technology,* 1960, **35,** 19–21.

Ráliš, H. M., Beesley, R. A., & Ráliš, Z. A. *Techniques in neurohistology.* London: Butterworth, 1973.

Scharenberg, K. Silver carbonate impregnation of the nervous system. In J. Winkler (Ed.), *Pathology of the nervous system.* New York: McGraw Hill, 1968. Pp. 168–174.

Steedman, H. F. *Section cutting in microscopy.* Oxford: Blackwell, 1960.

Weil, A. A rapid method for staining myelin sheaths. *Archives of Neurology and Psychiatry,* 1928, **20,** 392–393.

Chapter 3

Computer–Assisted Recording of Neuroanatomical Data

Thomas A. Woolsey and Michael L. Dierker

Department of Anatomy and Neurobiology
Washington University School of Medicine
St. Louis, Missouri

47

I. Introduction

The application of computers to biological problems has advanced enormously over the past two decades. Perhaps because of their background in the quantitative aspects of neurobiology and their need to be familiar with the electronic components necessary to record data from the nervous system, neurophysiologists were not reticent in applying computers to their research problems.

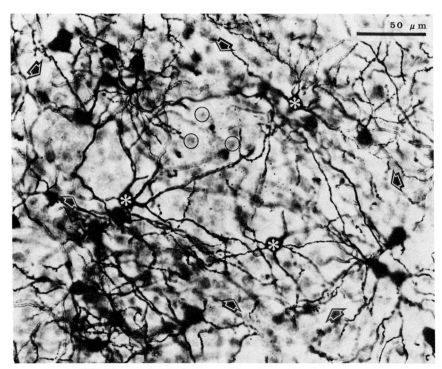

FIG. 1.   An illustration of the complexity of a neuroanatomical image. This is a 140-μm Golgi-Cox section, counterstained with thionine, through and in the plane of layer IV of the mouse SmI cortex. From the Nissl counterstain the boundaries of a barrel (see text, Section IV,A,1) can be recognized (note arrows). The somata of 3 Golgi impregnated neurons in the focal plane are indicated (*). Many other Golgi neurons are seen out of focus. Note the wealth of three-dimensional information available for any one Golgi impregnated neuron suggested by processes which pass out the focal plane. Approximately 98% of the neurons in this cortex have not taken the Golgi stain: three Nissl stained somata are indicated (○). The computer memory required to digitize this image assuming only 8 gray levels and a 512 × 512 matrix would be 786,432 bits. If additional focal planes are digitized at every 5 μm the required memory capacity would multiply 28-fold to 21 Mbits, a value which exceeds the core memory capabilities of all but the largest computers (672 Kwords; 32 bits/word).

Since the data already consisted of electrical events that could be readily fed to these computers, early emphasis in neurophysiology was to compute timing relationships among different action potentials (e.g., Rose, Gross, Geisler, & Hind, 1966). Although progress was initially slow, in part because new programs were developed with difficulty, the value of the small computer was quickly established. Now well-equipped laboratories of neurophysiology usually include a small computer. In contrast to the purely computational use for many problems, certain physiological problems lend themselves well to computer use for generation of stimuli. Thus, the computer has become important in both *generating* and *interpreting* neurophysiological data (e.g., Mountcastle, 1975). It has also been useful in testing mathematical models which describe physiological phenomena.

In contrast to neurophysiological problems, most neuroanatomical questions are inherently three-dimensional, nonelectrical, and not restricted to single events. The problem, then, was to convert anatomical data into a form that the computer could utilize. A first attempt was to fit a microscope with transducers, the voltages of which were subsequently digitized by the computer (Glaser & Van der Loos, 1965). The computational problems associated with the interpretation of three-dimensional structures are potentially very complex requiring the development of more sophisticated computers and programs for routine laboratory application. By 1970, relatively inexpensive yet powerful computers with more sophisticated programs had become available. This, coupled with technical progress in certain aspects of image processing concerning the digitization and subsequent analysis of three-dimensional information, made the application of computers to morphological problems feasible.

At present a number of groups including the authors' have experience in the use of computers to record anatomical data. This chapter will consider neuroanatomical data sources, observer interactions with the computer, and the appearance of results. In addition, certain aspects of computers in general, of systems design, and of component choices shall be discussed. As this is a rapidly changing field, the present perspective is likely to be outdated in the near future; nevertheless, the general principles should stand. The complexity of the neuroanatomical image is shown in Fig. 1.

II. Suitability of Computers to Recording Neuroanatomical Data

A judgment must be made upon the usefulness of applying a computer in every particular neuroanatomical data recording situation. This judgment can usually be made by providing answers to the following two questions. First, will the application of a computer increase the speed and/or accuracy of acquiring specific types of neuroanatomical data? Second, will the computational powers of the computer speed analysis, aid in data interpretation, or make experiments possible that are not otherwise feasible?

A. *Neuroanatomical Preparations*

Neurohistologists have two principal means to facilitate the analysis of a tissue: (1) to increase the contrast of particular cells or cell components by the introduction of staining and optical methods; and (2) to increase the resolution by mechanical or optical sectioning of the tissue. Most studies of the vertebrate nervous system utilize a combination of the two. On one extreme are the Golgi methods, in which relatively few neurons are examined in thick sections. On the other extreme are electron microscopic sections in which a number of standard stains reveal a wealth of high-resolution details about a fraction of a cell. The same factors which are of use in qualitative neurohistology are of value in computer-assisted neuroanatomy. These features assist either the operator or the computer in image recognition and processing. Most of the anatomical methods described elsewhere in this book can be sources of data for a computer.

Methods that produce high-contrast, uncomplicated images are best suited for more highly automated digitization. Selected examples are autoradiograms (high-contrast single-focal plane), electron microscopic (EM) stains for some synapses [phosphotungstic acid (PTA); high-contrast single-focal plane], and myelinated axons (high-contrast closed profiles). Methods that produce images which are easily recognized but may be complex or of low contrast are best dealt with by the human observer. Examples of this are synapses and synaptic vesicles (in EM images), cellular and nuclear outlines (stained with Nissl stain), and neurons of projection (histochemically labeled with horse-radish peroxidase (HRP). Thus, the properties of the particular technique used will dictate some of the decisions discussed below.

B. *Data Entry or Specimen Quantitation*

Neuroanatomical data are inherently three-dimensional and, in general, one deals with the quantization of three-dimensional structure in a specimen or a section of that structure. The objective of any data entry system is to somehow delineate this structure within the specimen, *quantitating* it for the computer. The three main approaches to this problem are (1) manual digitization and data entry, (2) semiautomatic or computer-assisted data entry, and (3) fully automatic data entry systems requiring no operator intervention.

1. MANUAL QUANTITATION

In this case, the operator assumes complete responsibility for delineating the structure of interest to the computer. The operator identifies and outlines the boundaries of structures (e.g., vesicular areas, nuclear and soma areas as repre-sented on electron micrographs and photomicrographs) with an instrument capa-ble of representing each point traced uniquely in a Cartesian coordinate system. Each structure traced is represented as a coordinate list within the computer.

After the data entry step, the computer can perform a variety of calculations designed to aid the operator in the task of interpretation. Since the delineation of the structure is left completely to the operator, a maximum amount of flexibility can be achieved with these systems.

2. SEMIAUTOMATIC QUANTITATION

These systems are characterized by the involvement of the computer in some of the structure quantization process. This requires a varying degree of operator intervention or monitoring. Of the various systems of this sort which have been developed, two classes may be distinguished, based on the degree of operator interaction.

The first type of semiautomatic system is really an extension of a manual system applied to more complex structures such as Golgi-stained neurons. In these systems the operator is still responsible for delineating structure and identifying specific important structural features, however, owing to the increased structural complexity, the computer is utilized to perform elementary bookkeeping functions, which allows the operator to concentrate primarily upon the task of identifying structural features. Since certain assumptions about the data to be entered are preprogrammed, these systems are less flexible and require more development than completely manual systems. Their obvious advantage is one of increasing the ease and speed of data entry.

A second type of semiautomatic system falls just short of being a truly automatic system. This class of systems is able to process the specimen image and extract the pertinent structural features mostly without operator intervention. However, concessions to the complexity of the image must be made, and the operator is usually required to monitor the computer's performance, aiding it where it is unable to completely delineate the desired structure. The advantage of these systems is that the data entry rate is often much higher than either the manual or the first type of semiautomatic systems mentioned above. The operator's job is particularly boring but the increase in the data entry rate is deemed sufficient compensation.

3. AUTOMATIC QUANTITATION

Systems of this nature should be able to process the specimen image for the desired features, extracting all of the relevant structural information entirely without human intervention. Many systems of this type have been proposed but difficulties caused by the inherent complexity of biological images have so far prevented implementation. This is not to say that completely automatic systems are impossible to implement, but a trade-off must usually be made between operating speed of system and the degree of automation. Complete automation requires considerable computation to resolve all possible ambiguities inherent in a complex biological image. However, a human operator, familiar with the material could make a decision in a matter of seconds.

4. SUMMARY

Increased automation requires increased *a priori* knowledge about the material under investigation and decreases the ease of applying the system to other materials. In general, if the researcher is familiar with the material and can conceive of a procedure to manually obtain the desired data, then a manual or some form of semiautomatic system probably could be implemented without great difficulty since the researcher defines the problem in terms that have meaning to the computer. Systems requiring great flexibility should be manual, and those restricted to a specific type of image input should be some form of semiautomatic system. Also, a person is often faster and more accurate in detecting the important aspects of a structure than is a computer utilizing the programming techniques and hardware technology available today.

C. Advantages of Computer Quantization

Obviously, some advantages must be accrued by the digitization of important structural aspects and subsequent computer storage. Advantages cited quite often are increased ease, speed, and flexibility of data gathering and interpretation. Consider the use of an automated tracing system which automatically calculates areas and perimeters (Cowan & Wann, 1973) versus the use of a planimeter (e.g., Kelly & Cowan, 1972) for the determination of nuclear areas as observed on a photomicrograph. A second example is the measurement and subsequent cataloging of true three-dimensional arc length of Golgi-stained neurons (Wann, Woolsey, Dierker, & Cowan, 1973) versus a manual computation and storing of the projected two-dimensional distances through the use of camera lucida techniques. In both of these cases the desired measurement upon the data is either made possible or made significantly faster and easier through the use of a small computer. Another benefit is increased reproducibility of measurements, owing to a minimization of errors attributable to operator fatigue (Brecher, Schneiderman, & Williams, 1956). This is an advantage in semiautomatic or automatic systems, where operator interaction is reduced to a minimum or eliminated completely.

Another important advantage is accrued from the increased speed and ease of data entry. Often it is very desirable to express a morphological change, in, say, a somal cross-sectional area, induced by some experiment in a quantitative manner (Loewy & Schader, 1977). To do so, a large sample may be required in order to express the results in a statistically rigorous fashion. Since time is limited, the increased data entry rate possible with the computer makes certain important experiments feasible.

The increased ease and speed of data entry also have implications beyond the ability to collect statistically significant data. When manual recording of data is laborious, there is a tendency to ignore so-called "abnormal" cells in favor of the "normal" cells (Tyner, 1975). These "abnormal"-cell types later may prove to be important exceptions to the rule and provide insight into the functioning of a specific system. Increasing the ease of the data-gathering process may encourage the experimenter to include these odd cells in the analysis. When enough data have been collected, these odd cells may sort themselves into separate and important groups. The second effect that increasing the ease and speed of data entry may have is that the researcher may be encouraged to include extra features along with his primary measurements (e.g., cell area and spine density included with a spatial and topological characterization of a neuron's processes). Recently, it has been illustrated that inclusion of such extra features can lead to a comprehensive classification of neurons in a local cortical area (Woolsey, Dierker, & Wann, 1975).

The most obvious advantage of the computer is its computational power. Further implications of this resource are discussed in Section IV, but one simple example will serve here for illustration. One basic problem in the sectioning of cortical material is that by physical necessity the image seen is that offered by one angle of view. By quantizing the topological and structural properties of a neuron and saving the resulting quantization in the computer, one is able to reconstruct a pseudo three-dimensional display of the quantized neuron on the screen of a cathode ray tube (CRT) which can be viewed from any spatial angle.

III. Methodology

When discussing the application of a new and changing technology to a multivariate field, it is not possible to provide a "cookbook" for the assembly of a functional system. First of all, specific requirements vary with respect to the specimen and the degree of automation sought in the final system. Second, as experience and commercially available equipment become more sophisticated, hardware configurations and operating software should become more standardized. Therefore, this section will primarily focus on (1) developing a conceptual framework for presenting the components and interactions in a neuroanatomical data collection system, and (2) illustrating these concepts for three specific systems.

A. Delineation of Components

We shall assume that the reader is relatively unfamiliar with computers and examine the operator's interaction with the system first in order to define four perceptual building blocks. Second, using these perceptual blocks, we shall briefly discuss the physical components which can go into the creation of each block. Finally, we shall consider requirements for the use of the computer in recording neuroanatomical data.

1. OPERATOR INTERACTION

A functioning system can be viewed by the operator as a black box comprised of subsystems which are perceptually delineated by the *interactions* possible with this black box. These interactions can be segregated into four modes: program interaction, data digitization, data analysis, and long-term data storage and retrieval. To illustrate the general organization of computer-assisted anatomical data recording systems and the perceptual blocks, a typical sequence of interactions leading to the digitization, analysis and subsequent storage will be outlined.

a. PROGRAM INTERACTION.   Generally, useful systems can be considered "turnkey systems". The system is designed so that the only computer expertise required of the operator is turning on the computer and starting the program; further interactions occur between the operator and the running computer program. Typically, use involves a preliminary interaction with the operating computer program through an interactive device (see Fig. 2, path 1). The scope of this interaction ranges from supplying name, date, and specimen identification to supplying information such as the image magnification and the kind of illumination. Thus, one perceptual block of the system is a device for these interactions.

b. DATA DIGITIZATION.   After the preliminary information has been supplied to the computer program, the operator is ready to quantize the specimen. Typically, quantization involves the use of some specialized equipment interfaced to the computer. The nature of this equipment is primarily dependent upon the physical properties of the specimen to be entered (e.g., contrast the different formats required for a tissue section, a photomicrograph, or an autoradiographic slide) and the degree of system automation. In Fig. 2, it can be seen that specimen quantization forms another perceptual block. In this case the primary operator interaction occurs with the special devices utilized for specimen data input. Path 3 of Fig. 2 illustrates one interaction, specifically a direct visual feedback path such as would exist in the manual tracing of cell boundaries with a pen. A second pathway (Fig. 2, path 4) involves responses by the operating computer program to various operator supplied inputs; thus the computer program is essential for the closing of the feedback loop. A transaction along this pathway might be moving a motorized microscope stage in response to an operator action, while the operator monitors the computer performance. In addition to the specialized input devices, a direct pathway to the operating program is also in operation (illustrated by path 2 in Fig. 2). This pathway allows the operator to

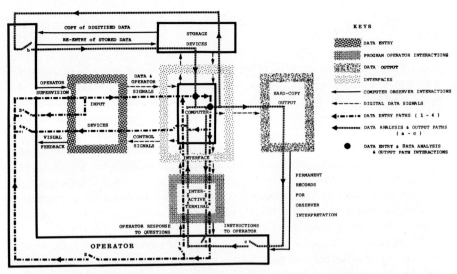

FIG. 2.   A generalized schema for illustrating the principal components and their interactions for computer-assisted neuroanatomy. All paths cross at the observer (whose size is proportional to importance in entering and interpreting data) and the computer. A variety of peripheral devices facilitate communication, between these two essential components of any system, through interfaces which surround the computer. To show the flow of information two kinds of paths have been illustrated: data entry and data analysis (see key at the right). While these paths freely intersect in the computer, to function properly each path must form a closed loop, which is done by the operator. These are indicated symbolically by switches. To enter data the operator must close switch 1 to start the program and depending on the nature of the data entry system he must interact by closing switches 3 (e.g., tablet system), 4 (e.g., Golgi system), or 2 (e.g., grain count system). Once the data are entered these may be analyzed: switch a; stored; switch b; and/or permanently recorded: switch c. For further description, see text.

interact with the digitization program during the process of specimen quantization. In this way, program operation or parameters may be changed.

c.   DATA ANALYSIS.   After specimen quantization is complete, the operator encounters a third subsystem which is primarily concerned with data analysis. This block corresponds to the pathway linked by a and c in Fig. 2. Once again the terminal is utilized for communication with the operating program, but various other devices are utilized in producing the results of the data analysis, using the computational power of the computer. Results may be printed, with a statistical breakdown of the measurements; they may be displayed as in Fig. 7c, d, and e to graphically illustrate a result; or they may be plotted as in Fig. 6a, b, e, and f. These devices may be incorporated into the previous blocks (e.g., the CRT and printer may also serve as an interactive terminal). It is useful, however, to group these devices in each perceptual block to visualize the functions of a particular device.

d. DATA STORAGE AND RETRIEVAL. This last block consists of devices which permit the saving and subsequent retrieval of the data. The advantages of this capability are twofold. First, the user is allowed to save quantized data for later analysis, without reentry. The second advantage has more impact upon future development. By accumulating data from many specimens, a data base is formed and further manipulations become possible, such as cluster analysis or measurements of uniformity among specimens. Path b illustrates this block in Fig. 2. Intersecting this operation is pathway a in Fig. 2, which allows specific operator commands to the computer program. These commands allow examination of previously collected specimen identifiers or saving and retrieval of previously quantized specimens or analyses.

To summarize, from the operator's viewpoint an anatomical data recording system can be dissected into four distinct blocks, composed of a set of physical devices not necessarily disjoint. These blocks can be identified with specific devices for: (1) communicating with the computer program, (2) quantizing the anatomical data, (3) furnishing some type of permanent output for the purpose of analysis, and (4) providing some medium to save and subsequently retrieve the data for later study or comparison with data from other specimens.

A final point is that all of these blocks rely upon the existence of some *controlling computer program* to make them operational. This component of the system is often the least noticeable to the operator (except in those cases where it malfunctions). Some idea of the complexity embodied in this component can be gleaned from Fig. 2 by noting that it is internal to the computer where all paths intersect.

2. DEVICE REQUIREMENTS

The hardware[1] necessary to implement the perceptual building blocks is considered in this section. Discussion will be limited to a listing of possible options and their advantages and disadvantages, and what equipment may be purchased directly from the manufacturer and what may need to be constructed.

a. INTERFACES. There is one aspect of physical device selection not immediately apparent to the casual user. In Fig. 2 it should be noted that communication between the computer and the various perceptual building blocks is mediated by devices referred to as *interfaces*. An interface between the computer

---

[1]The term *hardware* refers to physical devices which are *hard* wired such as a computer or CRT. *Software* refers to programs which can be written and changed at will without necessarily altering the conformation of any machine.

and other devices is necessary because of differing signal types (e.g., digital versus analog), differences in control signals, different voltage levels, or because some data need to be preprocessed before entering the computer. The complexity and cost of these interfaces varies considerably, but they usually account for a significant portion of the total system cost. An interface is frequently coupled to a specific type or family of computers and cannot usually be plugged easily into different computers. For many pieces of equipment the computer manufacturer or the device manufacturer will sell an appropriate interface to interconnect a specific computer and device, but interfaces are commercially available only for a limited selection of computers. For this reason it is wise to select one of the more popular computers, since this increases the likelihood that interfaces will be commercially available, reducing both the time and the amount of technical support needed to produce a working system.

A particular problem arises concerning the special input devices required for neuroanatomical data entry. In some cases the equipment has been standardized and can be purchased with an interface to many computers (e.g., The Scientific Accessories Graphic Sonic Digitizing Tablet), but in many other cases a specialized interface must be constructed. The later is the case for most video imaging systems, although some commercial interfaces do exist. However, prices are often too high to be considered feasible for most small laboratories. Accordingly, electronics support is necessary. For more details of interface description and design the following references should be of some assistance: Digital Equipment Corporation (1972, Ch. 5; 1973, Ch. 4; 1976a, Chs. 7 & 8; 1976b, Ch. 5), Kehl (1976), and Souček and Carlson (1976, Ch. 3).

b. INTERACTIVE TERMINAL AND HARD-COPY OUTPUT. The interactive terminal serves as a vital communication link between the operator and the operating computer program. Therefore, its selection should be made with some thought given to the nature of the communications between the operator and computer program. Program interaction is typically a series of queries and responses or specific operator commands. For this an input device (typically a standard typewriterlike keyboard) and a display device (a typewriterlike printer or CRT) are joined together as a single unit, the *interactive terminal*.

The signals to and from the terminal are standard and consequently may be purchased with proper interface for any computer; indeed, terminals for display and printing are often sold with the computer. Display systems are particularly useful, providing a high-speed interactive device which has no paper consumption. Often these systems are available with simple graphics plotting capability (e.g., Digital Equipment Corporation VT55 Graphics Terminal and the Tektronics Model 4010 Computer Display Terminal) and associated devices to produce hard-copy output directly from the display screen (Tektronics Model 4631 Hard Copy Unit and the Versatec Matrix Printer/Plotter with the Versatec C-Tex

Controller). These items are relatively cheap, easy to use, and provide an excellent means of rapid interactive communication between the operating program and the user.

c.   DATA STORAGE.   Digital storage devices are necessary for both program and data storage. Specification of these devices requires some understanding of their characteristics, defined as follows: (1) access time (i.e., the time needed to locate and read or write an item of data), (2) storage capacity, (3) portability (i.e., the ease with which the storage medium may be removed and utilized on the same or other systems), and (4) expense (i.e., cost per unit of information stored). Device requirements stem from (1) the limitations in the total system cost, (2) the necessity of providing a fast access device for program storage and possible data storage during analysis, (3) allowing individual users to save their quantized data on a portable storage media that is relatively inexpensive, and (4) supplying a compatible medium for the transportation of data from one computer to another. To satisfy the above requirements, two storage subsystems are usually chosen. A magnetic disk storage system is usually obtained from the computer manufacturer for program and temporary data storage. This type of storage system has the following characteristics: (1) short storage access times, (2) medium cost per unit of information to be stored, (3) low-to-medium storage capacity, and (4) low practical portability. If it is necessary to provide some type of large capacity, easily portable storage, then some form of tape system is chosen to complement the disk storage system. Typically, tapes: (1) are very portable, (2) are hardy (i.e., only ordinary care needs to be given the media to prevent damage), (3) have long access times, (4) offer low cost per unit of information to be stored, and (5) possess large storage capacities.

d.   INPUT DEVICES.   There are essentially four types of equipment which have been found to be useful in the quantization of anatomical data. They are categorized as follows: (1) manual tracing graphic digitization equipment, (2) equipment for mechanizing the movement of conventional microscope stages, (3) video equipment for visualizing transmitted and reflected light images and for signals originating from the electron microscope, and (4) specialized operator interface hardware. The last category is perhaps the least discussed in current literature, although proper operator interface greatly improves the effectiveness of semiautomatic systems. The equipment providing an operator interface usually needs to be specially constructed and requires competent technical assistance.

The degree of automation in a system is often the primary determinant of the types of feature quantization equipment which will be needed. For example, a manual area tracing system would need only a graphic digitizer in addition to the equipment mentioned in Sections III,A,2,a and b above. A system designed to trace these areas automatically, however, would require a video image system interfaced to the computer. The complexity of the task also influences the choice of equipment. For example, an on-line light microscope serial reconstruction system for the quantization of individually stained neurons requires more feature input equipment than a single section system.

This discussion of input devices has been necessarily brief. For the newcomer to the field the best advice that he can receive is to review the key articles describing systems currently in use and consult with the originators of these systems, who should have some knowledge on what new devices have become available from specific manufacturers and should also be able to provide helpful hints on what equipment will most likely be needed.

3. THE COMPUTER

A key piece of equipment is the computer and its associated software and hardware. From an examination of Fig. 2 it is seen that all paths cross within the computer. It is the controlling program which gives form to a specific anatomical data recording system and function to each specific peripheral device. A discussion of *all* the factors considered in the selection of a computer will not be presented but a summary of two important aspects of all computer systems—the hardware and the software—will be presented.

a. BACKGROUND. Our experience is largely with what are popularly known as mini-computers. Systems of this type should be differentiated from the large computers, such as the IBM 360 or 370 series or the CDC 6400 or 6600 series of computers, which are often encountered at university computing facilities. Mini-computer systems are much cheaper but their flexibility and power, both in hardware and software, are usually less than their large computer counterparts. Cost differences are typically greater than two orders of magnitude, while differences in available computing power are about tenfold.

A primary hardware difference between the mini- and so-called maxi-computers is a reduction in the intrinsic word length (i.e., the number of binary bits[2] which the machine is able to process simultaneously in arithmetic operations), typically a reduction from 32 bits to 16 or 12 bits in length. This reduction, coupled with a reduced instruction repertoire and increased instruction execution time, significantly reduces the size and cost of the mini-computers. The reduced word size also limits the amount of storage—both the main mem-

---

[2]A *bit* is the basic datum with which a computer deals and has two possible values, 0 or 1. This may be likened to the two possible states of a light switch; off = 0 or on = 1.

ory[3] and the mass storage memory[4]—which can be directly addressed by the computer. For a 32-bit-word versus a 16-bit-word machine this amounts to a difference between addressing 4096 million words and 64 thousand words directly. Although it is easy to show that large computers are capable of solving many problems significantly faster than the small mini-computers, it is just as easy to show that this power is not needed for the overwhelming majority of anatomical data collection and analysis problems. Indeed the cost difference can rarely be justified. Therefore, the following discussion will be limited to mini-computers.

b. HARDWARE. The hardware aspect of the computer refers to the physical computer, and includes such divergent specifications as a definition of the instruction set,[5] instruction execution speed,[6] word length, maximum memory size, and a definition of interface requirements (i.e., a digital description of the signals which provide communication with the devices described earlier). There are two aspects of computer hardware which we believe to be of importance in the selection of a computer in today's market place.

Since every physical device requires an interface for each specific computer, it is advantageous in terms of development and maintenance time to have available directly from a manufacturer the peripheral device and interface for the researcher's specific computer. This availability is largely determined by the popularity of a particular computer. Examples of computers for which interfaced equipment is readily available are (1) Digital Equipment Corporation's PDP-8 and PDP-11 line, (2) Data General's Nova line, (3) Hewlett-Packard's 2100 line, and (4) Interdata's line of 16-bit computers.

The second aspect of hardware which is of interest in choosing a computer is the breadth of the computer family to which the specific computer belongs. By breadth we are referring to the range of performance for various models of the same computer family. Typically these different models share the same basic machine language instruction set; the differences for the higher performance

[3]The *main memory* of the computer refers to that portion of storage which is directly addressable in a random manner (i.e., the time required to fetch or store a word is not a function of location). Access times are typically very short, less than 1 μsec.

[4]The *mass storage memory* of the computer refers to that portion of storage which is directly addressable at the block or record level (a block or record is a contiguous word grouping of fixed or variable length). Mass storage memory may be organized in the random access format described above, but most often is not. Access times are typically in the tens of milliseconds.

[5]The *instruction set* of a computer refers to the complete repertoire of basic commands, stored in main memory, which are executable by a particular computer. For example, the computer would be instructed to add two binary numbers stored at specified locations in main memory and store the result at a designated address in main memory.

[6]The *instruction execution speed* of a computer refers to the time necessary for the completion of a basic computer instruction as defined above. Typical speeds fall in the range 0.3–4 μsec.

models are shorter instruction execution time, higher price, and some extended machine language instructions. The members of a computer family typically utilize the same programs, hence it is simple to transfer programs written upon one machine to another. The great advantage gained by purchasing computers from a broad family line, is that the user need initially only pay for the amount of performance that is needed, yet if there is a future requirement for a computer with higher performance, there is no need to change the software. One example of a computer family with this breadth is the PDP-11 family from Digital Equipment Corporation, which has members ranging in performance from a microcomputer—the LSI-11—with an instruction execution time of about 3.7 μsec to the high-performance computer—the PDP-11/70—with an instruction execution time of about 300 nsec (i.e., about ten times faster).

c. SOFTWARE. The software supplied with a computer system is the second major aspect of the total system. The computer software, comprised of all of the programs[7] written for the specific computer, allows the computer hardware resources to be efficiently and easily utilized by the programmer. The total software package supplied with today's computer systems typically consists of two distinct packages, the operating system and the programming languages.

The operating system consists of a set of programs, written in the physical language of the computer, which handle the various management duties necessary to utilize the computer's features. These functions consist of a filing system for the storage and retrieval of various programs utilized by the operator; a means for the creation of new and modification of existing programs; a mechanism for the execution of other programs developed by the operator; and finally, a number of other items useful in program preparation and documentation, as well as for the support of the various peripherals attached to the computer. Various accessories are often provided on today's small computer systems, which are standard on large computer systems. One is the capability to run in a batch mode allowing the operator to run the computer by giving a predetermind string of commands, which the computer stores and then executes sequentially. Each of these commands may be very complex and may require some period of time for execution. Thus, running in the batch mode allows the operator to specify the computer operation for long periods of time without further operator intervention. This is the mode of operation typically used on most large university computers.

A second available option of mini-computer operating systems is more sophisticated and potentially more useful for neuroanatomical data recording. This is known as "time sharing" and allows several investigators to use the computer

[7]A *program* is the actual list of instructions in a particular computer language for accomplishing a task. An *algorithm* is a logic outline or block diagram describing the sequence of steps that are followed to do this task.

"simultaneously" by permitting several programs to run together. Most neuroanatomical data collection procedures have relatively low data entry rates when compared to the computer speed and thus make relatively inefficient use of the computer's powers. Therefore, *if* sufficient use can be generated for the system, a time-sharing or realtime operating system allowing concurrent use of two subsystems may be economically justifiable.

The programming languages supplied with the computer belong to three distinct categories: machine, interpreter, and compiler languages. A machine programming language allows the user to develop programs in the physical machine language of the specific computer at hand, optimizing both execution time and reducing required main memory size over programs developed in one of the other two classes of languages. Unfortunately, programs developed by this method are not easily transportable to another computer owing to differing machine languages. Also program development and documentation time is often increased over that required in one of the other two languages. Interpreters and compilers are actually just different methods of implementing various "high-level" languages. The term, high-level language, denotes in this context a programming language, which has the following properties: (1) problems may be easily stated in the language and as a consequence it is easily readable, (2) programs written in such a language are transferrable from machine to machine with a minimum of additional programming effort, and (3) a statement of the high-level language typically generates several physical machine language instructions. Examples of some of the typical high-level languages implemented on today's mini-computers are BASIC, an interpreter language, and FORTRAN IV, ALGOL, or COBOL, all compiler languages.

The primary differences in the operation of interpreters and compilers can be expressed very simply by the concept of binding. Binding refers to the association of the physical-machine-language representation with each statement in the high-level language. In a compiler, the binding is done early, thus the resultant program consists of instructions which can be directly executed by the computer. For efficient compilers, the execution speed of the resultant code (i.e., program statements) can be comparable to that of the same task programmed directly in machine language. The problem with compilers occurs in program development, since it is often difficult to relate program errors at the machine language level directly back to specific high-level-language statements. Nevertheless, compilers remain a popular method for the implementation of higher level languages, due in large part to the high execution speed of the resultant program.

An interpreter is distinguished by binding at the last possible moment. This means that the high-level-language statements are translated as the program is executed. Since the statements must be reinterpreted for each execution, the majority of the computer time is spent in the process of interpretation. Thus, it comes as no surprise that interpretative implementations of high-level languages tend to execute programs more slowly than a similar compiler implementation. The tremendous advantage of interpreters is in program development. Since the high-level language is always present in "raw" form, program errors may be easily traced to specific high-level-language statements, and the program may be easily and quickly modified in an interactive manner and executed without the necessity of recompilation. A suitable blend would be a dual implementation of the same high-level language by both interpretative and compilative techniques. This is occurring for some large computers and to some extent on smaller computers, and in the near future dual implementations may be commonplace.

The efficacy of complete system is determined by the gestalt of the computer hardware and the computer software. The principal problem encountered in the use of high-level languages, is some sacrifice in machine execution speed over comparable machine language programs. Thus, it is important to determine if programs will run fast enough for on-line, real-time utilization.

B. Three Specific Systems

To put the foregoing into a concrete context and to explain the data sources for the subsequent discussion an existing computer system which has been in use at the Department of Anatomy and Neurobiology at Washington University for several years shall be considered. A block diagram of this system is shown in Fig. 3. The various components have been grouped according to the general scheme shown in Fig. 2 but more specific details are given. Fuller descriptions of each of the three major systems are provided elsewhere (Cowan & Wann, 1973; Wann, Price, Cowan, & Agulnek, 1974; Wann et al., 1973). All systems use the same central processing unit (CPU) (DEC PDP-12) and many share some hardware and software as indicated in the upper left of the figure. The systems that will be discussed in turn are (1) the Tablet System, (2) the Golgi System, and (3) the Silver Grain Count System.

1. The Tablet System

The input to this system comes from drawings, photomicrographs, photographic negatives, and projected images, all of which are placed on a Science Accessories Corporation acoustical tablet. The tablet, which comes as a preassembled hardware package, consists of two strip microphones at right angles (X and Y axes), a spark-pen, and a digitizer which is interfaced to the computer.

When the spark-pen is pressed to the tablet surface, a switch is closed that is sensed by the computer which initiates the generation of sparks across a gap at the pen tip. The resultant high-frequency sound is picked up by the two strip microphones, and the conduction delay (between the emission and receipt of the sound) is measured. This is converted to an X–Y point in Cartesian coordinates. As the pen is moved, new coordinate points are recorded and stored until the

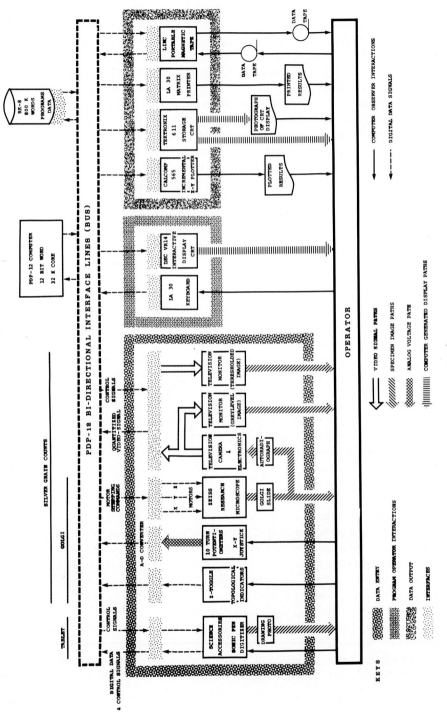

Fig. 3.

operator or the computer signals the end of the sample. To use this system the operator places the objects to be traced on the tablet and proceeds to trace them with the spark-pen. In this system, feedback to the operator is via a storage CRT on which all digitized points are shown (Fig. 3, right).

Cell bodies or axons as shown in Figs. 4 and 5 are closed profiles and an algorithm has been written to detect when the spark-pen has returned to the initial position. When this happens the computer stops the pen from sparking (via the control signal pathway in Fig. 3), and computes the requested parameters. These computations are printed out (on the right of Fig. 3) for the observer to examine before the next observation is entered. If the observer wishes to reject an erroneous observation a button is pushed which erases the observation from computer memory, or if the observation is "correct" it is stored and the observer can go on to the next observation and repeat the sequence. When a number of objects have been recorded (in our experience this can approach 200/hr) the observer can push another button to terminate data collection. In our system this calls for a program which performs elementary statistics, mean, standard deviation, range, etc., on the sample population. Next another program is entered which formats the results in bins, the size of which is entirely under observer control. These results are printed out as histograms (e.g., Figs. 4 and 5). These data may be stored on magnetic tape to which additional data from the same population can be added at a later time or from which statistical comparisons can be made with other preparations (see right of Fig. 3).

2. THE GOLGI SYSTEM

In this system the data are collected from a microscope slide which the observer views through a research microscope. The principal observer tasks are two: to center and focus portions of the Golgi-impregnated cell under cross hairs in the microscope occular and to indicate the topological identifiers in a prescribed sequence, i.e., from cell body out. The data consist of three-dimensional coordinates, available to the computer from stepping motors mounted upon the microscope. These motors are stepped in response to user commands via computer initiated pulses. It is then a simple matter to keep track of how many pulses (0.5 μm/pulse) were taken from left to right, up and down, etc., and thus store the relative three-dimensional position of one of a series of topological identifiers such as a branch, an end, or a sample point. As long as certain sequential rules

---

FIG. 3. A simplified schematic diagram of our computer system at Washington University. The components are organized in the same manner as indicated in Fig. 2 (see key below). The components of these principal systems described in Section III,B are indicated in the upper left. Note the hardware overlap for the Golgi and the grain count systems. The importance of the Operator is again emphasized, and the patterns of his interaction with the various components are schematically indicated in Fig. 2. The PDP-12 computer, RK8 Disc, and LA30 Teletype are Digital Equipment Corporation (DEC) products.

are followed, the three-dimensional topological structure of the cell may be easily entered and verified by the computer for errors during entry (Woolsey, Wann, Cowan, & Dierker, 1972).

To use the system the operator loads the program and uses the z toggle and x–y joy stick (Fig. 3, left) to center the soma under the cross hairs, signaling this to the computer by a pushbutton. This sets the x, y and z coordinates to zero, and centers the coordinate system at the cell body. Processes emanating from the cell are then sequentially centered under the cross hairs; their position is indicated by another pushbutton, and these points are stored in the computer for future reference. Then the observer tracks down a process, saving enough points to faithfully indicate the process curvature. When a branch is reached, it is indicated by a branch button, and the observer proceeds down one limb of the branch. When the end of the process is reached, it is indicated by stepping on a foot peddle. The program has been written to perform a series of simple topological checks (i.e., a terminal branch point should have two limbs with ends), and if the branch point has an unsaturated limb the stage is moved so that the branch point is recentered under the cross hairs and the observer can follow the unrecorded limb. The user systematically records data for the entire cell in this way, until all of the processes have been entered. During the data entry the observer has the opportunity to correct errors at will via special controls and keyboard commands.

The data for an individual cell thus consist of a series of coordinate triplets and topological identifiers. Such a sequence of coordinate triplets can be connected by vectors to "draw" the neuron on the face of a CRT or X–Y plotter (Fig. 3, right). The computational power of the computer permits easy and rapid coordinate transformations so that projected views of a neuron can be generated for any spatial angle (Dierker, 1976b) (see Fig. 6). These displays can be called at any time during the logging sequence for data verification. Other forms of feedback consist of a "beeper," which sounds every time a new point is entered, and statements that are printed on the teletype to signal observer errors or to request verification of the observations. In practice, the time required to "draw" a cell in this way is about one-half to one-third that required to draw it by hand with a camera lucida.

When the observations on a particular cell have been completed the user has several options. He can go on to enter data from additional cells or examine the data. The true three-dimensional lengths of the processes can be computed with hard copy presented by teletype or by scaled length diagrams on the x–y plotter (Figs. 6c and g). Finally, these data can be stored on magnetic tapes for future examination (Fig. 3, right).

3. The Grain Count System

The data are taken directly from autoradiographic slides placed on the research microscope. The optical image is converted to an electrical image by a television camera attached to the microscope. This "gray"-level image is converted to a black-white image (data reduction) by a thresholding procedure that is implemented in the computer interface. The silver grains appear as black objects against a white background in bright-field illumination and *vice versa* under dark-field illumination. In principal, the computer simply scans the thresholded image counting the black or white objects, depending upon the illumination.

To use the system the observer places the specimen on the microscope and loads the program. At this point the observer usually specifies the illumination (dark- or bright-field) and the size of the microscope fields to be examined. Next the observer uses the X–Y joystick to move relevant boundaries in the specimen (e.g., blood vessels or the barrel outlines in Fig. 7) under the occular cross hairs, just as in the Golgi System described above. The boundary points are indicated by pressing a button and their coordinates are stored in the computer. These anatomical boundaries are then displayed as a series of points on the CRT, and by controlling a number of potentiometers the observer can superimpose a rectangle of any dimension, defining the region(s) to be counted, over the anatomical landmarks. The computer stores the coordinates of this rectangle and then systematically scans the bounded area in a checkerboard of optical fields, the size of which is determined largely by the optics of the microscope (i.e. 25× vs 100× objectives). As each optical field is counted, the results are printed out in the appropriate spatial organization, and these values can be precisely correlated with relevant features in the specimen.

The principal function of the observer in this data collection is to quickly judge the television image of the microscopic field for correct focus and artifacts such as dirt or blood vessels. The observer can press a button to bypass the artifacts and can interrupt the program to correct the focus using the z toggle switch.

When an area has been counted, the results are printed. We have recently added programs in which the data can be placed in bins, the size of which are entirely under operator control. Thus, the results can be represented as histograms (counts/per field vs position of the field in the rectangle) made on the X–Y plotter or on the CRT as dots sized in proportion to the various bin values (see Fig. 7). Again the data can be stored on magnetic tapes for scaling (to compare different experiments) or subsequent manipulation (to subtract background counts or examine details of the projection—Fig. 7).

4. GENERAL COMMENTS

It is clear that the Tablet System, the Golgi System, and the Grain Count System represent systems that are progressively more automated. The input to the Tablet System is almost entirely manual while the Grain Counting System requires only occasional operator intervention during data collection. It is also evident that given suitable material the Tablet System can be used for an enormous variety of applications which are not necessarily restricted to anatomy.

Indeed the system has been used to integrate areas under dilution curves and from optical scans of polyacrylamide gels. Of the subsystems described, the Tablet System is the least expensive in terms of hardware and technical backup. By contrast the present Silver Grain Counting System is most restricted in its application and most expensive in terms of hardware and technical backup.

A second general point that can be made is that the layout of our system as shown in Fig. 3 is by no means unique. Other hardware in other configurations could be used to accomplish the same tasks in many cases more efficiently and less expensively. However, once one system has been implemented it is often possible to utilize components of both the hardware and software directly in a new system designed for a different task. In Fig. 3 it is obvious that many of the components originally developed for the Golgi system are common to the Grain Count System as well. Historically, the Golgi System components and software formed an essential core of the Silver Grain Counting System and their existence saved time and effort in the development of this system. Now, certain aspects of the Grain Counting System are being integrated into the Golgi System to allow the reconstruction of neurons from serial sections.

Finally, as outlined in Fig. 3, the details of our system have been oversimplified. There are a large number of peripheral control knobs and switches which the observer uses in the collection and manipulation of the data. These have functions which may differ depending upon the particular task being performed. In our "turnkey" systems, explanations of various operations are readily available. For example, the user need know only the correct keyboard command to cause an explanation of the controls for a particular segment of the program to be displayed on the CRT. Also explanation "frames" allow the user to freely go from one program to another without the cumbersome execution of irrelevant intermediate steps which is typical of a number of commercially available systems.

IV. Appearance of the Data, Analysis, and Interpretation

To give a feel for the appearance of the data and their use we shall take four examples from our own work which have been chosen to illustrate the application of computers to some of the neuroanatomical methods discussed in subsequent chapters.

A. *Examples*

1. CYTOARCHITECTURE AND CELL SIZES

The various criteria and bases for the study of cytoarchitectonics are known (e.g., Lorente de Nó, 1938; Rose & Woolsey, 1958). We applied the computer to an unusual cytoarchitectonic arrangement of cells seen in the somatosensory cortex of the mouse (Pasternak & Woolsey, 1975). As seen in sections that are taken parallel to the overlying pia, the neurons of layer IV are arranged in discrete cytoarchitecture units called barrels (Woolsey & Van der Loos, 1970) (Fig. 4a). The question to be resolved was "Are the barrels recognizable because of (a) a spatial segregation of neurons on the basis of neuronal size; (b) a regional variation in packing density; or (c) some combination of the above factors"?

The individual neurons were drawn with a camera lucida and the positions of the cell somata noted in the arbitrary scheme shown in Fig. 4b. The data were entered into the computer utilizing the Tablet System described in the previous section. Figure 4d shows the results taken from computer-printed histograms in which the cross-sectional areas of cells lying in the periphery or side (Fig. 4b) of the barrel were compared with those in the center or hollow (Fig. 4b). When the histograms were scaled they were identical as supported by the statistics. With the same Tablet System the areas of the rings shown in Fig. 4b were measured and as the thickness of the sections from optical measurements and the *number* of cells in each ring from the statistical computations was known, it was possible to compute a regional cellular density. Thus, with the aid of many measurements taken rapidly with the computer it was possible to show that the cytoarchitectonic appearance of layer IV was simply a neuronal packing phenomenon (Fig. 4c).

2. DIAMETERS OF PERIPHERAL NERVE FIBERS

Another class of applications of the Tablet System is measurement of profiles which may be recognized in the electron microscope. Although subcellular organelles, synapses, and synaptic vesicle profiles have been measured with this system (Johnson, Ross, Meyers, & Bunge, 1978), our own application has been made to certain peripheral nerve fibers (Lee & Woolsey, 1975a, 1975b). In this case we were interested in the fiber spectrum of myelinated fibers. As in the cell-size measurements, fiber cross-sectional areas could be obtained from which the equivalent diameter can be computed. From a glance at a preparation of a peripheral nerve it is obvious that distortion is introduced in the tissue processing. A better parameter to extract is the perimeter or circumference of the fiber from which an equivalent diameter may be determined by dividing by π.

A comparison of these two measures is shown in Fig. 5. As a result of the perimeter measurement the distribution of the fiber diameters becomes more sharply focused in at least three peaks. Here the computer has been utilized to perform a relatively straightforward correction for a processing artifact, which is tedious and difficult by hand. This example illustrates one of the chief advantages in this application of the small computer; the data are stored and can be manipulated through appropriate programs to examine various possibilities, in this case tissue distortion, to produce a result which is closer to the situation in the living state.

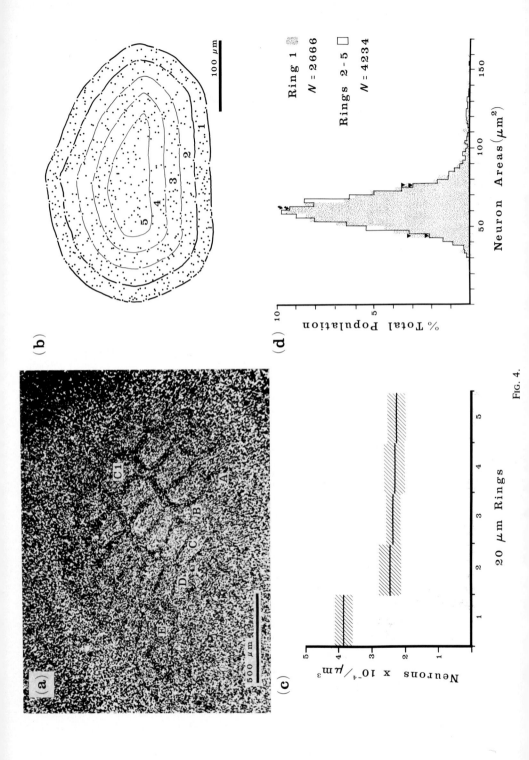

Fig. 4.

3. APPLICATION TO NEURONS SELECTIVELY STAINED IN THEIR ENTIRETY

Anyone familiar with neuroanatomy will recognize that one of the most distinctive features of neurons as a class of cells is extensive processes which typically are highly branched and distributed in three dimensions (Fig. 1). The application of computers has made practical the collection of large quantities of three-dimensional information on the morphology of neurons which are individually stained (for example, by one of the variants of the Golgi method, Chapter 4 by the Scheibels).

In the present case we examined the three-dimensional structure of cortical neurons comprising the barrels of the mouse (Woolsey *et al.*, 1975). The problems were to correlate views of these cells obtained in different planes of section and to use the data as an aid to classification of the members of the neuronal population. To correlate the views of neurons seen in sections taken tangential to the pia with those seen in the more customary plane normal to the pia, the cell projections were rotated 90° as shown for 2 cells in Figs. 6a,b and 6e,f. The three-dimensional length data, computed by the Pythagorean theorem can be conveniently displayed by the computer as scaled length diagrams (Figs. 6c and g). From these it is clear that the two neurons shown in Figs. 6a and e have different dendritic branching patterns which can be used to distinguish them from one another and help to allocate them to one class of cells or another. These results from a larger sample are shown in the histograms of Figs. 6d and h. These histograms emphasize the distinctions between the two classes of neurons (we have called them Classes I and II) and help give an indication of the variability in these parameters. In fact, when these two parameters, distance of branch points from soma and distances of ends from soma, as measured along the processes, are correlated with another parameter, somal cross-sectional area, it is possible to successfully classify over 90% of the cells on the basis of numerical data alone. The data have been stored on tapes. As more is learned about these cells it is

---

FIG. 4. To show the application of the computer system to a cytoarchitectonic problem. (a) This is a Nissl-stained section 75-μm thick taken parallel to the pia to show the cytoarchitectonic organization of layer IV of mouse SmI cerebral cortex. The rings of cells outline structures called barrels which can be recognized consistently from one brain to the next. One of these, C1, is identified in the left hemisphere. (b) To study the distribution of neurons the position of each cell was marked (·) and barrel C1 was arbitrarily divided into a series of 20-μm rings (1–5) labeled radially inward. (c) Using the tablet system the areas of each of the rings were measured, and knowing the thickness of the sections a volume could be computed. From the cell counts, the density of neurons for each ring was computed. Clearly, there are more neurons/unit volume in ring 1 than the other rings. The hatching indicates the standard deviations about the means (bars). (d) The cytoarchitectonic appearance of the barrels is the result of a neuronal packing phenomenon, since there is no difference in cell somal size, as measured with the tablet system, between the ring 1 and rings 2–5. The means (➡) and standard deviations (▼) for the two populations are identical. Adapted from Pasternak and Woolsey (1975).

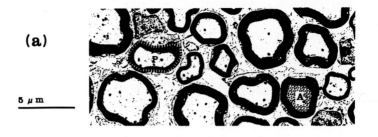

(a)

5 μm

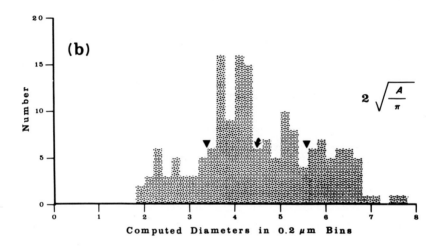

(b)

$$2\sqrt{\frac{A}{\pi}}$$

Number

Computed Diameters in 0.2 μm Bins

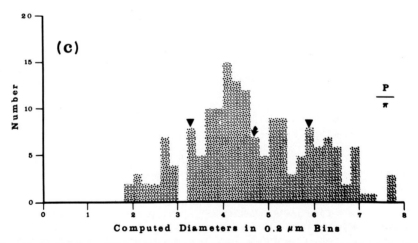

(c)

$$\frac{P}{\pi}$$

Number

Computed Diameters in 0.2 μm Bins

FIG. 5. To show the application of the computer to the measurement of myelinated fibers inervating vibrissa C1 of the mouse face. (a) A low-power electron micrograph of a portion of a vibrissal nerve. This nerve contained 191 myelinated fibers and 67 unmyelinated fibers. Notice that many of the profiles are compressed and not circular, possibly a reflection of shrinkage and handling in process-

possible to develop appropriate algorithms to search for, compute, collate, and display the parameters of interest. A number of statistical classification techniques can also be applied to serve as useful data compression techniques (i.e., classification by cell morphology) and which may have predictive value in that they suggest hypotheses about the biology of the neurons so studied.

4. QUANTITATIVE AUTORADIOGRAPHY IN TRACING OF AXONAL PATHWAYS

The final illustration is taken from the Grain Counting System. Here the problem is quantifying the variation in grain densities as a function of spatial distribution. Counting silver grains has revealed regional differences within particular specimens which in one case suggested a developmental hypothesis (Gottlieb & Cowan, 1972) and in many others has shown graphically the distribution of silver grains with respect to certain anatomical landmarks (e.g., Cowan, Gottlieb, Hendrickson, Price, & Woolsey, 1972).

The present example is again taken from the somatosensory cortex of the mouse (Woolsey, 1978). In this case the somatosensory thalamus [ventrobasal complex (V.B.)] was injected with a mixture of tritiated proline and leucine. After a 48-hr survival the brain was sectioned tangentially to the pia and the brain was processed for autoradiography (see Chapter 9 by Hendrickson). The questions "How uniform is the projection to the center or hollow of the barrel?" and "Is there a projection to the spaces—septa—between the barrels?" were asked. From grain counts, the projection was not found to be uniform, and there was no evidence of projections to the septum (Fig. 7b). By comparison to background levels it is clear that the VB complex projects to the barrels via sharply delineated terminal fields. Furthermore, the data can be compressed and more clearly illustrated in the resultant dot displays. By altering the threshold of these displays and the bin sizes it is possible to eliminate background counts (also determined by the device) and to appreciate certain details within the projection (Figs. 7c–e). These data can be used to compare projections from a single source to several brain target areas and as a result one can begin to speak of quantitative

---

ing. From the data the area (A, stipple) or the perimeter (P, dashed line) can be computed. (b) Histogram of fiber diameters computed from the areal measurements. The mean (➤) and standard deviation (▼) are indicated. The distribution taken directly from the computer printouts (see Section III,B,1) is roughly trimodal. A similar distribution would be obtained by devices in which a fiber perimeter is approximated by a circle. (c) Histogram of fiber diameters computed from the perimeters (i.e., circumference). The means and standard deviations are slightly shifted with respect to (b) by about 1 bin, however, the distribution appears more clearly trimodal. Presumably this calculation will diminish some of the artifacts owing to handling and processing the tissues which are evident in (a). A comparison of (b) and (c) shows the power of the computer in that the same data can be treated in several ways to approximate the natural condition (from Lee & Woolsey, 1975b).

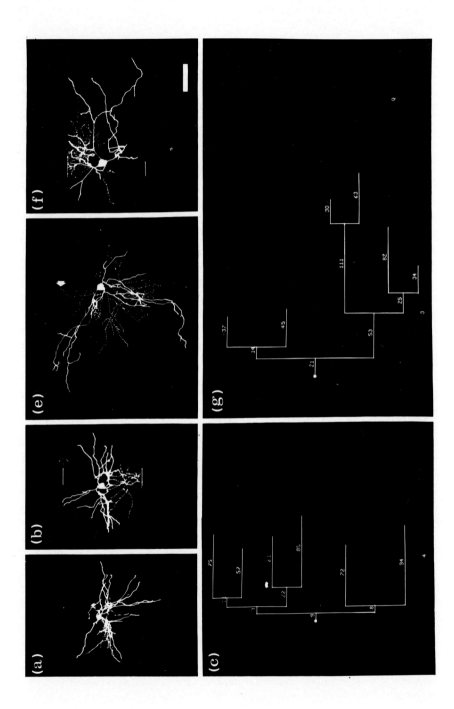

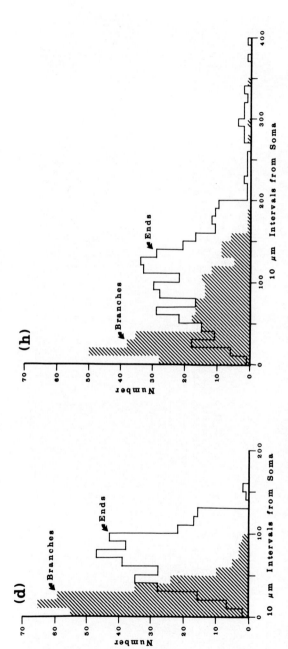

FIG. 6. To show the use of the computer in the analysis of Golgi-impregnated neurons. The processes of a neuron are generally distributed in space which makes them hard to draw faithfully and particularly difficult to measure (see Fig. 1). On the upper and lower left are shown data from cells in layer IV of mouse SmI cortex which we have named Class I (a–d) and on the upper and lower right those which we have named Class II (e–h). (a) CRT display of a cell as it was "drawn," i.e., in its observed orientation. (b) The cell in (a) has been rotated 90° about the Y axis and is viewed as if the brain's were sectioned perpendicular to the pia. The lines (⍦) indicate the surfaces of the section observed while data from the neuron were being recorded. (c) A dendrite from this cell was analyzed for the three-dimensional length of between nodal points (i.e., branches and ends). A computer display of this length diagram (Sholl plot) indicates that the bifurcations of dendrites and their ends occur at prescribed distances along the process from the cell body. (e) A class II neuron viewed as it was "tracked." The solid arrow (➤) indicates that special processes such as axons can be distinguished by a variety of methods such as CRT (axon is dotted vs dendrite solid) or x–y plotted on which the processes can be drawn in different colors. (f) The cell in (e) has been rotated 90°. Observe that the processes are truncated at the plane of section. The bar which applies to (a), (b), (e), and (f) is 100μm. (g) The Sholl (Sholl, 1956) plot of dendrite 3 of the class II cell in (e) and (f) shows the branches and ends of the cells (c). A comparison of the histograms of the distances of dendritic branches and ends for the (d) class I and (b) II cells are consistent and statistically significant. The histograms were compiled from approximately 40 cells in each class and when combined with somal-size data (from the tablet system) provided a basis for the *numerical classification* of these cells (from Woolsey *et al.*, 1975).

75

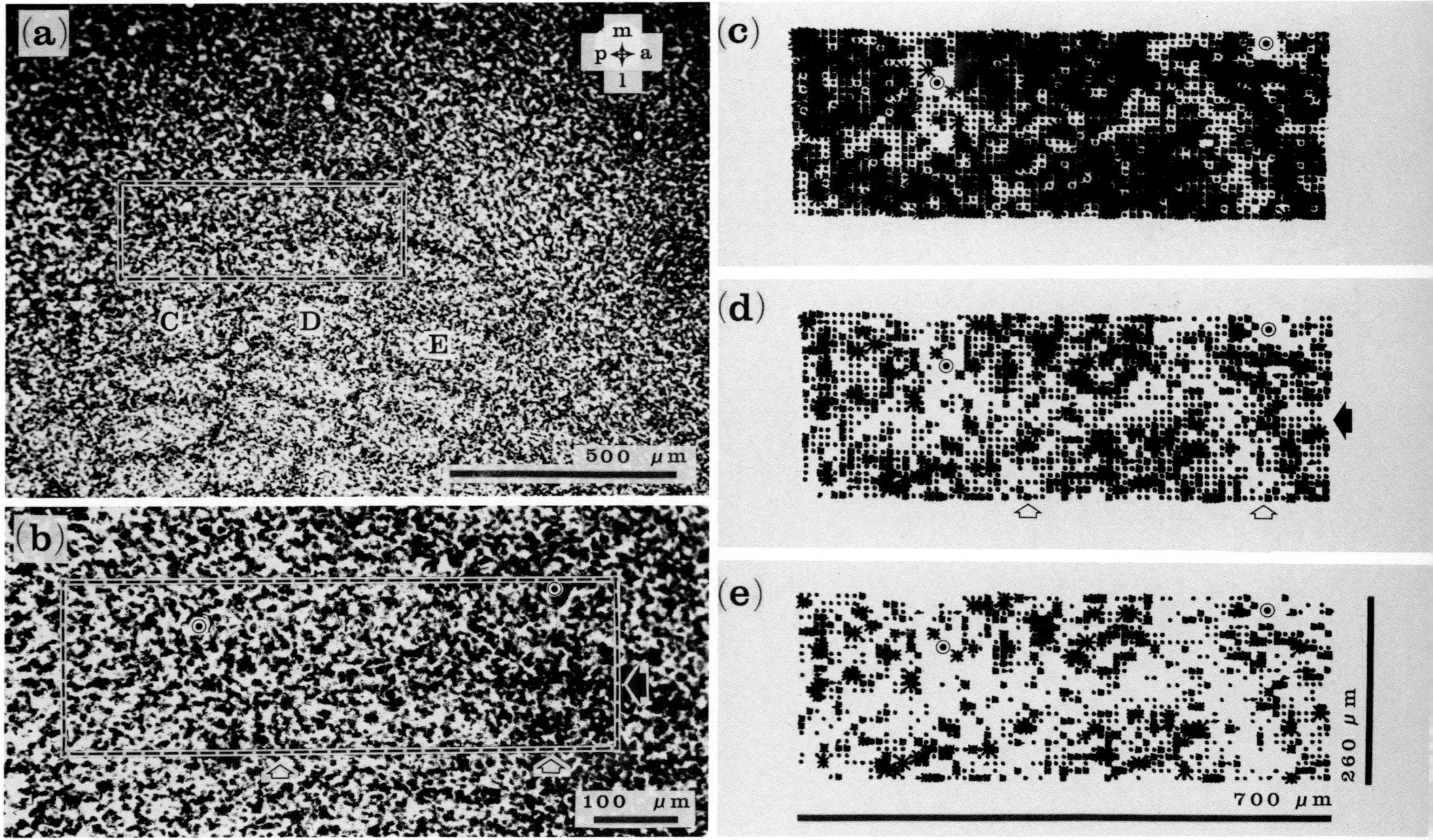

FIG. 7.

relationships between various projection fields for which qualitative descriptions are less satisfactory.

B. Other Applications

There are now a number of research groups actively involved in developing and using computer systems for anatomical research. In addition to these, most microscope manufacturers now market computer–microscope packages which are becoming generally available. Most of these commercially available systems, while convenient for some morphological applications, are probably of limited use to neuroanatomists as they assume a certain uniformity of the specimens. Thus, while stereological principals are of use in metallurgy, general histology, and cytology, their value in the study of the decidedly nonhomogeneous nervous system is generally restricted. However, as these systems become more widely used, they may be modified for certain problems in neuroanatomy.

1. TABLET SYSTEMS

Tabletlike (planimeter) systems have been widely applied in a wide range of studies. Examples are (a) measures of synaptic vesicle size and flattening (e.g., major vs minor axis) in electron microscopy (Johnson *et al.*, 1978; ch. 7); (b) measures of cells which remain after retrograde degeneration (Loewy & Schader, 1977; Ch. 12); (c) somal cross-sectional areas of neurons labeled retrogradely by HRP (Jones, 1975; Ch. 13); (d) assessment of the area of cytoarchitectonic fields and neocortical surface area (Welker & Woolsey, 1974); and (e) evaluation

---

FIG. 7. To show the results obtained from the counts of silver grains in autoradiograms of the somatosensory thalamocortical projection in mice. Two days prior to the sacrifice of the animal the ventrobasal (VB) neurons were labeled with tritiated leucine and proline. The projection of these cells to layer IV of the SmI cortex cut tangential to the pia are revealed by the distribution of the silver grains. (a) This is a survey of the organization of the barrel field of the right cerebral hemisphere of a mouse (frozen section, 80μm thick, stained with thionine). The orientation with respect to the animal is given: m, medial; a, anterior; l, lateral; and p, posterior. The field examined by the computer is outlined by the dashed lines. (b) A higher power view of the field shown in (a). Two prominent blood vessels are indicated (⊙) and the barrel outlines are clearly seen. The open arrows mark the septa between adjacent rows of barrels C and D and D and E. The solid arrows mark the septum between barrels in the same row [see (d)]. (c) The data from the raw counts are presented. In this figure the silver grain counts/100μm^2 field are presented as dots in which the size is proportional to the silver grain areas. Note that the densest projection is to the hollows and that the position of the blood vessels in (b) is shown. (d) By altering the threshold it is possible to subtract background levels of silver grains which were determined elsewhere on the same section. Clearly, there is little or no projection of VB to the space between the barrels. The arrows (open and solid) correlate with (b). (e) If the threshold is raised further it can be seen that the pattern of projection of VB to the barrels is not homogeneous. The terminals from the thalamus are apparently clustered in a way that was not appreciated by simple inspection of the specimen. The time necessary to count the 1820 fields shown was less than 2 hr.

of normal and experimentally altered developmental phenomena (Guillery, 1972; Woolsey & Wann, 1976).

2. GOLGI SYSTEMS

Several systems have been assembled to obtain quantitation data from stained single neurons, principally those revealed by Golgi's method. Indeed one of the first computer-aided systems was designed for this (Glaser & Van der Loos, 1965). Subsequently developed systems have been used with varying success. One relies on a more fully automated tracking routine using a gray level image (Garvey, Young, Coleman, & Simon, 1973), and another depends upon assembly of several quantified video images taken at planes of focus through a Golgi impregnated neuron (Lliñas & Hillman, 1975).

3. SERIAL RECONSTRUCTIONS

An area that we did not illustrate but which has received a great deal of attention and effort has been serial reconstruction principally from serial electron micrographs. Levinthal and his colleagues (Levinthal & Ware, 1972), Brenner's group (Ward, Thomson, White, & Brenner, 1975), Russell and co-workers (Ware, Clark, Crossland, & Russell, 1975), Selverston (1973), etc., have been successful in reconstructing substantial portions of nervous systems in small invertebrates in this way. A great deal of effort has gone into the development of these systems, which must be capable of storage and manipulation of vast quantities of data.

Serial reconstructions have addressed the question of whether a migrating neuroblast in the primate brain is attached at the pia and ependyma (Rakic, Stensaas, Sayre, & Sidman, 1974). Computer-assisted serial reconstructions have been used to evaluate the variability of human striate cortex to determine the suitability of prosthetic devices (Stensaas, Eddington, & Dobelle, 1974). In spite of some of the limitations which are principally technical, the value of the computer-assisted serial reconstruction is considerable. These involve easily manipulated reconstructions of nuclear groups in the brain on which have been plotted the extent of degeneration or silver grains, to aid three-dimensional interpretation in following processes of Golgi-stained neurons through many sections with which we now have some experience (Dierker, 1976a).

C. Computer Analyses of Quantized Data

Such features as counts per area, object area, planar length, true three-dimensional length measures, or structural identifications (such as presence/absence and location data of spines and/or varicosities on neuronal dendritic processes) represent direct measurements. These measurements may be conceived as forming a primary data base describing the structure. From these data further features may be derived, utilizing the computational power of the computer. For example, given a spatial description of a neuronal process and the identification and location of such dendritic specialization as spines and/or varicosities, plots of spine density versus radial distance from the cell soma may be easily obtained. Such measures (obtained by manual computation) have already proved useful in visual deprivation studies (Valverde, 1971), illustrating that such computer-extracted features can prove to be directly applicable to experimental situations.

One possible application of computer analysis is used in the study of neuronal classification by attempting to identify classes of similar neurons on the basis of their morphology (Woolsey *et al.*, 1975). The computational power of the computer allows comparison of cells on an objective basis from all the features gathered, and generates cell classifications based upon quantitative repeatable rules rather than the qualitative, albeit useful, classification systems developed in the past (e.g., Lorente de Nó, 1922; Ramón-Moliner, 1968). Studies of this nature appear to be very promising in developmental, genetic, or deprivation experiments providing quantitative measures of cell changes between animals and indicating which neuronal features were most affected by the experiment.

V. Technical Problems, Methodological Limitations, and Interpretive Difficulties

The use of computers in neuroanatomy is no different than any other technique. There are certain limitations, some of which are considerable. The results can be no better than the quality of the preparations which have been examined, the accuracy of the instruments which have been employed, the soundness of the programs which have been used, and the intelligence with which the data are interpreted.

A. *Specimens*

The importance of careful preparation of biological specimens cannot be over-emphasized. Some examples of oversights which can be devastating follow. Failure to scrupulously clean glass slides and cover slips in autoradiography can introduce spurious results in grain counts which are nothing more than particles of dirt and other debris. This is particularly true for observations made under dark-field illumination. The failure to choose the appropriate section thickness, cover slip thickness, or objective lenses can be the source of a variety of difficulties. In sections that are too thin, the process can be excessively truncated and lead to an underestimate of the true extent of a cell's territory. A cover slip that is too thick, on the other hand, can result in a loss of data which are otherwise available from the depths of a particular section. If the numerical aperture of a lens is low the focal plane will be deep, and in situations where measurement along the optical axis are critical there will be a loss of resolution. If the mountant is insufficiently hardened, additional errors can be introduced, particularly when observations are being made under oil. Friction between the objective and the cover slip can generate shear on the section underneath which will be propelled by the microscope stage unpredictably.

These are factors which, while a nuisance, can usually be solved directly. More difficult to deal with are phenomena which have as their basis a biological property of the system or particular method. Ignoring these questions will frequently lead to errors in interpretation of the results. An example is that the stoichiometry of the uptake and incorporation of radioactive precursors is still unknown. A direct comparison of numerical results from different experiments is unjustified. Other relatively uncontrollable factors include variable tissue shrinkage in processing, individual and genetic variations in different specimens, and variability and/or errors in the observer decisions which may be necessary in entering the data.

Finally, one must consider the appropriateness of a particular histological technique for the computer methodology employed. A system which has been designed to consider preparations of high optical contrast may fail if the preparations do not meet some minimum criterion. Particularly for highly automated systems spurious results may be obtained if care is not taken to make sure that the preparation can be effectively analyzed.

B. Hardware and Software

A number of significant limitations are imposed by hardware. On one extreme is the consideration that all mechanical systems, i.e., microscope stages, have limitations. These become particularly important in the microscopic range. The principal considerations are what are the errors that can be expected from the mechanical system and to what extent are they predictable. Some consideration must also be given to compatibility with existing computers that the investigator may wish to use or adapt for his own investigations. These are somewhat akin to video taping systems a few years ago; there is not yet a standardization of small computers.

A second problem is that input devices are not yet standardized. Since the input sources often determine the data-gathering algorithm, different input sources may require extensive or complete revision of existing programs. In addition, programs written for a particular machine or system are not transferable without considerable or complete revision. The later is being rectified by the introduction of software systems which utilize flexible and generally understood languages. This restriction further complicates the standardization and interchange of programs. A third problem which to some extent is dependent upon the previous two is that data storage formats differ considerably. Conceivably, the data from one system can be "decoded" for another by appropriate programs but this can be a time-consuming proposition.

C. Technical Assistance

In our own experience the success which we have had in applying computer technology to neuroanatomical problems has been due to extensive cooperation and interaction among individuals whose backgrounds are in engineering and biology. The technical support for the development and implementation of new hardware and software can be a considerable expense and for ambitious or novel applications constitutes a significant limitation.

However, as various systems become perfected and generally available the necessity for full-time technical backup may be diminished. That is, given a standard setup and a convenient programming language, an anatomist can learn to modify existing programs to suit his own needs. A wide variety of commercial programs are available for various statistical and display purposes. If an investigator wishes to have a larger and more varied capability even for proven systems, some form of permanent and readily available technical assistance is advisable in order to avoid costly delays from hardware and software failures. Finally, for an active program of development of new applications which requires skilled machinists, electrical engineers, and computer programmers, a long-range commitment must be made in terms of staff and funds. This notion has long been recognized as a necessity in physiology departments but for anatomists the concept is relatively new.

D. Investigator Responsibility

It may seem a trivial point, but the success or failure in computer applications to neuroanatomy ultimately depend upon the neurobiologist. Too frequently the availability of numbers in convenient hard copy is considered the end in itself. The numbers that a computer generates are solely dependent on the quality of the input and the correctness and appropriateness of the software packages. Careful formulation of the biological question and the technical innovations that aid in its solution are of utmost importance. Attention to details at each point from specimen to hard copy is critical. At the end one must always ask, what does this tell me about the system under study and is it biologically significant?

VI. Summary and Comments

The various problems in computer-assisted recording of anatomical data have been reviewed and are briefly summarized. (1) Almost any anatomical preparation is suitable for computer-assisted treatment. The exact nature of these preparations will determine a number of decisions in the implementation of a computer-based data-gathering system. (2) There are three basic methods of recording the data: manual, semiautomated, and fully automated. The least-automated systems offer greater flexibility at lower cost. (3) From the user's point of view, the interaction with the computer system can be considered as consisting of four perceptual blocks: (a) program interactions, (b) data digitization, (c) data analysis, and (d) permanent data storage and retrieval. (4) The general features of equipment necessary to perform these tasks are discussed. In addition some general considerations about various computer hardware and software provide useful background for selecting equipment. (5) The operation of three specific computer-based systems is described. (6) Examples of data collected and analyzed with these specfic systems are given to illustrate some of the potential uses for computer-assisted recording of neuroanatomical data. (7) Some problems associated with system implementation, data gathering, and data interpretation are reviewed.

To conclude this chapter, some general comments seem warranted. It is hoped that the general topics outlined will provide a suitable introduction to the basic principles for the building of a computer-assisted data recording system. It is advisable to assess the technical resources within one's own environment so as to identify potential sources of equipment and hardware and software expertise as these are frequently the ultimate limitations on what can or cannot be accomplished. When a particular project is undertaken it is extremely valuable to examine similar systems which exist in other laboratories and to obtain technical advice from equipment manufacturers.

Cost is of particular importance. The Tablet System which was described can be implemented for about $10,000, while the grain counting system can be approximately five times as much. There is reason to believe that the prices of many of the electronic components, including computers, will continue to fall over the next few years. Thus, simple systems may be implemented for less in the future. When a system incorporates optical devices, the trend of rising prices for optics may offset the economics in the electronics field.

There is little doubt that future improvements in hardware and software will make a large variety of system packages generally available. These improvements will be attended by greater ease of usage. A proposal, now under consideration, which should help investigators wishing to enter this field, is the establishment of a central registry which will provide technical information, programs, and advice to neuroanatomists.

Many of the initial technical problems in applying computers to morphology have now been overcome. A number of groups are now routinely collecting data, which in several cases has already improved our understanding of the nervous system. The present status of computers in neuroanatomy may be likened to the situation in neurophysiology about a decade ago. Scientists are beginning to appreciate the computer's power in solving problems. And, just as in neurophysiology and electron microscopy, the solution of the initial problem may be expected to pave the way for exciting future discoveries, particularly as experience with and new insights about the underlying biological phenomena are gained.

Acknowledgments

We are grateful to Janet Hake for some of the histology and illustrations, Dana Grellner for photography, and Mary Murphy for typing. This work was supported by National Institutes of Health Grants NS 10244 and EY 01255. M.L.D. was supported by National Institutes of Health Training Grant 5 TO1GM01827-08.

References

Brecher, G., Schneiderman, N., & Williams, G.Z. Evaluation of electronic red blood cell counter. *American Journal of Clinical Pathology*, 1956, **26**, 1439-1449.

Cowan, W.M., Gottlieb, D.I., Hendrickson, A.E., Price, J.L., & Woolsey, T.A. The autoradiographic demonstration of axonal connections in the central nervous system. *Brain Research*, 1972, **37**, 21-51.

Cowan, W.M., & Wann, D.F. A computer system for the measurement of cell and nuclear sizes. *Journal of Microscopy (Oxford)*, 1973, **99**, 331-348.

Dierker, M.L. An algorithm for the alignment of serial sections. In P. B. Brown (Ed.), *Computer technology in neuroscience*. New York: Halsted Press, 1976. Pp. 131-133. (a)

Dierker, M.L. An algorithm for the display and manipulation of lines in three dimensions. In P. B. Brown (Ed.), *Computer technology in neuroscience*. New York: Halsted Press, 1976. Pp. 139-151. (b)

Digital Equipment Corporation. *PDP-12 system reference manual*. Maynard, Massachusetts. Author, 1972.

Digital Equipment Corporation. *Small computer handbook*. Maynard, Massachusetts: Author, 1973.

Digital Equipment Corporation. *PDP-8/a mini computer handbook*. Maynard, Massachusetts: Author, 1976. (a)

Digital Equipment Corporation. *PDP-11 peripherals handbook*. Maynard, Massachusetts: Author, 1976. (b)

Garvey, C.F., Young, J.H., Coleman, P.D., & Simon, W. An automated three-dimensional dendrite tracking system. *Electroencephalography and Clinical Neurophysiology*, 1973, **35**, 199-204.

Glaser, E.M., & Van der Loos, H. A semi-automatic computer-microscope for the analysis of neuronal morphology. *IEEE Transactions on Biomedical Engineering*, 1965, **12**, 22-31.

Gottlieb, D.I., & Cowan, W.M. Evidence for a temporal factor in the occupation of available synaptic sites during the development of the dentate gyrus. *Brain Research*, 1972, **41**, 452-456.

Guillery, R.W. Binocular competition in the control of geniculate cell growth. *Journal of Comparative Neurology*, 1972, **144**, 117-127.

Johnson, M.O., Ross, D., Meyers, M., & Bunge, R.P. Changing synaptic cyto-chemistry in cultured sympathetic neurons, 1978, in preparation.

Jones, E.G. Some possible determinants of the degree of retrograde neuronal labelling with horse-radish peroxidase. *Brain Research,* 1975, **85,** 249–253.

Kehl, T.H. On uniformity of digital computer interface design. In P. B. Brown (Ed.), *Computer technology in neuroscience.* New York: Halsted Press, 1976. Pp. 601–611.

Kelly, J.P., & Cowan, W.M. Studies on the development of the chick optic tectum. III. Effects of early eye removal. *Brain Research,* 1972, **42,** 263–288.

Lee, K.J., & Woolsey, T.A. A proportional relationship between peripheral innervation density and cortical neuron number in the somatosensory system of the mouse. *Brain Research,* 1975, **99,** 349–353. (a)

Lee, K.J., & Woolsey, T.A. The relationship of peripheral innervation density (vibrissae) to cortical neuron number (barrels) in the mouse. *Anatomical Record,* 1975, **181,** 408. (b)

Levinthal, C., & Ware, R. Three dimensional reconstruction from serial sections. *Nature (London),* 1972, **236,** 207–210.

Lliñas, R., & Hillman, D.E. A multipurpose tridimensional reconstruction system for neuroanatomy. In M. Santini (Ed.), *Golgi centennial symposium proceedings.* New York: Raven Press, 1975. Pp. 71–79.

Loewy, A.D., & Schader, R.E. A quantitative study of retrograde neuronal changes in Clarke's column. *Journal of Comparative Neurology,* 1977, **171,** 65–82.

Lorente de Nó, R. La corteza cerebral del ratón. *Trabajos del Laboratorio de Investigaciones Biologicas de la Universidad de Madrid* 1922, **20,** 41–78.

Lorente de Nó, R. Architectonics and structure of the cerebral cortex. In *Physiology of the nervous system,* J. Fulton (ed.). London: Oxford University Press, 1938, Pp. 291–327.

Mountcastle, V.B. The view from within: Pathways to the study of perception. *Johns Hopkins Medical Journal,* 1975, **136,** 109–131.

Pasternak, J.F., & Woolsey, T.A. The number, size and spatial distribution of neurons in lamina IV of the mouse SmI cortex. *Journal of Comparative Neurology,* 1975, **160,** 291–306.

Rakic, P., Stensaas, L.J., Sayre, E.P., & Sidman, R.L. Computer-aided three-dimensional recon-struction and qualitative analysis of cells from serial electron microscopic montages of foetal monkey brains. *Nature (London),* 1974, **250,** 31–34.

Ramón-Moliner, E. The morphology of dendrites. In G. Bourne (Ed.), *The structure and function of nervous tissue.* Vol. 1. New York: Academic Press, 1968. Pp. 205–267.

Rose, J.E., Gross, N.B., Geisler, C.D., & Hind, J.E. Some neural mechanisms in the inferior colliculus of the cat which may be relevant to localization of a sound source. *Journal of Neurophysiology,* 1966, **29,** 288–314.

Rose, J.E., and Woolsey, C.N. Cortical connections and functional organization of the thalamic auditory system of the cat. In *Biological and biochemical bases of behavior,* H. Harlow and C. Woolsey (eds.). Madison: University of Wisconsin Press, 1958, Pp. 127–150.

Selverston, A.I. The use of intracellular dye injections in the study of small neural networks. In S.B. Kater & C. Nicholson (Eds.), *Intracellular staining in neurobiology.* Berlin & New York: Springer-Verlag, 1973. Pp. 255–280.

Sholl, D.A. *The organization of the cerebral cortex.* London: Methuen, 1956.

Souček, B., & Carlson, A.C. *Computers in neurobiology and behavior.* New York: Wiley, 1976.

Stensaas, S.S., Eddington, D.K., & Dobelle, W.H. The topography and variability of the primary visual cortex in man. *Journal of Neurosurgery,* 1974, **40,** 747–755.

Tyner, C.F. The naming of neurons: Application of taxonomic theory to the study of cellular populations. *Brain, Behavior and Evolution,* 1975, **12,** 75–96.

Valverde, F. Rate and extent of recovery from dark rearing in the visual cortex of the mouse. *Brain Research,* 1971, **33,** 1–11.

Wann, D.F., Price, J.L., Cowan, W.M., & Agulnek, M.A. An automated system for counting silver grains in autoradiographs. *Brain Research,* 1974, **81,** 31–58.

Wann, D.F., Woolsey, T.A., Dierker, M.L., & Cowan, W.M. An on-line digital computer system for the semi-automatic analysis of Golgi-impregnated neurons. *IEEE Transactions on Biomedical Engineering,* 1973, **20,** 233–247.

Ward, S., Thomson, N., White, J.G., & Brenner, S. Electron microscopical reconstruction of the anterior sensory anatomy of the nematode *Caenorhabditis elegans. Journal of Comparative Neurology,* 1975, **160,** 313–338.

Ware, R., Clark, D., Crossland, K., & Russell, R. L. The nerve ring of the nematode *Caenorhabditis elegans:* Sensory input and motor output. *Journal of Comparative Neurology,* 1975, **162,** 71–110.

Welker, C., & Woolsey, T.A. Structure of layer IV in the somatosensory neocortex of the rat: Description and comparison with the mouse. *Journal of Comparative Neurology,* 1974, **158,** 437–454.

Woolsey, T.A. Some anatomical bases of cortical somatotopic organization. *Brain, Behavior, and Evolution.* 1978 (in press).

Woolsey, T.A., Dierker, M.L., & Wann, D.F. Mouse SmI cortex: Qualitative and quantitative classification of Golgi-impregnated barrel neurons. *Proceedings of the National Academy of Sciences of the U.S.A.,* 1975, **72,** 2165–2169.

Woolsey, T.A., & Van der Loos, H. The structural organization of layer IV in the somatosensory region (SI) of the mouse cerebral cortex. *Brain Research,* 1970, **17,** 205–242.

Woolsey, T.A., & Wann, J.R. Areal changes in mouse cortical barrels following vibrissal damage at different postnatal ages. *Journal of Comparative Neurology,* 1976, **170,** 53–66.

Woolsey, T.A., Wann, D.F., Cowan, W.M., & Dierker, M.L. Computer analysis of Golgi impregnated neurons. A sixteen millimeter sound and color motion picture, Copyright: Washington University, 1972 (copies available on loan from the authors).

Techniques for the Study of Normal Tissue

Chapter 4

The Methods of Golgi

Madge E. Scheibel* and Arnold B. Scheibel

Departments of Anatomy and Psychiatry, and
Brain Research Institute
University of California
Los Angeles, California

*Deceased.

89

"... provando e riprovando ..."
Camillo Golgi

I. Introduction

It has been said that the methods of Golgi are the only histological techniques with personality. At their best, they are not difficult to execute and make relatively few demands on the resources of the investigator. They may be run in a well-equipped laboratory or in a mountain cabin (Cajal surreptitiously processed tissue for several weeks while he was a guest in the Sherrington home), and they may offer, in one fortunate impregnation, sufficient data on axon–dendrite relations to galvanize an entire institute of electrophysiologists. At their worst they may appear as cranky, mystique-ridden methods, unclear as to rationale, uncertain as to result, and unacceptable to those whose scientific training demands that a technique, if properly applied and attended, produce a predictable result.

Although we have etched the contrasts with a heavy hand, no Golgi practitioner would deny this Janus-like quality of the methods. The fact remains that 100 years after its discovery, the technique is in active and increasing use, not only in its primary role in qualitative histology, but as a keystone in the new quantitative neurobiology, experimental neurology, and neuropathology, and as the essential handmaiden of the electron microscope. Surrounded by precocious neophyte methods such as fluorescence microscopy, autoradiography, and retrograde transport techniques with capacities to track and identify that are almost uncanny, the Golgi methods continue to provide a view of the nervous system both unique and indispensable. It seems appropriate, therefore, to take another look at the technique, with respect and affection—an old friend whose tasks never run out.

The term "Golgi" has become generic for a group of impregnation methods that rely on the preparation of neural tissue with dichromates (or chromate–dichromate salts in equilibrium) followed by exposure to heavy metal ions, either silver (as $AgNO_3$) or mercury (as $HgCl_2$). The basic method was initially discovered by Camillo Golgi while working as a resident physician in the home for incurables at Abbiategrasso, and described almost casually in one paragraph in a short paper on cerebral cortex (Golgi, 1873). Legend claims that the first Golgi impregnation occurred when an overzealous cleaning woman threw some pieces of dichromate-fixed tissue into a refuse jar containing discarded silver nitrate solution. However, Ramón-Moliner (1970) has called attention to a little known work of Retzius (1953) containing a more respectable version of the discovery. The method is said to have been discovered unexpectedly while Golgi was trying to impregnate with silver the innermost lamina of pial membrane on dichromate-hardened brain tissue. The chance observation of a few brown-stained neurons in adjacent, underlying gray matter led to a series of experiments with the results that are now a matter of historic record. In 1954 we met at Pavia Professor Antonio Pensa, the last of Golgi's great pupils and a living storehouse of Golgiana. When asked about the circumstances surrounding the discovery, he smiled but could offer no alternative version. We prefer to think of that apocryphal cleaning lady, immersed in the urgent minutiae of her life, hurriedly clearning that work desk for the thousandth time, throwing the young master's tissue specimens into the refuse bucket—and history.

II. Why Use the Golgi Methods?

The main strengths of the Golgi methods lie in their capacity to reveal all components of the nervous system; neuronal, glial, and vascular. The revelation is tinct with restraint, so much so that only a small percentage of the elements in any one area (1–10%) are ever impregnated in a single preparation. Of the neurons rendered visible, however, virtually all portions can be seen, as illustrated in Figs. 1 and 2 and in Plate I in the Color frontispiece. The cell body, dendrites, dendritic substructures if any, and at least part of the axon are visible, providing a panoramic view of the entire neural element silhouetted in black on a yellow or pale orange background. Quite apart from the enormous informational content resident in this fragment of neural circuitry, many see a unique biological work of art as impressive, in its own way, as the X-ray revealed structure of a crystal, or a high-resolution photograph of a distant galaxy.

It follows that no single Golgi impregnation provides a complete roster of the elements available at any one point and in any one section. Instead, there is provided a partial picture—a vignette—revealing one aspect of connectivity patterns. If the same area were stained in 100 consecutive brains, no two impregnations would ever show precisely the same elements in the same relation to each other. It is here that an essential and highly personal ingredient must be added—the investigator's capacity to conceptually reconstruct the totality of the neuropil field which has been glimpsed so many times, and in so many guises (see Fig. 3). It is one thing to describe faithfully the size and shape of dendrite systems or the total number and course of axon collaterals and the minutiae of terminal axonal structures. It is another thing to see, however dimly, the general model of interconnections, and the putative consequences of such linkage patterns. For the Golgi is, above all other light microscope techniques, a methodology conducive to physiological interpretation and predictive of physiological consequences.

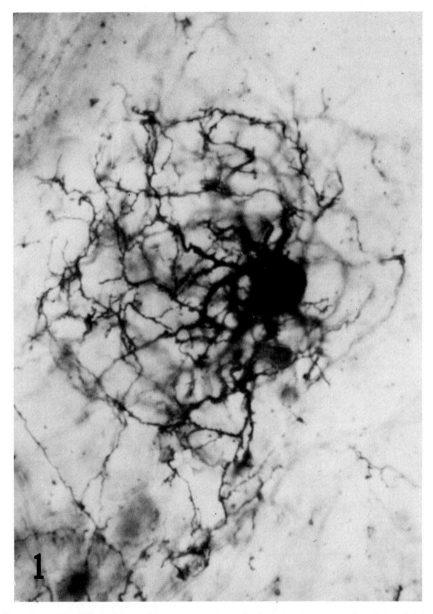

FIGS. 1 AND 2. These figures illustrate one of the basic and unique strengths of the Golgi methods, the visualization of the dendritic domain and details of dendritic surface structure.

FIG. 1. A rapid Golgi impregnation of an inferior olive cell from the brainstem of a 3-week-old cat. Because of the spherical shape of this type of domain, many dendrite branches are out of focus or invisible. The relative thinness of each dendritic element and the irregular excrescences are typical of the immature olive cell. Note how the stain seems to selectively impregnate every element of the domain although there is no impregnation in the immediate surroundings. Calibration bar: 20 μm.

92

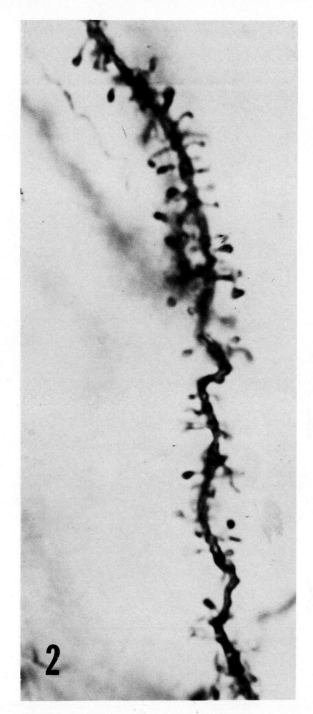

FIG. 2. A relatively high-power (original magnification: ×800) photograph of a branch of an adult cat cortical neuron. The rather sparse and pleomorphic spines are characteristic of more remote (secondary or tertiary) pyramidal cell dendrite branches compared to the more regular arrays on more proximal shafts. Calibration bar: 10 μm.

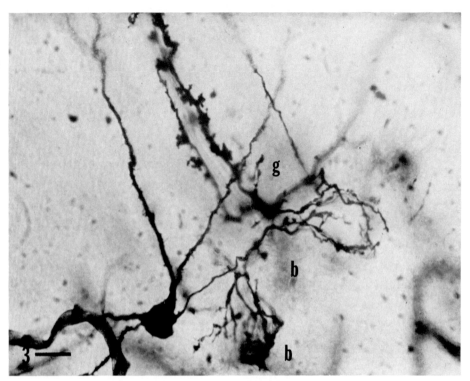

FIG. 3.   Rapid Golgi impregnation of 2-month-old cat cerebellum showing sensitive staining of a very small number of elements. The basket cell axon generates basket formations, b, around two Purkinje cell bodies which are unstained. An epithelial glial cell of Golgi, g, is also stained showing the characteristic leafy excrecences which cover the surfaces of both adjacent Purkinje cell dendrites and cerebellar blood vessels (both unstained). This type of sparse impregnation epitomizes the typical Golgi-stained fragments of circuitry which must gradually be resynthesized into an overall connectional schema by the investigator. Calibration bar: 10 μm.

III.   Where Is the Golgi Useful?

The number of problem areas in which the Golgi methods have been found useful are numerous, especially when it is remembered that until 1950, its one widely recognized use was in qualitative descriptive neurohistology. Developmental and comparative neurohistology and neurocytology, already appreciated by Ramón y Cajal (1952, actually a reissue of the classic 2-volume set originally published in 1909–1911) as potential grist for his mill have been increasingly explored with the Golgi methods where the capacity to compare size, shape, and patterning of dendrite domains and of the synaptic connections which result, provide important clues on developmental trend and specialization of function.

In similar fashion, the entire realm of neuropathology has remained virtually untouched with a very few exceptions (De Moor, 1898) until recently. Here the Golgi methods are rapidly proving to be enormously productive in revealing changes in dendrites, dendrite spines, and axons which cannot be inferred from the entire panoply of neuropathological stains in use for many decades. Significant new information has already been revealed as to the nature of pathological processes at work in various types of mental deficiency (Huttenlocher, 1974; Marin-Padilla, 1972; Purpura, 1975), temporal lobe epilepsy (Scheibel, Crandall, & Scheibel, 1974), and senility (Scheibel, Lindsay, Tomiyasu, & Scheibel, 1975), and the surface has barely been scratched.

Quantitative neurobiology [or histonomy as Bok (1959) would have it] is becoming an engaging reality, approximately two decades after the seminal work of Sholl (1953). Modern high-speed computer graphics technology crossfertilized with Golgi histology already provides the capability for enormously detailed descriptions of cell laminae, dendrite field structure, and axonal elaborations, with only minimal participation of the investigator, once the appropriate programs are written (Lindsay & Scheibel, 1974). The prospect of totally automated, high-speed operations is now imminent, promising detailed quantitative descriptions of large numbers of neurons with almost incalculable consequences for future research, clinical diagnosis, and treatment.

The coupling of Golgi and electron microscope techniques has become a powerful and permanent element of neurohistological research. The strengths of each technique complement the deficiencies of the other and their intelligent joint use probably constitutes the most effective neurocytological approach to the nervous system yet employed. Here are combined the panoramic viewing qualities of the Golgi with the high resolving capabilities of the electron microscope. Pioneering work by Blackstad (1975), Stell (1965), and Chan-Palay and Palay (1972) has made it feasible to utilize the power of the electron microscope directly on the tissue specimen already studied with Golgi techniques. Aside from the obvious relevance of this joint method to definitive identifications and descriptions of individual synapses and associated structures, it has also thrown revealing light on the nature of the Golgi impregnation itself.

We have certainly not exhausted the list of problems and approaches amenable to Golgi analysis, but these suffice to suggest the range, and indicate that the intrinsic power of the method will continue to ensure its attractiveness to the venturesome investigator.

IV.   How to Use the Golgi Methods

The basic variants of the Golgi methods have come down to us from Golgi and Cajal. The original or slow method was, of course, developed by Golgi, the rapid

brown or black. On the other hand, long impregnations of 30 to 60 days often led to very light impregnations of great delicacy in which only a few axons with their collateral systems showed up against a very pale yellow translucent background.

3. IMPREGNATION

At the end of the fixation period, the fixative is poured off and the tissue drained briefly on absorbent paper. A 0.75% silver nitrate (AgNO₃) solution is used as the impregnation medium. Specimens are washed in successive baths of the fluid until no further red silver chromate precipitate appears, then are placed in a relatively large volume of fresh silver nitrate solution (50–100 ml/tissue blocks), and are stored in the dark, or at least in brown glass bottles. It is stated generally that impregnation times vary from 24 to 48 hr and that tissue may be kept in silver without deleterious effects for days or even weeks. In our experience, the duration of silvering is almost as critical as the time of fixation in that it gives the worker one of his few opportunities to control the depth and intensity of impregnation. Somewhere between 4 and 8 hr after the beginning of silvering, impregnation begins to take place, depending on the nature of the tissue and the length of previous fixation (i.e., the longer the fixation, the later the onset of impregnation). If this point can be determined through experiment, and 30 to 60 min extra allowed for gradual development of the impregnation, extremely sensitive stains can be achieved in which a small number of individual elements are rather completely stained. This endpoint is rather critical and is passed quickly so that often, by 12–14 hr, the impregnation is already rather heavy and too complex for the unravelling of individual elements, as illustrated in Figs. 4–7. On the other hand, when a general overall view of the area is desired or where the appearance and orientation of dendrite masses and axon plexuses is indicated, longer impregnation of up to 48 hours may be necessary. In our experience there is little further development of the staining picture with still longer impregnations, with or without change of silver baths. On the other hand, after several days of immersion, the impregnation begins to deteriorate, culminating in diffuse precipitation of silver salts throughout the tissue.

Ramón y Cajal (1952) and Valverde (1970) have recommended double and triple immersion procedures to intensify or enhance the result. Following silvering, the tissue is returned to the original chromating solution for a few days and then into fresh silver solution. This procedure may then be repeated once more before mounting and cutting. In our hands, this variant has not proven advantageous over the usual single immersion and, consequently, is not a regular part of our staining routines.

4. SECTIONING AND MOUNTING

All further operations on the tissue are dependent on one limiting factor. The silver impregnation appears to be in some type of nonstable suspension within the

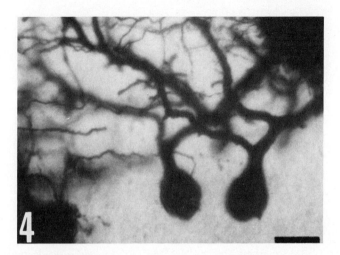

FIGS. 4 AND 5. These figures illustrate the importance of the characteristic of selective partial staining by the rapid Golgi method.

FIG. 4. Two Purkinje neurons from a 3-month-old cat cerebellar vermis. Only primary and secondary dendrites are impregnated (no tertiary dendrites) and a few basket axons. Calibration: 20 μm.

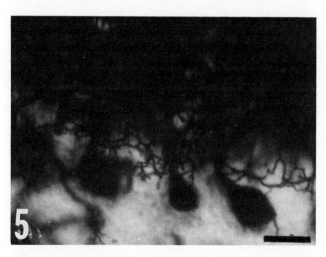

FIG. 5. The result of complete impregnation of five Purkinje neurons with their total complement of dendritic branches and ramifications. Aside from a few isolated ventral branches, the dendrite systems are visually inpenetrable, and this portion of the impregnation is useless. Calibration: 20 μm.

FIGS. 6 AND 7. The approximate upper limits of "legibility" for multielement-stained Golgi preparations.

FIG. 6. A rapid Golgi stain of a 2-month-old cat hippocampus. The heavy dendritic neuropil is generated by approximately six hippocampal pyramids. Many cell bodies are out of focus, and an even greater number remain unstained. Calibration bar: 25 μm.

FIG. 7. A cluster of pyramids in the fifth layer of motor cortex of a 35-year-old man, showing the intricacy of the dendritic neuropil and the problem of maintaining selective focus. Arrows point to three breaks in apical shafts. The artifact is usually due either to tissue shrinkage during fixation and/or preparation for cutting, or to compression during the cutting process. Calibration bar: 30 μm.

tissue (see Section VI), and careless handling at this stage can ruin the result. Immersion of the block in aqueous solutions (including the original silver bath) or in hot paraffin, or even in the high alcohols for appreciable lengths of time can cause diffusion of the silver, so the usual paraffin or celloidin embedding techniques are not advisable. Sectioning on the freezing microtome is possible if the tissue has not been hardened too long. Usually, however, this technique makes rapid Golgi material too brittle. Sectioning is often possible as the tissue begins to thaw but limitations of the technique may force the worker to cut at thicknesses which are not optimal for the result he wishes to obtain.

The paraffin shell technique is very simple and the least destructive to the impregnation. Because relatively little support is provided for the specimen however, it is useful mainly for small pieces of tissue, or for specimens of fairly homogenous density where compression artifacts or cracking are less likely (see Fig. 7). We have used this technique successfully in sectioning the lower brainstem of small laboratory animals, and small or moderate-sized pieces of cerebral and cerebellar cortex.

The tissue is removed from the silver solution, drained briefly on absorbent paper, and then laid on a wooden or fiber mounting block with the correct orientation. Low-melting-point paraffin (48–54°C) heated over a burner just to the melting point is dropped slowly over the tissue until it is completely covered. The block and preparation are then inverted and, with a heated spoon or spatula, the excess paraffin is worked in and around the point of junction of tissue and block, and made to flow over the ''top'' of the specimen, thereby strengthening the shell and sealing off any cracks or bubbles that have developed.

Celloidin (or parlodian) impregnation is most likely to provide a continuous series of undistorted sections, but unless carried out carefully, may impair the impregnation. In general, we have found that the shorter the dehydration process and subsequent immersion in celloidin or parlodian, the better the final result. In the rapid dehydration process, we avoid the lower alcohols entirely and place the tissue immediately in 95% alcohol for 10–15 min. The blocks are then advanced into several changes of absolute alcohol totaling 2–4 hr at the maximum, with an equal time in 2:1 ether–alcohol mixture. Four hours in a 3% solution of celloidin or parlodian (made up in 2:1 ether–alcohol solvent) and a similar time in 10% solution is usually sufficient.

The specimen is then mounted on a fiber block and a few drops of the celloidin or parlodian solution are allowed to run over the tissue. Specimen and block are then carefully lowered into a container of chloroform. Within an hour, the mountant is usually hard and ready for cutting. Small amounts of silver are lost from the specimen block during the process, expecially while in the celloidin or parlodian bath. These solutions, therefore, become cloudy and should be discarded after a few uses. Additionally, these solutions should be kept and used only in dark glass bottles and shielded as much as possible from light. There is some evidence that this material is light sensitive and that when silver-containing tissue is immersed in a solution previously exposed to light, severe fading of the stain occurs.

Other methods of embedding have been reported for preparing Golgi tissue for the microtome including embedding procedures in paraffin, in water soluble wax, and in celloidin followed by paraffin (double embedding). In our experience, all of these have interfered with the quality of silver impregnation.

5. PROCESSING

Tissue slices should be processed as soon as possible, because leaving blocks embedded and ready for sectioning for any length of time again tends to produce gradual deterioration of the impregnation. Sections are cut on the sliding microtome at thicknesses varying from 50–200 μm depending on the tissue and the needs of the investigator. Sections are immediately removed from the blade with a heavy camel's hair brush, and placed in absolute alcohol. We have found that it is necessary to keep the knife, specimen, and brush continuously wet with alcohol to facilitate manipulation of the thick sections without cracking or fragmenting them. The sections are moved serially through two to three changes of absolute alcohol over a 10-min period, then transferred to oil of cloves, wintergreen, or other aromatic heavy oil for clearing. As soon as each section becomes translucent and sinks of its own weight to the bottom of the dish, it can be transferred to reagent grade toluene or xylene; one or more baths totaling not more than 2–3 min should remove all traces of alcohol or aromatic oil.

Sections are mounted serially on glass slides previously treated with egg albumen in glycerol. Sections are pressed against the glass with blotting or filter paper, then covered with a thin layer of mounting medium. Commercially available synthetic resins such as Permount, Clarite, or Technicon mounting medium are used exclusively since solutions of natural balsam resins are somewhat acidic and may lead to fading of the sections. The classical schools of Golgi and Cajal advised against covering the sections with cover slips. Some modern investigators have suggested either coverage after the mounting medium is partly set, or the use of plastic cover slips. If study with oil immersion objectives appears impossible because of the thickness or unevenness of the medium, it may be soaked off by immersion in toluene or some xylene for some hours; the sections then are recovered with cover slips. The material must then be studied, drawn, or photographed immediately since a fair proportion of such sections deteriorate rapidly.

Slides should be allowed to dry for at least a week in a horizontal position, shielded from dust and currents of air. Preparations not so protected usually develop rippled or "washboard" surfaces that make study of the sections impossible. Such material may be reclaimed by painting over the surface with xylene or toluene which usually dissolves enough of the superficial layers of the mounting medium to reproduce a smooth surface.

6.   MODIFICATION OF THE RAPID GOLGI METHOD FOR ADULT TISSUE

It has become increasingly clear over the past decade that the rapid Golgi method can be rendered more effective in staining adult and myelin-rich tissue by means of short periods of prefixation in buffered formalin solution. This is usually accomplished by perfusing the deeply anesthetized animal with buffered (pH 7.0–7.2) formalin (7–10%) following a quick perfusion wash with physiological saline or Ringers solution. If the tissue is also to be used for parallel studies with the electron microscope, a mixture of 3–4% formalin and 1% glutaraldehyde is preferred. Following perfusion for about 4 min in the case of an adult rat and 6–8 min in the case of an adult cat, the brain is allowed to remain *in situ* for about 30 min, is dissected free, is cut into appropriate sized blocks, and is immersed in fresh buffered formalin solution for another 6–12 hr. After this, the blocks are removed, blotted, and started through osmic-dichromate solution. Fixation times average 4–6 days, and the results after 1–2 days of silvering are often surprisingly good.

B.   The Cox Modification of the Golgi Method

The Cox modification of the Golgi method allows simultaneous fixation and impregnation of the tissue, fixation being primarily the function of a group of chromate salts, while the heavy metal responsible for impregnation, in this case, is mercury. The method remains popular among contemporary investigators primarily because its use increases the probability that an appreciable number of nerve cells will be stained. Indeed, the Cox modification is excellent for studying dendrite masses with respect to length, position, and orientation. On the other hand, the impregnation is seldom sufficiently delicate to allow adequate staining of axons and their plexuses. For these reasons, the Golgi–Cox has been used frequently in histological studies of dendrite patterns in cerebral cortex (Conel, 1939; Sholl, 1953), but other techniques have been used to reveal data on axonal detail. The technique as described by Cox (1891) has shown remarkably little variation in 85 years. Those variations that have developed are discussed in detail by Ramón-Moliner (1970).

1.   FIXATION AND IMPREGNATION

Tissue fixation and impregnation are carried out in a solution made up of 20 parts of 5% potassium dichromate and 20 parts of 5% mercuric chloride, which are mixed together and then added to 40 parts of distilled water. This mixture should then be warmed in a water bath (to decrease precipitation following the next step) after which 8 parts of 5% potassium chromate are added with continuous stirring. Blocks of brain tissue removed as in the manner already described for the rapid Golgi method are placed in the mixture, are sealed well, and are stored in the dark, at room temperature for 6–10 weeks. We have found that it is useful to change the solution once or twice during the course of impregnation to keep tissue precipitate to a minimum.

The end point of the impregnation must be determined empirically. This is done most easily by slicing off very thin sections by hand with a razor, rapidly clearing and mounting them on glass, and looking for impregnated elements. When an appropriate point in the impregnation has been attained, tissue blocks are removed from solution, drained on absorbent paper, and prepared for cutting and processing as with the rapid Golgi. Like the former, celloidin or parlodian appear to give the best result in sectioning, and there is relatively less chance of damaging the somewhat cruder Cox impregnation. Sections are cut on the sliding microtome at 50–200 μm and are transferred into 70% alcohol.

2. Processing

Processing of Cox–Golgi material differs only in one detail from the rapid Golgi: the Cox mercury impregnation needs intensification and darkening before microscopic study is possible. The classical technique for accomplishing this was to bring the section from 70% alcohol to distilled water, then to a weak ammonia solution for a few minutes until gross darkening occurred, then back to distilled water, and finally through dehydration and clearing. Sholl (1953) has devised an alternative method that not only adequately darkens the cellular elements, but simultaneously bleaches out the unstained background, thereby providing a clearer field and somewhat better contrast. After removal from 70% alcohol, sections are transferred to distilled water with a few drops of a wetting agent for 15 min. They are then placed in 5% potassium sulfide solution with 2–3 drops of 5% oxalic acid for 30–60 min, and then rinsed again. Sholl follows a somewhat different method for clearing and fixing the specimens to the slide. Sections are transferred to equal parts of absolute alcohol and chloroform, then placed on the slide, flattened with filter paper, and coated with 1% celloidin. The whole slide is then transferred to equal parts of alcohol and chloroform, cleared in terpineol or essential oil mixture, and then mounted in a thick warm mounting medium.

3. Counterstaining

Counterstaining of the nonimpregnated neurons is readily achieved with Cox material and considerably adds to the informational content of the tissue. Any one of a number of aniline stains, either in aqueous or alcoholic solution, can be used. For example, following the darkening procedure and return to absolute alcohol, sections are immersed in a 2% mixture of Thionine (or cresyl violet) and methylene blue in 95% alcohol, where they are allowed to remain from 15 min to 1 hr depending on the thickness of the section. When the tissue sections appear grossly blue to inspection, they are washed and differentiated in several baths of 95 and 100% alcohol until all of the background color is judged to be gone. The sections are then cleared and mounted as before. Preparations such as this are useful for studying the relationship between dendrite masses and nerve cell bodies and, if the counterstain is successful, should enable the derivation of at least semiquantitative data. Ramón-Moliner, Vane, and Fletcher (1964) have described a variant of this procedure using cresyl violet, neutral red, or Darrow red. The results appear to be at least equally good.

Cox preparations are, on the whole, more durable than rapid Golgi material and often may be kept under coverslips with little loss of detail over a period of 3–5 years or more. Our personal preference is to leave them uncovered, adding slips only when specifically indicated.

C. Modifications Using Formalin

A number of modifications of the rapid Golgi method are built around the substitution of formalin for the more expensive osmic acid. One of the better known of these is the method of Strong (1895) and Kopsch (1896) which is, in every other respect, a rapid Golgi method. The procedure presented below, originally described by Hortega del Rio (1928) and recently redescribed by Klatzo (1952), is essentially the formol–dichromate method of Kopsch (1896) and Strong (1895) with the addition of chloral hydrate. Although its role in the reaction is not clear, it is possible that chloral hydrate functions as a buffer. This method has given the authors satisfactory results in cerebellar cortex, and to a somewhat lesser degree in cerebral cortex and basal ganglia. Adult brainstem and spinal cord are, as always, more difficult to impregnate but the method often gives surprisingly adequate results even here where the classical rapid Golgi is usually impotent. While it is possible that this method is the best of the Golgi modifications for studying all types of neuroglia in the central nervous system, the authors have been especially interested in its ability to delineate small granule and Golgi type II cells in cerebral and cerebellar cortices (see Color Plate I, frontispiece). Although the method is infrequently used today, its strengths suggest that increased use might well be made of it.

1. FIXATION

Fixing the tissue is achieved by adding 10 ml of neutral reagent grade formalin, 4 gm of potassium dichromate, and 4 gm of chloral hydrate to each 90 ml of distilled water. Suitably sized blocks of brain tissue (4–8 mm in thickness) are removed quickly from the brain and immersed in the solution at room temperature. High ambient temperatures do not appear to be as destructive to this type of preparation as to rapid Golgi material, although speed of hardening will, of course, increase as temperature of the mixture rises. Heavy brown sludges form in the fixative by the second or third day, and it has seemed helpful to change the solution several times depending on the length of fixation. The fixation period depends on the neural components that the worker wishes to impregnate. In general, protoplasmic astroglia, oligodendroglia, and microglia are likely to be stained within the first 12–24 hr. After this, nerve cell bodies and dendrites may be stained and, somewhat later, their axons. In the authors' work on the granule cell layer of the cerebellum, it has been found that fixation periods of from 1–3 weeks are optimal in cat, monkey, and man.

2. IMPREGNATION

At the end of this time the fixative is decanted, the tissue washed briefly in distilled water then immersed in a number of baths of 0.75% silver nitrate solution until the reddish-brown silver chromates cease to appear. The tissue blocks are then immersed in an excess of fresh 0.75% silver nitrate and stored in the dark at room temperature. Once more, the duration of the silvering process depends on the intensity of impregnation which the worker hopes to achieve. However, the end point appears to be less critical than that with the rapid Golgi, and the stain appears to develop in intensity over a much longer period. For example, it may take from 2 to 10 days for complete impregnation to develop in oligodendroglia or in granule and Golgi type II cells.

Following termination of impregnation, all further steps are substantially the same as those for the rapid Golgi. Of all the Golgi modifications, this one is probably the least durable, with or without coverslip. For this reason, it is usually wise to study valuable preparations as soon as possible after mounting. They have their greatest clarity and brilliance at this time, qualities which will have usually disappeared in several weeks' time. Despite this unfortunate feature of the method, it remains the best one for simultaneous visualization of neuronal and neuroglial elements and is, in our opinion, the best available light microscope method for the study of the morphology of neuroglia.

The authors, drawing on their experience with chloral hydrate as an additive, have experimented with a considerable number of other drugs and chemicals added to the basic osmic or formalin-based Golgi stain. These have included sympathomimetic and parasympathomimetic drugs, hypnotics, analeptics, narcotics, and psychotropic agents. As might be expected in the case of a pragmatic exercise of this sort, the results have been highly variable. Certain classes of agents appear capable of producing changes which are, in some cases, beneficial to the stain. Problems in establishing optimal parameters make it premature to describe these at present, but it is hoped that this material may be reviewed and published in the future.

D. Golgi-Type Impregnations of Completely Formalin-Fixed Tissue

The modification described by Fox, Ubeda Purkiss, Ihrig, and Biagioli (1951) was one of the first to stain formalin-fixed tissue with some degree of success. Although our own results with the method have not been as satisfactory as we could have hoped, we have seen the method in the hands of Fox and his students produce excellent and sensitive impregnations in many portions of the central nervous system. This serves to underlie, once more, the individual peculiarities of each of the Golgi techniques and the necessity of thorough familarity with a method before passing judgment on its capabilities. In addition to its affinity for older formalin-fixed material, a second strength of the technique lies in its capability, like the Strong–Kopsch method, of staining myelin-rich tissue. Thus, the technique is potentially capable of use in any portion of the mature brain, a quality notoriously lacking in the rapid Golgi.

1. FIXATION

Brains are perfused and fixed in neutral 10% formalin, but "old" formalin-fixed tissue, including human brain tissue can also be used. It is reported that tissue fixed for at least 2 years gives results clearly superior to that fixed for a month or two. Slices of 2–4 mm are cut from suitably formalinized material and immersed in chromating solution consisting of 4–5 gm (and up to 8 gm in some cases) of zinc chromate dissolved in 98 ml of distilled water and 2–4 ml of formic acid. Two days of fixation is considered adequate.

2. IMPREGATION

After the chromated slices are removed from the fixative and blotted dry, threads are passed through them, and they are suspended in a large volume of 0.75% silver nitrate solution. A similar effect can be obtained by placing glass beads or glass wool in the bottom of the bottle. In each case, the aim is to ensure contact of all sides of the tissue block with the silver. After 24 hr the material is transferred to fresh silver solution, and the process is considered complete in 48 hr. The authors suggest that before and after the final silvering, all adhering crystals be brushed away with a camel's hair brush. If the crystals are soft and pliable and resemble gold foil, a successful impregnation has usually resulted.

3. SECTIONING AND MOUNTING

Fox *et al.* (1951) suggest that their thin tissue slices are best prepared for sectioning by rapid dehydration in 95–100% alcohol (totaling 1 hr) followed by 10 min in xylene and an equal time in low-melting-point paraffin, after which the specimens are embedded and mounted. Sections are cut at 100 μm, passed through absolute alcohol and several changes of xylene, and mounted on glass slides under Permount or other synthetic resin. The authors later apply coverslips after moistening the surface of the mounting medium with xylene, then apply flat lead weights while warming the slide to hasten drying. In the authors' limited experience with the method, gradual deterioration occurs in some of the preparations as with most Golgi material. The technique is less successful with young and fetal material.

Since the method of Fox *et al.* (1951) was described, there have been a number of other acidified chromate methods published in the literature. Of these, the authors have found most useful a method described by Davenport and Combs (1954). As in the case of the Fox modification, this technique is maximally useful on old formalin-fixed adult tissue from most mammalian species. It has also been found that it works well on human brain tissue.

The fixative consists of 1.5 gm of cadmium or zinc chromate, 25 ml distilled water, and 6 ml glacial acetic acid. Sixty-five milliliters of 5% potassium chromate solution are added when solution of the first mixture is complete. Optimal immersion time is 48 hr followed by the same period in 0.75% silver nitrate solution. Tissue blocks may then be cut in paraffin shells or after rapid parlodian impregnation and processed as previously described. Under some as yet unspecified conditions, this modification may stain as densely as the Golgi–Cox and as delicately as the best rapid Golgi. However, it is not reliable, and the investigator must expect to come up with a number of "blanks," no matter how faultless his technique.

V.   What Are the Artifacts and Pitfalls?

The problem of artifacts has long been considered one of the more significant drawbacks to the Golgi methods. However, such stories are, like the report of Mark Twain's death, greatly exaggerated. As in the case of other methods, aberrant or neurologically meaningless structures are visualized but, considering the extraordinarily high information content of Golgi material, a fair amount of background noise can easily be tolerated. As a matter of fact, the range of artifacts is quickly run through and their means of recognition easily established.

Precipitates of flat orange chromate crystals are possible in any of the chrome–silver methods. It is almost impossible to mistake these for neural structures. In the chrome–mercury (Golgi–Cox) modification, there may be areas of generalized black stippling or larger black spherules, occurring individually or in clusters (see Fig. 8). Once again, it is almost impossible to mistake these for biological structures, and they probably represent small amounts of excess mercury which have precipitated out of solution for one of a number of reasons.

Figure 9 illustrates the larger, rather amorphous dark brown or black masses which are seen in Golgi impregnations from time to time and probably represent an aberration of the impregnation process. It is assumed that these represent excesses of the Golgi impregnation deposited in a neuronal-glial stroma in which, for some reason, the impregnation was not "turned" off at the appropriate time. It might be added in passing that this represents part of a fascinating question about the Golgi methods as a whole, i.e., what triggers the process of

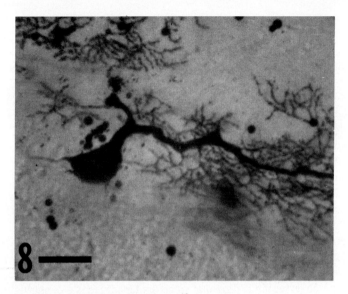

Figs. 8 and 9. Two common types of staining artifact.

Fig. 8. A partially stained Purkinje neuron from adult cat cerebellum. This Golgi–Cox stained section had been kept under coverslip for over 10 years. Only a portion of the dendrite system visible is still heavily stained. The finer tertiary system of branches has undergone considerable fading. A number of small round dark shotlike bodies above and below the cell body are typical Cox artifacts—possibly tiny globules of mercury. Calibration bar: 30 μm.

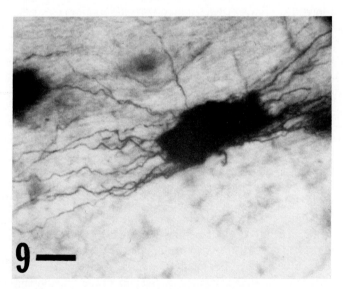

Fig. 9. A photograph of 3-month-old cat medulla. It shows a presumed deposit of silver salts in a neuronal-glial matrix of uncertain composition. This constitutes a very common type of artifact characterized by a large amorphous central dark zone with fibrous elements streaming away in all directions. Calibration bar: 30 μm.

impregnation, what guides it, and what turns it off. While there are no direct answers to these questions, there are some educated guesses as to the nature of the reaction, and a little spectrographic data, as will be noted in Section VI. In any case, these amorphous dark masses are usually easy to distinguish from the more precisely shaped neurobiological entities around them.

A third artifact appears based more exclusively in fibrous atroglia and usually results in sunburst types of structures with long straight radiating spicules often many hundreds of micrometers in length. These may, at times, be mistaken for large glial colonies (which, in a sense, they are) but a little practice usually serves to differentiate these structures from the more structurally specific astrocytes.

It seems to us that there are two major pitfalls of which the reader and potential practitioner should be aware. The first is the unrealistic expectation that a complete picture be available from one or two series of sections. The other is, in a sense the obverse; the feeling that since the technique is a century old and still not fully capable of being controlled, it has no place in the modern high-technology laboratory. With regard to the first, it should be clear by now that the Golgi methods are all, by nature, partial methods. In fact, if they stained every element in a field in the same sense that a Nissl or reduced silver type of stain does, the field would appear as an impenetrable black mass . . . so great is the density of neuropil (see Fig. 5). The "empty" background is as much a part of the Golgi picture as the occasional elements silhouetted against it, and the more the investigator works with the system, the more grateful he is that the method supplies a set of built in filters or windows to allow some degree of recognition and focusing on a few paradigm elements. It follows that the most successful impregnation can supply at best a fragment or caricature of the elements and the connectivity patterns that are actually present (see Fig. 3). Development of the complete picture becomes a work of synthesis, and this is the real Golgi "work" from the point of view of the investigator. The completeness and validity of the model reconstructed are entirely a function of the powers to visually process and conceptually redistribute a host of seemingly unrelated neural elements. It might be said that every neural schema built up from Golgi studies is as much a work of art as a fragment of science.

It is in part because of this qualitative, intensely personal component in Golgi work, that its curiously maverick nature need not be a source of embarrassment in an increasingly predictable world. Undoubtedly, a concerted attack on the physicochemical nature of the reactions involved, using modern methodology would soon reveal the basis for the reaction, and conceivably, the reasons for its curious neuronal selectivity. For intellectual reasons, all would be gratified to see this. Indeed the initial studies reported in Section VI are exciting and satisfying. Whether the method might yield more handsomely after it is "understood" is of course another question. Does the lark sing sweeter when the innervation of its larynx is understood?

It might also be pointed out that all methodologies have their idiosyncracies, and it is not surprising that a technique as poorly understood from the methodological point of view as the Golgi might have its share. But to reiterate a former point, the informational content of each section is usually so high that the signal-to-noise ratio can be considered remarkably favorable. As experimental methods go, it must be considered docile and easy to use, a victim more of its own mystique than of any intrinsic and unsolvable problems.

VI.   Rationale of the Golgi Methods

It can be said that the genre of Golgi methods was probably conceived in serendipity and developed pragmatically. After 100 years there is still no clear understanding of the mechanisms substrate to this remarkable group of methods. However, data are available which may throw at least some light on the subject. Virtually all Golgi modifications rely on the use of dichromates which are recognized as vehicles for stabilizing lipoprotein membranes (Fregerslev, Blackstad, Fredens, Holm, & Ramón-Moliner, 1971b). All Golgi techniques, whether relying on silver or mercury impregnation, are characterized by the formation of microcrystals within spaces delimited by lipoprotein membranes. Electron microscopy shows clearly that the impregnation is variable and often spotty, but occurs largely within the cytoplasmic envelop of the cell, dendrite, and axon. The actual staining material has been thought to be a lipoprotein–chromesilver complex maintained in some type of loose association and not "bound in" like the silver in a reduced silver method. There has also been a good deal of speculation on the nature of the reactions involving chromates, osmic acid, silver ions, and protein. Valverde (1970) has suggested that osmium tetroxide enters a series of reactions which carry it to a diol and then a diketone form, the latter being able to link to chromium and react with silver nitrate.

On the other hand, Golgi-impregnated material recently has been studied by tbe X-ray powder diffraction method and the black precipitate identified as AgCrO$_4$ (Fregerslev, Blackstad, Fredens, & Holm, 1971a) with the possibility of a secondary formation of metallic silver. Using similar methods on Golgi–Cox impregnations, the same group found that, in this case, the precipitate appears to be Hg$_2$Cl$_2$ prior to alkalinization (darkening) and following this procedure, apparently metallic liquid mercury (Fregerslev et al., 1971b). What is clear in any case is that the Golgi impregnation does not simply provide metal-plated relics of neuronal elements, but, rather, introduces metallic salts directly into the cyton, molded and limited by the membrane complexes making up the living system. It is equally clear that much still remains to be learned before one can hope to sort out critical from uncritical operations in the techniques described. Among the items of obvious relevance in any total approach to operational principles would be the role of pH control, the interactions of the heavy metal salts with membranes, and the significance of the immediately preceding vital state of the neuron just before fixation.

VII.   Interpretation of the Data

We have always encouraged our students and colleagues to do their own Golgi technical work personally and have, so far as academic and personal responsbilities allowed, followed this dictum ourselves. Each block that is processed is totally unlike any other, and each section coming off the knife carries in it the potential for exciting and unexpected insights. This unusual state of affairs is quite as true today as it was in the final heroic decades of the nineteenth century when the optical microscopy of the nervous system mined its richest lodes. The investigator who is midwife to the birth of each section and who gives it its brief initial check as soon as the mounting medium has covered the tissue, is in an optimal position to "know" his material and to build up impressions while he still has the kinesthetic feeling of that particular block. Although this may seem an exaggeration to some readers, the fact remains that "second-hand" Golgi material produced by even the most fastidious laboratory technician can never mean quite the same thing to the investigator as the material he has personally ushered through.

The study and interpretation of this material is an even more personal process, no aspect of which can be relegated to another. For one thing, it takes a great deal of looking before one can begin to see critically. Even an average and not particularly distinguished section reveals potentially enormous amounts of information about organization of cell groups, neuronal circuitry, interrelations and structure of dendritic domains, axonal trajectory, and cytology. It is a rich visual feast and one cannot seek answers until the questions are formulated. Our own experience teaches us that inevitably, when we begin study of a batch of material, particularly if the project is new or the part of the brain is less familiar to us, we must expect to work back and forth over the sections, often for several days or weeks, before we begin to reformulate our questions. (Our initial formulations usually seem irrelevant after the study of the tissue.) We assume that a new set of personal filters has to be generated for purposes of recognition and screening before the process of "looking" can truly be said to have become "seeing."

We conclude that doing Golgi work is a peculiarly personal scientific experience where the rigor of objective description of many parital images is enriched by coneptual synthesis and tipped by aesthetic judgment. Golgi himself feared what he called "... fantasy which leads only to an appearance of progress," a point of view reminiscent of the echoing "hypothesis non fingo" of Newton more than two centuries earlier. But the fact remains that both deduced magnificant consequences from their observations—which were often less than complete and sometimes less than objective.

The enormous power of the Golgi methods is essentially a latent power. Perhaps more than any other histological technique, it depends on the interpretative gifts of its particular master for what it will teach. Unlike virtually all other optical microscopic methods, it is almost endlessly adaptable, as much at home on the kitchen table laboratory of the nineteenth century as in the computerized edifice of the twenty-first. Its remarkable capacity for showing all elements of the nervous system as they "really are" ensures its uniquely central position for as long as investigators have the ingenuity to put new questions to it.

Acknowledgments

The studies upon which this review is based were supported by the U.S. Public Health Service via Grant Nos. NB 1063, NS 10567, and NS 11468.

References

Blackstad, T. Golgi preparations for electron microscopy: Controlled reduction of the silver chromate by ultraviolet illumination. *In* M. Santini (Ed.), *Golgi centennial symposium.* New York: Plenum, 1975. Pp. 123–133.

Bok, S. T. *Histonomy of the cerebral cortex.* Amsterdam: Elsevier, 1959.

Chan-Palay, V., & Palay, S. L. High voltage electron microscopy of rapid Golgi preparations. Neurons and their processes in the cerebellar cortex of monkey and rat. *Zeitschrift fer Anatomie und Entwicklungsgeschichte,* 1972, **137,** 125–152.

Conel, J. L. *The postnatal development of the human cerebral cortex.* Vol. I *et seq.* Cambridge, Massachusetts: Harvard University Press, 1939.

Conn, H. J., Darrow, M. A., & Emmel, V. M. *Staining procedures used by the biological stain commission.* (2nd ed.) Baltimore, Maryland: Williams & Wilkins, 1960.

Cox, W. Impregnation des centralen Nervensystems mit Quecksilbersalzem. *Archiv fer Mikroskopische Anatomie,* 1891, **37,** 16–21.

Davenport, H. A., & Combs, C. M. Golgi's dichromate-silver method. 3. Chromating fluids. *Stain Technology,* 1954, **29,** 165–173.

De Moor, J. La mécanisme et la signification de l'état moniliforme des neurones. *Annales de la Societe des Sciences Medicales at Naturelles de Bruxelles,* 1898, **7,** 205–250.

Fox, C.A., Ubeda Purkiss, M., Ihrig, H. K., & Biagioli, D. Zinc chromate modification of the Golgi technique. *Stain Technology,* 1951, **26,** 109–114.

Fregerslev, S., Blackstad, T., Fredens, K., & Holm, M. J. Golgi potassium dichromate silver nitrate impregnation: Nature of the precipitate studied by X ray powder diffraction methods. *Histochemie,* 1971, **25,** 63–71. (a)

Fregerslev, S., Blackstad, T., Fredens, K., Holm, M. J., & Ramón-Moliner, E. Golgi impregnation with mercuric chloride: Studies on the precipitate by X ray diffraction and selected area electron diffraction. *Histochemie,* 1971, **26,** 298–304. (b)

Golgi, C. Sulla struttura della sostanza grigia dell cervello. *Gazzetta Medica Lombarda,* 1873, **33,** 244–246.

Golgi, C. Di una nuova reasione apparentemente nera delle cellule nervose cerebrali ottenuta col bichloruro de mercurio. *Archivio per le Scienze Mediche,* 1879, **3,** 1–7.

Hortega del Rio, P. Contribucion al conocimiento citologico de los tumores del nervio y quiasma optico. *Memorias Sociedad Espanola de Historia Natural,* 1928, **14,** 7–41.

Huttenlocher, F. R. Dendritic development in neocortex of children with mental defect and infantile spasms. *Neurology,* 1974, **24,** 203–210.

Klatzo, I. A study of glioblastoma multiforms by the Golgi method. *American Journal of Pathology,* 1952, **28,** 357–362.

Kopsch, F. Erfahrungen über die Verwendung des Formaldehyde bei der Chromsilber-Impregnation. *Anatomischer Anzeiger,* 1896, **11,** 727.

Lindsay, R. D., & Scheibel, A. B. Quantitative analysis of the dendritic branching pattern of small pyramidal cells from adult rat somasthetic and visual cortex. *Experimental Neurology,* 1974, **45,** 424–434.

Marin-Padilla, M. Structural abnormalities of the cerebral cortex in human chromosomal aberrations. A Golgi study. *Brain Research,* 1972, **44,** 625–629.

Porter, R. W., & Davenport, H. A. Golgi's dichromate-silver method. 1. Effects of embedding. 2. Experiments with modifications. *Stain Technology,* 1949, **24,** 117–126.

Purpura, D. Normal and aberrant neuronal development in the cerebral cortex of human fetus and young infant. In N. A. Buchwald & M. A. B. Brazier (Eds.), *Brain mechanisms in mental retardation.* New York: Academic Press, 1975. Pp. 141–166.

Ramón-Moliner, E. The Golgi–Cox technique. *In* W. J. H. Nanta and S. O. Ebbesson (eds.). *Contemporary Research Methods in Neuroanatomy.* New York, Springer. Verlag, 1970, pp. 32–55.

Ramón-Moliner, E., Vane, M. A., & Fletcher, G. V. Basic dye counterstaining of sections impregnated by the Golgi-Cox method. *Stain Technology,* 1964, **39,** 65–70. Pp. 32–55.

Ramón y Cajal, S. *Histologie du système nerveux de l'homme et des vertébrés.* Madrid: Consejo Superior de Investigaciones Scientificas, 1952. 2 vols.

Retzius, G. *Biografiska Anteckninger och Minven.* Vol. II. Stockholm: Almkvist & Wiksell, 1953. [*Biographical notes and recollections*].

Scheibel, M. E., Crandall, P. H., & Scheibel, A. B. The hippocampal dentate complex in temporal lobe epilepsy. *Epilepsia,* 1974, **15,** 55–80.

Scheibel, M. E., Lindsay, R. D., Tomiyasu, U., & Scheibel, A. B. Progressive dendritic changes in aging human cortex. *Experimental Neurology,* 1975, **47,** 392–403.

Scheibel, M. E., & Scheibel, A. B. Observations on the intracortical relations of the climbing fibers of the cerebellum. A Golgi study. *Journal of Comparative Neurology,* 1954, **101,** 733–764.

Sholl, D. A. Dendritic organization in the neurons of the visual and motor cortices of the cat. *Journal of Anatomy,* 1953, **87,** 387–406.

Stell, W. K. Correlation of retinal cytoarchitecture and ultrastructure in Golgi preparations. *Anatomical Record,* 1965, **153,** 389–397.

Strong, O. S. Notes on neurological methods and exhibition of photomicrographs. *Anatomischer Anzeiger,* 1895, **10,** 494.

Valverde, R. The Golgi method. A tool for comparative structural analysis. *In* W. J. H. Nauta & S. O. E. Ebbeson (Eds.), *Contemporary research methods in neuroanatomy.* Berlin & New York: Springer-Verlag, 1970. Pp. 12–28.

Chapter 5

Fluorescence Histochemistry

Robert Y. Moore and Rebekah Loy

Department of Neurosciences
University of California
La Jolla, California

I. Introduction

The purpose of this chapter is to describe a group of fluorescence histochemical methods which permit the visualization of certain groups of neurons on the basis of their neurotransmitter content. At the present time these methods can only be applied to neurons which utilize certain biogenic amines, the catecholamines and indoleamines, as neurotransmitter agents. Thus, these histochemical methods are unique as neuroanatomical methods in that they use a neurotransmitter substance for identification of the neurons to be studied, and they are highly selective in the neurotransmitter substances which will undergo the histochemical reaction. Despite these apparent limitations, however, these fluorescence histochemical methods are now widely applied. The intensive investigation of central catecholamine and indoleamine neurons, which followed the introduction of these methods 15 years ago, has contributed substantively to our understanding of the organization and function of the central nervous system.

The history of the development of these methods has been reviewed (cf. Björklund & Moore, in press; Corrodi & Jonsson, 1967) and will not be repeated here. Briefly, the essential background for the development of the methods was the establishment of the concept of chemical neurotransmission and the identification of putative central neurotransmitter agents. Although the catecholamine, noradrenaline, was known to be the peripheral sympathetic neurotransmitter from the work of von Euler, (1947), it was not until 1954 that M. Vogt unequivocally demonstrated the presence of noradrenaline in brain, in a regional distribution and independent of its sympathetic innervation (Vogt, 1954). At approximately the same time, Twarog and Page (1953) and Amin, Crawford, and Gaddum (1954) identified an indoleamine, 5-hydroxytryptamine or serotonin, in brain. Subsequently, Bertler and Rosengren (1959) demonstrated the striking regional localization of dopamine in the neostriatum, and this was followed shortly by the important clinicopathologic observation that neostriatal dopamine content is markedly reduced in the brains of parkinsonian patients (Ehringer & Hornykiewicz, 1960).

During the late 1950's and early 1960's the regional distribution of biogenic amines in brain was studied intensively, and it became evident that this information was most easily interpreted by the assumption that the compounds were associated with specific groups of central neurons. To establish this association a histochemical method obviously was required. For biogenic amines it is necessary that the neurotransmitter remains in its *in vivo* position and that the histochemical product be microscopically visible.

Several investigators attempted to develop such methods, but the introduction by Falck, Hillarp and their co-workers (Falck, 1962; Falck, Hillarp, Thieme, & Torp, 1962; Falck & Torp, 1961) of a method based on the condensation of a biogenic amine with formaldehyde in a gas phase reaction met these qualifications better than prior methods. The use of the gas phase reaction on dried or freeze-dried tissue prevented the diffusion of the water-soluble biogenic amines, which had vitiated previous methods and, in the Falck–Hillarp method, produced an intense fluorophor which could readily be identified in the fluorescent microscope. The application of the original Falck–Hillarp method (cf. Björklund, Falck, & Owman, 1972a; Falck, 1962, Falck & Owman, 1965) led to extensive advances in our knowledge of the organization of central monoamine neuron systems (cf. Andén, Dahlström, Fuxe, Larsson, Olson, & Ungerstedt, 1966;

Dahlström & Fuxe, 1964, 1965; Fuxe, 1965; Hökfelt, & Ungerstedt, 1968; Ungerstedt, 1971). Subsequent improvements in the fluorescence histochemical methodology such as the Vibratome–formaldehyde method (Hökfelt and Ljungdahl, 1972a) and, particularly, the glyoxylic acid method (Björklund, Lindvall, & Svensson, 1972b; Lindvall & Björklund, 1974a; Lindvall, Björklund, Hökfelt, & Ljungdahl, 1973) have led to further advances so that our knowledge of the organization of central catecholamine neuron systems is probably nearly as great as that for any nonsensory, central system (cf. Björklund & Moore, in press; Lindvall & Björklund, 1974b).

The striking advantages of the fluorescence histochemical techniques are their selectivity and their sensitivity. At the present time it is generally agreed that the methods demonstrate only noradrenaline, dopamine, 5-hydroxytryptamine, and adrenaline neurons in the mammalian central nervous system. The sensitivity of the methods has been reviewed by Björklund, Falck, and Lindvall (1975), who conclude from information in the literature and their own observations that as little as 5×10^{-6} pmol of noradrenaline or dopamine can be detected in one varicosity with the Falck–Hillarp technique. The glyoxylic acid technique is even more sensitive and for dopamine, for example, they calculate that 10^{-7} pmol can be readily detected. The methods are less sensitive for adrenaline and the indoleamines, and this will be discussed in regard to application of the individual methods.

In the body of this chapter two major variations of the fluorescence histochemical methods shall be presented and comment made upon their special uses, advantages, disadvantages, and pitfalls. The first will be the original Falck–Hillarp technique and its modification for the Vibratome; the second will be the glyoxylic acid methods. Each variant has special uses and advantages, but the reader should be warned that this is a period of intense activity in this field, and very rapid technological advances are being made in neurotransmitter histochemistry. These include the development of new methods as well as the improvement of existing ones. Thus, anyone who becomes seriously involved with the application of these methods should keep a close eye on the literature.

II. Fluorescence Histochemical Methods for Demonstrating Biogenic Amines in Neurons

A. The Chemical Basis of the Methods

1. THE FALCK–HILLARP METHOD (FORMALDEHYDE CONDENSATION METHOD)

The chemical basis of the Falck–Hillarp method was worked out principally by Corrodi and Hillarp (cf., for reviews, Björklund et al., 1975; Corrodi & Jonsson,

1967) and based on the foundation of many years of work by numerous investigators in organic chemistry of cyclization reactions. The principal reaction in the Falck–Hillarp method is the condensation or cyclization of formaldehyde with a primary catecholamine or indoleamine. This reaction, called a Pictet–Spengler reaction, is well known in organic chemistry and has been studied extensively. For the catecholamines it is a general condensation method for reacting primary or secondary β-arylethylamines with a carbonyl compound to yield a tetrahydroisoquinoline. A 5-hydroxyindolylethylamine (e.g., 5-hydroxytryptamine) undergoes the condensation reaction to form a 6-hydroxy-1,2,3,4-tetrahydro-β-carboline (see Fig. 1). In each case, the product of the condensation reaction, which is only weakly fluorescent, undergoes a secondary dehydrogenation in a protein-promoted reaction which is not understood. The reaction will not take place in the absence of protein, but the substitution of peptides or even amino acids (only alanine and glycine are effective) will also catalyze the reaction. The reaction products, a 3,4-dihydroisoquinoline in the case of the catecholamines and a 6-hydroxy-3,4-dihydro-β-carboline in the case of the indolamine, are intensely fluorescent. At neutral pH, as occurs in tissue, these fluorophors are in their quinoidal forms, whereas the nonquinoidal forms predominate at lower pH values. This pH-dependent tautomerism is reflected in characteristic changes in the spectral properties of the fluorophor and

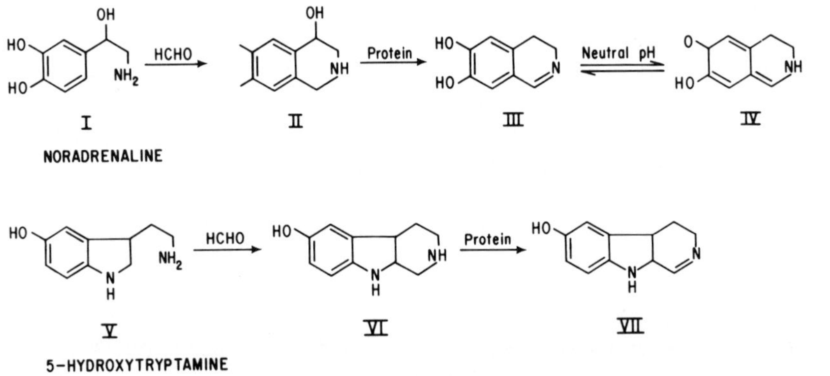

FIG. 1. Chemical basis of the Falck–Hillarp method. At the top of the figure, noradrenaline (I), as a representative phenylethylamine, undergoes a cyclization reaction with formaldehyde to form a 6,7-dihydroxy-1,2,3,4-tetrahydroisoquinoline (II) which is converted to a 3,4-dihydroisoquinoline (III) in a protein-promoted reaction. At neutral pH, as in tissue, the predominant fluorophor will be the tautomeric quinoidal form (IV). At the bottom of the figure, 5-hydroxytryptamine (V), as a representative indolylethylamine, undergoes a similar condensation reaction with formaldehyde to form a 6-hydroxy-1,2,3,4-tetrahydro-β-carboline (VI). This is dehydrogenated in a protein-promoted reaction to an intense fluorophor, 6-hydroxy-3,4-dihydro-β-carboline (VII). (Based on the work of Corrodi and Hillarp, reviewed by Björklund et al., 1975, and Corrodi and Jonsson, 1967).

forms the basis by which the dopamine fluorophor can be distinguished from the noradrenaline fluorophor by acid treatment (Björklund et al., 1975).

2. THE GLYOXYLIC ACID METHOD

The fluorophor formation from primary and secondary phenylethylamines and indolyethylamines is quite similar in the glyoxylic acid method to that in the formaldehyde condensation method (Björklund et al., 1975). The reaction proceeds in two steps, as shown in Fig. 2. In the first a phenylethylamine or indolylethylamine reacts with glyoxylic acid in an acid-catalyzed Pictet–Spengler reaction to yield the 1,2,3,4-tetrahydroisoquinoline-1-carboxylic acid or 1,2,3,4-tetrahydro-β-carboline-1-carboxylic acid, respectively, via a Schiff's base. In the second step these very weakly fluorescent compounds may be transformed into strongly fluorescent products in two alternative ways: via autoxidative decarboxylation to the 3,4-dihydroisoquinoline or 3,4-dihydro-β-carboline; or through an intramolecularly acid-catalyzed reaction with glyoxylic acid to the 2-carboxymethyl-3,4-dihydroisoquinolinium or 2-carboxymethyl-3,4-dihydro-βcarbolinium compounds. A further decarboxylation to the 2-methylated compounds may occur. These glyoxylic acid-induced fluorophors will exhibit a pH-dependent tautomerism, as shown by the formaldehyde-induced fluorophors. At neutral pH they will be in the quinoidal forms and, at acid pH, in the nonquinoidal forms.

Glyoxylic acid yields a stronger fluorophor than formaldehyde (Björklund et al., 1975). This can be attributed to several phenomena: the formation of strongly fluorescent products by the intramolecular acid catalysis exerted by the carboxyl group on the 1-carbon of the tetrahydroisoquinoline or tetrahydro-β-carboline molecules; the promotion of the initial cyclization step due to the acid catalysis participating in both steps of the fluorophor formation; and acidification, which is known to increase the fluorescence intensity of some amine fluorophors.

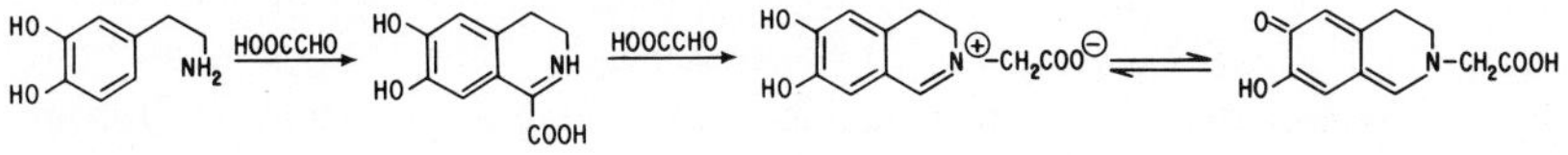

FIG. 2. Chemical basis of the glyoxylic acid method. Dopamine (I), as a representative phenylethylamine, undergoes an acid-promoted cyclization reaction with glyoxylic acid to form a weakly fluorescent 6,7-dihydroxy-1,2,3,4-tetrahydroisoquinoline-1-carboxylic acid derivative (II). The fluorophor formation then proceeds through an intramolecularly acid-catalyzed reaction with another glyoxylic acid molecule to form an intense fluorophor, a 2-carboxymethyl-6,7-dihydroxy-3,4-dihydroisoquinolinium compound (III), which is in a pH-dependent tautomeric equilibrium with its quinoidal form (IV). (Based on the work of Lindvall et al., 1974.)

It should be noted, however, that fluorophor formation occurs with a reaction between glyoxylic acid and the indolylethylamines in model experiments but this has not, as yet, been applied successfully to tissue as a histochemical method for the indolylethylamines. Consequently, the glyoxylic acid methods to be described are only applicable to the study of catecholamine neuron systems in living tissue.

B. Equipment

1. THE FLUORESCENCE MICROSCOPE AND FLUORESCENCE MICROSCOPY

The fluorescence microscope is the basic and essential unit of equipment common to all fluorescence histochemical methods. Microscopes suitable for use with these methods are produced by virtually all major manufacturers, and standard laboratory microscopes can often be modified by adding attachments so that they may serve both in routine microscopy and fluorescence microscopy. Two forms of illumination, by incident light or by transmitted light, can be used for fluorescence microscopy in these methods. Transmitted light is usually advantageous for work with low-power objectives, whereas incident light illumination is particularly advantageous for use with high-power immersion (oil or water) objectives with a numerical aperture greater than 0.75. The differences are not sufficiently great, however, that one would necessarily recommend one form of illumination over the other. Transmitted light illumination has been used much longer than incident light illumination and for that reason, and because we use transmitted light almost exclusively in our laboratory, we will describe a transmitted light system and its application to the fluorescence histochemical methods. The microscope is an ordinary light microscope which is fixed to a stand that has attached to it a lamp housing and a camera holder. In our work we use a 200-W mercury vapor lamp (Osram HBO 200) which requires a special power supply. The light emitted from the lamp passes through a heat absorption filter (Schott KG 1) and through suitable primary (excitation) filters which are thick enough to minimize unwanted excitation light and which select activation light in wavelengths as close as possible to the activation maximum (absorption maximum) of the fluorophors. The excitation maximum for the biogenic amines which can be visualized in the nervous system using the fluorescence histochemical methods is at 390–410 m, and the excitation, or primary filter, commonly used is a Schott BG 12. In our microscope we have four containers for excitation filters which can be placed in or out of the light path, in which we have BG 12 filters 2, 3, and 4 mm (2) thick, respectively. By careful selection of filters it is possible to obtain optimal visualization of the fluorophor with minimum background fluorescence. The secondary, or barrier, filter should have a high absorption below 500 mμ; an example of such a filter is a Schott OG 4. This excludes the blue component of the excitation light and, since the catecholamine fluorophors have an emission maximum about 475 mμ, the combination of a BG 12 primary filter and an OG 4 secondary filter results in catecholamines appearing green to greenish-yellow in the fluorescence microscope. 5-Hydroxytryptamine, which has an emission maximum about 525 mμ, will appear yellow with this combination of filters. The fluorescence microscope should be set up so that the light goes through the heat absorbtion and primary filters to a dark-field condenser that directs the light on the specimen. The fluorescence emitted by the specimen then goes through the objective lens of the microscope through the secondary filter to the eyepieces. It should be emphasized that the fluorescence is of low intensity and no beam-splitter or prism should be in the light path, as it will sufficiently reduce the light to make even good specimens unusable. For this reason, and because the electronics of the instruments are inadequate to assess the low levels of illumination obtained as fluorescence in the fluorescence histochemical methods, we have not found any of the automatic photomicroscopes we have tried to be usuable for these methods.

In summary, then, the fluorescence microscope contains a light source (200-W mercury vapor lamp), a cooling filter, primary filters, a dark-field condenser (we use only condensers that require oil; the dry dark-field condensers do not provide sufficient illumination), specimen (see below), objective lenses, a secondary filter, and eyepieces.

A trinocular head of our microscope contains a removable mirror, which directs all of the light to the eyepiece or, when removed, to the camera. The camera consists of a simple back, to hold and transport 35-mm film, attached to a connector between the microscope and the camera. The connector has a viewing lens, and the light can be directed either into the viewing lens or into the camera but, again, all of the light goes one direction or the other. The connector also has another position which will direct the light to a photometer, but this is not used as the standard commercial models of photometer will not measure the low levels of illumination obtained in the fluorescence histochemical methods. To give the reader some idea of the levels of illumination, it is useful to make a few comments on photomicrography. If a record of the appearance of material only is needed that will not be published, Kodak Tri-X film and exposures of 15–30 sec are used. The exposure times are determined empirically. If the photographs are for publication, either Kodak Panatomic-X or Kodak SO-410 is used. Exposure times for Panatomic-X are from 2 to 10 min, depending upon the specimen and the objective lens, and for SO-410, which is less sensitive in the green ranges of the spectrum, exposures are of 6–12 min duration. The latter film has the advantage, for the patient microscopist, of offering superb contrast and definition of the material. For color photographs Kodak High-Speed Ektachrome with exposures of 3–8 min are used.

One further practical point that should be made about fluorescence microscopy concerns the mercury vapor lamps. In the authors' experience these commonly have a useful life of about 200 hr. They do not "burn out" often in the conventional sense, but, rather, slowly become less and less intense. This can be very difficult to perceive, but is usually evident when exposure times for photography that had routinely been satisfactory result in underexposed film. Consequently exposure times are monitored carefully and a running record is kept of how long the lamp has been in use. Lamps have failed as early as 100 hr of use and, as a routine measure, any lamp which has been in use more than 200 hr is replaced.

2. FREEZE-DRYERS

The object of freeze-drying is to obtain the rapid removal of water from frozen tissues by sublimation at a temperature below the freezing point of the tissue. This is carried out by keeping the tissue under vacuum to promote vaporization of the water and providing a trap for the water vapor so that residual water is removed and complete drying is accomplished. Two types of freeze-dryer are currently in use for fluorescence histochemical techniques, one in which the water vapor trap is provided by a cold-finger and one in which the trap is a desiccant material (cf., for reviews, Björklund et al., 1972a; Eränkö, 1967; Falck, & Owman, 1965). Commercial models are available but, as Björklund et al. (1972a) point out, simple and inexpensive freeze-dryers can be constructed almost anywhere without difficulty. In this laboratory a commercial desiccant-type freeze-dryer designed by Olson and Ungerstedt (1970b) and produced by the Bergman and Beving Co., Stockholm, Sweden is used. The tissue is held on a flat metal plate which rests in the bottom of a cylindrical metal well immersed in an alcohol solution. Thus, the temperature of the plate will equilibrate rapidly to the temperature of the solution. Above the tissue-containing plate are placed metal mesh baskets containing the desiccant, granular phosphorus pentoxide (Granusec-Baker). The top of the cylinder is covered by a glass plate which rests on an O ring. The metal cylinder has a sidearm attached to an outlet to a vacuum guage, an outlet to a valve to allow air to be let in at the end of the freeze-drying, and an outlet to the vacuum pump which is a two-stage, mechanical and oil diffusion pump. This generates a vacuum of 10^{-3} torr in the freeze-drying chamber. The alcohol bath is contained in a standard compressor with a well and cooling coils. It is arranged so that the alcohol is continuously circulated and can be maintained at a constant temperature ranging from $-35°$ to $+30°C$. The major advantage of this freeze-dryer, and those described by Björklund et al. (1972a), is that they have a high capacity and can handle as many as 50–100 tissue specimens at one time depending upon their size.

3. VIBRATOME

The Vibratome is produced by the Oxford Instrument Co., San Mateo, California. Tissue is placed on a chuck in a cold solution in a trough and fixed in place as in any microtome. The Vibratome operates by moving a sharp blade forward in short rapid horizontal strokes. This allows small blocks of relatively soft tissue to be cut into sections of reasonably good histologic quality. The instrument is set up so that the rate of horizontal movement, the rate of forward movement, and the thickness of the sections can be controlled independently. These parameters will vary for different tissues, and the use of the Vibratome is empirical in respect to these parameters. Successful use of the Vibratome takes considerable practice, and cutting good sections from fresh tissue or glyoxylic acid-perfused tissue is often difficult even for the expert.

4. CRYOSTAT

Two cryostat methods will be described in Sections II,C and D. These can be done using any good, commercially available cryostat. A freezing-microtome is not an adequate substitute, however, because the cooling chamber of the cryostat keeps the sections frozen until mounted, and this permits the easily diffusable amines to be maintained in their original location.

C. Methods Using Formaldehyde Condensation: The Falck–Hillarp Technique and Variants

1. THE FALCK–HILLARP TECHNIQUE

a. STRETCH AND SMEAR PREPARATIONS. The Falck–Hillarp method was originally applied either to fresh tissue, which could be rapidly dried and exposed to formaldehyde vapor (Falck, 1962), or to model systems in which amines in a dried protein film were exposed to formaldehyde vapor (Falck et al., 1962). The stretch preparation continues to be a useful one for two purposes. First, it is performed simply and can allow the investigator to assess whether the formaldehyde treatment is working properly and to validate the function of the fluorescence microscope and photographic exposure times. Second, certain experimental situations only require stretch or smear preparations or can be adapted to use these techniques. The tissue most often used for a stretch preparation is the iris of the albino rat. This procedure is carried out by removing the eye, incising the entire circumference of the globe with an iris scissors, removing the lens, and placing the anterior portion of the globe, cornea down, on a clean glass slide. Using a small forceps the iris then can be removed by gently teasing away its attachment to the ciliary body, using a dissecting microscope. The iris is then moved to an adjacent portion of the glass slide where it is gently stretched to approximate its original shape. The remaining tissue is discarded, the excess fluid around the iris blotted away, and the slide placed overnight in an evacuated desiccator containing phosphorus pentoxide. The next day it is treated with formaldehyde vapor (this will be discussed in greater detail below) at 80°C for 1 hr, covered with immersion oil, coverslipped, and viewed in the fluorescence

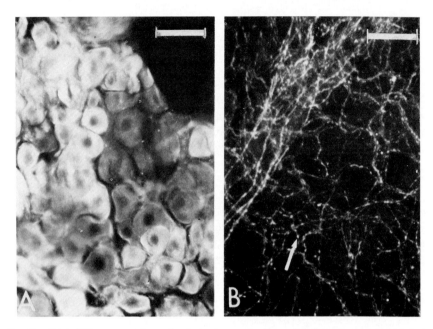

FIG. 3.   Falck–Hillarp preparations. A. photomicrograph of a section from the superior cervical ganglion showing intense fluorescence of the cytoplasm of the ganglion neurons. The centrally placed nucleus of the neuron is not fluorescent. Calibration bar = 40 μm. B. Photomicrograph of an iris stretch preparation. The band of several straight, fluorescent axons along the left-hand border of the photograph is peripheral, sympathetic innervation along a blood vessel. The adjacent plexus of fluorescent, varicose axons is the sympathetic component of the autonomic ground plexus of the iris. The varicosities (one is designated by an arrow) represent the vesicle-containing terminals of the plexus. Calibration bar = 25 μm.

microscope. This technique is applicable not only to iris but to any thin tissue (e.g., mesentery) which can be stretched and is sufficiently thin to dry rapidly and allow visualization of discrete structures in the microscope. An iris stretch preparation is shown in Fig. 3B.

An alternative to the stretch preparation is the smear preparation (Olson & Ungerstedt, 1970a), which can be applied to the central nervous system. It is identical to the stretch preparation except that the tissue to be examined is dissected so that a small piece, approximately 1 × 2 mm or less, is removed and placed on a clean glass slide. A second slide is placed on edge at an acute angle to the first slide and pulled across the tissue with gentle pressure so that a thin, even smear of tissue is obtained. This is identical to the technique of obtaining blood smears. The slide is placed overnight in an evacuated desiccator over phosphorus pentoxide and treated the next day with formaldehyde vapor, generated from paraformaldehyde equilibrated to a relative humidity of 60%, for 1 hr at 80°C.

Coverslips are placed on the slides, and then the slides are examined in the fluorescence microscope. For most areas of brain this method demonstrates fluorescent varicosities, which are present in number roughly in proportion to the innervation of the smeared area; this is assessed by chemical analysis of amine content or by the more conventional histochemical methods to be described below. Thus, since the tissue organization is disrupted by the smearing, the method provides a semiquantitative assessment of catecholamine content (serotonin terminals are not demonstrated by the smear technique) in the area being studied. In this respect it has a useful function as a screening procedure for rapid assessment of the effects of pharmacologic manipulation or lesions.

b. THE FALCK–HILLARP FREEZE-DRY METHOD. The conventional Falck–Hillarp freeze-dry method has been described in detail on a number of occasions (cf., for reviews, Björklund et al., 1972a; Eränkö, 1967; Falck and Owman, 1965; Fuxe, Hökfelt, Jonsson, & Ungerstedt, 1970). Although the basic procedure is similar in all laboratories, there are great individual variations in detail, and no attempt will be made to encompass these here. Rather, the procedure as it is used in this laboratory shall be described; the reader interested in variations and further details should consult the reviews cited.

The technique as used for catecholamines in this laboratory is as follows. The animal is sacrificed and the brain removed as rapidly as possible. For rats decapitation with a guillotine is used, but larger animals are decapitated after anesthesia with pentobarbital (40 mg/kg). The brain is dissected into pieces which may range in size from 2 mm in diameter (or smaller) to as large as 1×2 cm in diameter. The tissue is placed on a rigid piece of paper (an index card works well) which is slightly larger than the specimen. It is our practice to place the face of the block to be cut against the paper, as this both prevents confusion in orientation and makes a flat surface. The back of the paper is numbered so that each specimen frozen can be recorded. After the block has been placed on the paper, it is covered with a fine-mesh cotton gauze. Since the tissue is fresh, the gauze readily adheres to it. The gauze should just cover the tissue. The tissue is then placed in a liquid propane-propylene mixture (a commercial propane-propylene tank is used in this laboratory) cooled to the temperature of liquid nitrogen. In practice, this mixture is prepared by running the propane through a copper tubing coiled in liquid nitrogen in a Dewar flask. The propane goes into a liquid phase, and the end of the tubing is placed in a cylindrical copper container (about 10×4 cm) immersed in liquid nitrogen in a second Dewar flask. Pure propane freezes at the temperature of liquid nitrogen and sufficient propylene is added to keep the mixture liquid. The copper container contains a copper-mesh basket with a handle, and the tissue pieces to be frozen are placed in this and then immersed rapidly in the propane–propylene mixture. They freeze immediately and after a few seconds are transferred to the tissue plate of the freeze-dryer which is submerged in another Dewar flask containing liquid nitrogen. The tissue can be stored for long periods in liquid nitrogen and, consequently, tissue may be collected for several days before a freeze-drying run is begun. The use of the propane-propylene mixture is to allow the most rapid possible quenching of the tissue. The critical aspect of the freezing is to pass rapidly through the temperatures between $-30°$ and $-40°C$, and thus minimize ice-crystal artifact to provide the best tissue preservation. Liquid nitrogen itself does not give good preparations because of the rapid gas formation adjacent to the tissue which slows freezing; and isopentane, though used widely, is less satisfactory because it, too, freezes at the temperature of liquid nitrogen. Rapid freezing of tissue, of the type described above, results in the formation of tissue cracks, and the gauze covering prevents loss of any tissue pieces during freezing.

Following freezing of the tissue, the specimens (blocks) collected in the tissue plate are transferred to the freeze-dryer. The alcohol bath has been cooled to $-35°C$, and the plate is rapidly lowered into the drying chamber, the phosphorus pentoxide baskets placed over it, the glass plate cover placed over the O ring (lightly covered with vacuum grease), and the mechanical vacuum pump turned on. Evacuation of the chamber should continue to a vacuum of 10^{-1} torr, when the oil diffusion pump can be turned on. For very small pieces of tissue, 4–5 days in the freeze-dryer are sufficient to assure good drying. Larger pieces (in the range of 1×2 cm) require 10–14 days. Beginning at about 2 days, or with the longer runs 3–4 days, the temperature in the alcohol bath is gradually raised to room temperature. At the conclusion of the period of time in the freeze-dryer, the oil diffusion pump is turned off, and air is admitted to the chamber containing the tissue plate. If the humidity of the room air is high, this air should be passed through a trap immersed in liquid nitrogen so that it is completely dry when it is allowed into the chamber. Pure nitrogen gas can be used as a substitute. The tissue basket is then removed and either placed in a desiccator over phosphorus pentoxide, under vacuum, or treated immediately with formaldehyde vapor. The tissue kept in a desiccator prior to treatment can be held for several days without noticeable depreciation of subsequent fluorescence.

The formaldehyde vapor treatment is carried out in a 1-liter glass vessel containing about 5 gm of paraformaldehyde in an oven at $80°C$ for 1–2 hr. Paraformaldehyde is depolymerized during heating to form formaldehyde gas which undergoes the specific condensation reaction with primary and secondary amines. It should be noted that the reaction with secondary amines (e.g., adrenaline) is slower, and a 3-hr exposure is optimal for their demonstration. The formaldehyde should be discarded after use. An extremely critical point is the water content of the paraformaldehyde (Björklund et al., 1972a; Falck & Owman, 1965; Hamberger, Malmfors, & Sachs, 1965). If the water content is too low, the fluorescence yield is low, and few amine-containing structures are visualized. On the other hand, if the water content is too high, the fluorescence yield is high but so diffused that structural analysis is difficult. Indeed, the diffusion may be so great as to make the morphological analysis impossible. The ideal is to have a paraformaldehyde with sufficient water content to produce an optimal yield of fluorescence with minimal diffusion. For reasons not entirely understood at this time the optimal water content appears to vary from time to time within the same laboratory. [Standardization of the water content of paraformaldehyde was carried out by Hamberger *et al.* (1965), who introduced the technique of equilibrating the paraformaldehyde with solutions of sulfuric acid.] In practice, paraformaldehyde is kept in a desiccator with varied solutions of concentrated sulfuric acid and water. These solutions produce, in a closed space, a very constant relative humidity and, hence, a constant water content of the paraformaldehyde. In most laboratories a relative humidity between 50 and 90% will produce optimal results, and this can only be determined empirically. For catecholamines we routinely use paraformaldehyde equilibrated to a relative humidity between 50 and 70%. Figure 3A shows noradrenaline-containing neurons of the superior cervical ganglion which exhibit a high fluorescence yield and, at the same time, good morphological detail. This is shown by the lack of fluorescence over the cell nuclei and the discrete cell borders. In contrast to this, Fig. 4B shows a very good fluorescence yield but with considerable diffusion, as exhibited by the fluorescence over the nuclei, the indistinct cell borders, and the diffuseness of the terminals within the arcuate nucleus (compare with the terminals in Color Plates A, B, and C of the frontispiece (and other figures).

Following the formaldehyde vapor treatment, the sections should be embedded as rapidly as possible in paraffin. The embedding should be performed *in vacuo* to ensure rapid and complete penetration of the dry tissue. The embedding is carried out with the tissue still attached to the paper strips and covered with gauze. After the initial infiltration the paper is removed, the gauze stripped off the specimen using a warm forceps, and the block embedded by any standard technique for sectioning (see Chapter 2 by Clark). Sections are cut on a rotary microtome at 6–15 μm; are mounted on clean glass slides in either Entellan (E. Merck, Darmstadt, West Germany), Fluoromount (E. Gurr, London, England), liquid paraffin, or immersion oil; and then cover-slipped. The sections are then warmed gently for a short time on a slide warmer at about 60°C to dissolve the paraffin into the mounting medium and are then visualized in the fluorescence microscope. Central nervous system tissue shows stable fluorescence in the paraffin blocks for 3–6 months but the deparaffinized sections show in a few days marked fading associated with a significant increase in background fluorescence. Consequently, once sections are made, the material must be analyzed and photographed immediately. Photographs are the only permanent record and, as noted above, these can be made rapidly using Tri-X film.

The above methodology is principally useful for catecholamines and is much less satisfactory for 5-hydroxytryptamine-containing neurons. Indeed, with that

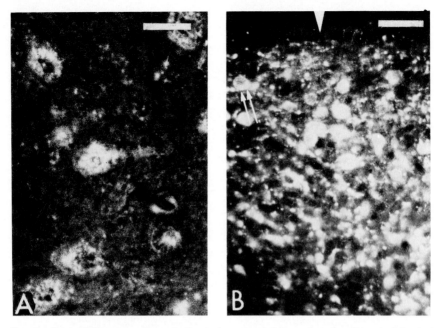

Fig. 4.   Falck–Hillarp preparations. A. Serotonin neurons of the nucleus dorsalis raphe lying just dorsal to the median longitudinal fasciculus in the periaqueductal gray of the midbrain. Calibration bar = 25 μm. B. Arcuate nucleus of the hypothalamus, adjacent to the third ventricle (the ventricular surface is designated by a single arrow). The nucleus contains numerous round to oval-shaped neurons (one is designated by the small double arrows). Interspersed among the neurons are numerous varicosities ranging from small to large. These represent the terminal plexus of catecholamine neurons whose cell bodies are located in the brainstem. Calibration bar = 25 μm.

methodology most investigators have only been able to demonstrate 5-hydroxytryptamine cell bodies and a few terminal areas (cf. Kuhar, Aghajanian, & Roth, 1972). The visualization of cell bodies and terminals is improved by pretreatment of the animal with a monoamine oxidase (MAO) inhibitor (e.g., nialamide, pargyline). In addition, Fuxe and Jonsson (1967) have reported that increased 5-hydroxytryptamine fluorescence is obtained using successive 1-hr exposures to formaldehyde vapor generated from paraformaldehyde equilibrated to a low relative humidity (60%) and then to one equilibrated to a high (90%) relative humidity. In the author's experience, the most successful demonstration of brain 5-hydroxytryptamine neurons is obtained with the method of Aghajanian, Kuhar, and Roth (1973). The animals are first given an MAO inhibitor (pargyline, 100mg/kg, i.p.), followed 15 min later by the 5-hydroxytryptamine precursor L-tryptophan (100mg/kg, i.p.). Two hours later they are decapitated and the freeze-drying carried out as detailed above; the formaldehyde vapor treatment is carried out for 1½ hr at 80°C using paraformaldehyde equilibrated to a relative humidity of 60%.

c. INTERPRETATION.   With the procedure outlined above, and the fluorescence microscope set up with primary and secondary filters as detailed in Section II,B, catecholamine neurons exhibit a green to green-yellow fluorescence. The method does not distinguish between noradrenaline, adrenaline, and dopamine fluorophors, but this can be done by special treatments and microspectrofluorometry (cf. Björklund *et al.*, 1975). These methods are cumbersome, however, and require very specialized equipment. With variants of the Falck–Hillarp technique, to be described below, some distinctions can be made on morphology alone. The 5-hydroxytryptamine fluorophor has a distinctive yellow appearance. In contrast to the catecholamine fluorophors, which are quite stable and fade slowly under illumination by the light transmitted from the primary filter, the 5-hydroxytryptamine fluorophor undergoes a very rapid photodecomposition and, consequently, is difficult to photograph. Nevertheless, the conventional Falck–Hillarp method is the only one now available which reliably demonstrates 5-hydroxytryptamine in central nervous system.

The major difficulties in interpretation of Falck–Hillarp material come from two sources. The first is the limited sensitivity of the method. In the normal situation only cell bodies and terminals of catecholamine or 5-hydroxytryptamine neurons are shown in central nervous system. Occasionally autofluorescent material in neuronal cell bodies may be misinterpreted as 5-hydroxytryptamine fluorescence, but the distinction is readily made. The autofluorescent material appears as orange-yellow granules in the neuronal cytoplasm, does not fade with exposure to light from the primary filter, and is much more common in older than in young animals. In addition, when present it tends to be ubiquitously distributed, and it appears independent of the formaldehyde vapor treatment. A further problem with sensitivity is that only when special procedures, such as lesions, are employed does the preterminal axon become visible. Further, the terminal plexus is shown in much less detail than with the more sensitive methods (see below). The second difficulty in interpretation comes with the vagaries of the condensation reaction. The line between an optimal reaction and one which produces little fluorescence on the one hand, or excessive diffusion on the other, are fine ones, and even the most experienced laboratories have periodic, inexplicable failures. Last, the inevitable cracking associated with the freeze-drying often results in an unsightly gap through a beautifully fluorescent area which would otherwise make an ideal illustration for publication.

Nonetheless, the method continues to be valuable and extensively used for several reasons. It offers the simplicity of the stretch and smear techniques. Once established, it is fairly easy to perform and, despite the failures, reasonably reliable. It is the only method currently available which is applicable to demonstration of 5-hydroxytryptamine neurons and for examination of large tissue blocks. If one is interested in large brains, it is possible to take quite large blocks for processing and, in addition, to obtain serial sections through them.

2. THE VIBRATOME–FORMALDEHYDE TECHNIQUE

The original Falck–Hillarp method was modified by Hökfelt and Ljungdahl (1972b) in two significant ways. First, they eliminated freeze-drying and introduced the use of perfused material. Second, they introduced the Vibratome to fluorescence histochemical methodology. The instrument has been briefly described above.

The Vibratome–formaldehyde method utilizes two solutions. The first is a buffered formaldehyde solution made by mixing 83 ml of a solution of monobasic sodium phosphate (2.26 gm/100 ml) with 17 ml of a sodium hydroxide solution (2.52 gm/100 ml), adding 4 gm paraformaldehyde, and heating to 60°C to dissolve the paraformaldehyde. The pH is then adjusted to 7.2–7.4. The second is a calcium-free Tyrode solution (sodium chloride, 0.8 gm; potassium chloride, 0.02 gm; magnesium chloride, 0.01 gm; dibasic sodium phosphate, 0.005 gm; monobasic sodium phosphate, 0.1 gm; dextrose, 0.1 gm; added to 100 ml distilled water; pH adjusted to 7.2–7.4).

The method is carried out as follows. The animal is anesthetized with pentobarbital (40 mg/kg, i.p.) and perfused through the left ventricle. The perfusion may be carried out at room temperature, but results are sometimes improved by chilling the animal in an ice bath prior to perfusion. As with other perfusion techniques the solution should go all the way to the tip of the cannula at the start of the perfusion as to avoid air emboli occluding components of the cerebral circulation. The perfusion is carried out over 10 min with 100–200 ml ice-cold buffered formaldehyde solution, depending upon the size of the rat. The brain is

---

FIG. 5.   A. Plexus of large and small fluorescent axons of the noradrenaline type in the lateral septal nucleus. The large varicose axons (large arrow) represent innervation from caudal brainstem noradrenaline neurons, whereas the small, varicose axons (small arrow) represent innervation from the locus coeruleus. Vibratome–formaldehyde method. Calibration bar = 50 μm. B. Caudate–putamen complex adjacent to the lateral ventricle at the level of the septum. The arrow designates the ependymal surface of the caudate–putamen complex. The innervation pattern is made up of a meshwork of extremely fine, varicose fibers. The density of the innervation is so great that individual fibers are difficult to demonstrate even at high magnification. The open, nonfluorescent areas represent groups of fibers of the internal capsule traversing the caudate–putamen complex. Vibratome–glyoxylic acid method. Calibration bar = 80 μm. C. Plexus of fluorescent varicose fibers within the stratum radiatum of the hippocampal CA3 zone. These fibers, with regularly spaced varicosities and thin intervaricose segments, are characteristic of noradrenaline axons of locus coeruleus origin. Vibratome–formaldehyde method. Calibration bar = 50 μm. D. Bundles of intensely fluorescent, distorted axons in the medulla caudal to a hemisection of the brainstem. These fibers would not normally be seen in material prepared by the Vibratome–formaldehyde method, but the hemisection has caused accumulation of noradrenaline in the axon proximal to the cut. Calibration bar = 50 μm.

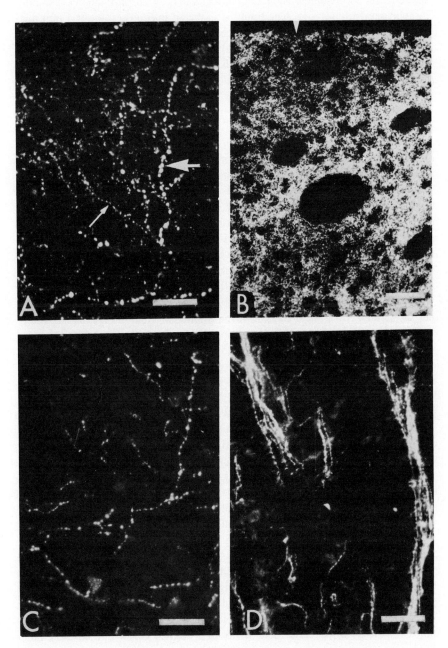

FIG. 5.

then removed immediately and stored in ice-cold buffered formaldehyde solution for approximately 20 min. The brain is dissected into pieces about 4 mm thick which are mounted for sectioning. Other pieces may be maintained for several hours in ice-cold buffered formaldehyde solution and give satisfactory results when subsequently sectioned. In this laboratory sections are usually cut at 10–25 μm as measured by the Vibratome scale. This is generally performed with the vibration rate at 6–9 scale units and a feeding speed of 1–3 scale units. For sectioning, the trough of the Vibratome is filled with the Tyrode solution kept at or below 5°C.

The sections are transferred to a clean glass slide immediately after being cut, blotted around the edges to remove excess Tyrode solution, dried under the warm air stream of a hair dryer for 15 min, and placed overnight in a desiccator under vacuum and over phosphorus pentoxide. The next day they are treated for 1 hr at 80°C in an oven with paraformaldehyde equilibrated to 70–80% relative humidity. The sections are mounted in immersion oil, Entellan (E. Merck, Darmstadt) or xylene; coverslipped; and examined in the fluorescence microscope. Once sections are prepared by the Vibratome–formaldehyde method, they are more durable than the Falck–Hillarp material. If kept refrigerated between use, the sections may be examined without significant loss of fluorescence quality for 1–2 weeks. Examples of material prepared by the Vibratome–formaldehyde method are shown in Color Plates A and B of the frontispiece and in Fig. 5A, C, and D.

The advantages of the Vibratome–formaldehyde method are (a) it requires only 2 days to obtain usable sections, (b) the sections are of very good quality, and (c) the method is quite consistent in providing good fluorescence with minimal diffusion. As in the original Falck–Hillarp technique, cell bodies and axon terminals (varicosities) are more readily demonstrated than preterminal axons. The Vibratome–formaldehyde method does show some preterminal axons when it is working optimally or when the axons are accentuated by a lesion, producing amine accumulation proximal to the lesion (Fig. 5D).

Parenthetically the examination of central nervous system material by dark-field microscopy is not an easy task for the individual inexperienced in neuroanatomy. Landmarks are hard to visualize, and the fluorescent axons and cell bodies may be difficult to localize. In this respect the Vibratome–formaldehyde method offers a significant advantage. The sections adhere well to the slide. Following visualization in the fluorescence microscope, the coverslip may be removed; the section may be cleared through xylene and graded alcohols to water; then they may be stained with cresyl violet, dehydrated, and again coverslipped for viewing by routine light microscopy. Thus, if drawings or photographs are made from the material in the fluorescence microscope, exact verification of their location can be obtained in the stained section. This also holds for the glyoxylic acid methods to be described below but, in these latter methods, the fixation of the material is much poorer than in the Vibratome–formaldehyde method and less satisfactory anatomical correlations are possible.

At the present time the Vibratome–formaldehyde method is the routine method used in this laboratory because of its reliability, the ease of cutting Vibratome sections from formaldehyde-fixed material, and the quality of the histological material. It does have several disadvantages, however. 5-Hydroxytryptamine neurons are not well shown. The axonal morphology of the Vibratome–formaldehyde method is not comparable in quality to that of good sections prepared by the glyoxylic acid method (Lindvall & Björklund, 1974a), and last, the necessity of cutting Vibratome sections limits the amount of material that can be prepared from a single brain in 1 day. As with all of the Vibratome methods, only small tissue blocks can be cut. Thus, if rapid screening of a large area is important, the original Falck–Hillarp freeze-dryer method may be preferable.

D. Methods Using Glyoxylic Acid

1. THE VIBRATOME–GLYOXYLIC ACID METHOD

The use of glyoxylic acid (GA) as a reagent to form a fluorophor with biogenic amines in a histochemical method was introduced by Björklund, Lindvall, Falck, and Svensson in a series of papers (cf., for reviews, Björklund et al., 1972a; 1975; Lindvall, Björklund, & Svensson, 1974). Although the reactions of amines with glyoxylic acid were worked out through these studies, the application of the Vibratome provided the basis for a viable histochemical technique (Lindvall & Björklund, 1974a; Lindvall et al., 1973). This technique will be described in detail, as it offers the most powerful histochemical method for the analysis of the organization of catecholamine neuron systems in brain. The technique is essentially that described by Lindvall and Björklund (1974a).

Two solutions are used in the Vibratome–GA method. The first is a 2% GA solution in Krebs–Ringer bicarbonate buffer. [The buffer composition in grams per liter is as follows: sodium chloride, 6.923; potassium chloride, 0.354; calcium chloride ($CaCl_2 \cdot 2H_2O$), 0.165; monobasic potassium phosphate, 0.162; magnesium sulfate ($MgSO_4 \cdot 7H_2O$), 0.294; sodium bicarbonate, 2.100; glucose, 1.8 1.800.] The buffer is approximately pH 7.0. After it is made up, it is saturated with a mixture of 95% O_2 and 5% CO_2. A concentration of 2% glyoxylic acid is then added (20 gm/liter). In this laboratory the monohydrate of glyoxylic acid is routinely used. This compound is somewhat unstable, and when received in the laboratory in 100 gm lots, individual containers (scintillation vials work well) are filled with small amounts of glyoxylic acid approximating what might be used in a day (5–10 gm). These are sealed, placed in a desiccator over phosphorus pentoxide, and kept in a freezer. They are then removed for use individually.

After the glyoxylic acid is added to the buffer the solution is acid, and the pH is adjusted to 7.0 with 10 N sodium hydroxide. The second solution is Krebs–Ringer bicarbonate buffer, pH 7.0, prepared as described above, including the saturation with 95% O_2 and 5% CO_2.

The procedure is essentially the same as that for the Vibratome–formaldehyde method. The animal is anesthetized and the head perfused with ice-cold glyoxylic acid solution through a cannula placed in the ascending aorta. For the GA method, 150 ml of 2% GA solution is perfused rapidly (in less than 2 min). This amount is for adult rats (about 200–250 gm); larger amounts of perfusion solution would be used for larger animals and smaller amounts for smaller ones.

Following perfusion the brain is rapidly removed, cooled in buffer, and dissected to provide the desired specimen (no greater than 5 mm thick) in the appropriate orientation. This is then glued to the Vibratome specimen holder with a rapidly hardening adhesive (Loctite Quickset Adhesive 404, Loctite Corp., Newington, Connecticut). The specimen and holder are immediately immersed in cold buffer, and then transferred as soon as possible to the Vibratome where they are submerged in ice-cold Krebs–Ringer bicarbonate buffer in the Vibratome trough. The buffer is kept at 0°–5°C throughout the sectioning, which is performed at a vibration rate usually in the range of 6–7 units and at a feeding speed of 1–3 scale units. Unlike the specimens prepared by formaldehyde perfusion, GA-perfused specimens are very soft, and it is extremely difficult to cut good sections. With experience, however, most individuals can cut acceptable sections at 25–35 μm. The tissue piece will usually be maintained in good condition for 3–4 hr, during which an experienced individual can obtain 15–20 acceptable sections. Some areas are much easier to cut than others. The brainstem, in particular, offers great difficulty in obtaining good sections, whereas septum, caudate, thalamus, and hypothalamus are much easier to cut.

When a section is cut, it is transferred with a blunt glass rod or a fine brush to a container of ice-cold GA solution (a small beaker or staining dish in an ice-bath will suffice) for 3–5 min. During this time the section must be kept submerged below the surface of the solution. It is then floated onto a clean glass slide, excess solution is blotted away, and the section is dried under the warm airstream of a hair dryer for 15 min. At this point the section is transferred to a desiccator where all sections from the day are kept in a slide rack over phosphorus pentoxide. When the slides have been collected, the desiccator is evacuated and kept in the dark overnight. For development of the fluorophor the sections are treated by one of two methods—glyoxylic acid vapor treatment or heating.

Glyoxylic acid treatment requires two vessels of about 1-liter capacity. In this laboratory small desiccators are used. One vessel, the reaction vessel, consists of a desiccator which has three outlets, one to a manometer, one attached to a piece of tubing from the second vessel, and one which can be opened by a valve to the outside atmosphere. The second, the GA vessel, is another small desiccator equipped with a valve and a connection to the reaction vessel. The treatment is carried out as follows. Two grams of GA are dried in a desiccator over phosphorus pentoxide for 24 hr, placed in the GA vessel and heated at 100°C for 1 hr. The two vessels are then placed together in the oven at 100°C. The microscope slides containing the sections are placed in the oven for about 3 min and then transferred to the reaction vessel. The reaction vessel is evacuated with a vacuum pump and hot GA-saturated air from the GA vessel is introduced into the hot reaction vessel to a pressure of 300 torr. Hot air from the oven is then allowed into the reaction vessel until atmospheric pressure is attained. The sections are treated in the reaction vessel for 2 min then removed.

For the reaction by heating, the slides containing the sections are placed in a slide rack and heated at 100°C for 6 min. In general, the fluorescence yield from the heating treatment is less than that from the GA vapor treatment, but heating alone produces less background fluorescence.

Following either the heating or GA vapor treatment, the sections are mounted in liquid paraffin or immersion oil, coverslipped, and examined in the fluorescence microscope. When not being examined, the slides should be kept at −20°C. Over time GA sections tend to have increasing background fluorescence but, when kept cold, this is decreased while the specific fluorescence is maintained, so that the sections are usable for several months.

Examples of GA method material are shown in Fig. 5B and in Color Plate II,C of the frontispiece. The advantages of the GA method are the extremely discrete morphology revealed in good preparations and the marked sensitivity of the method as compared to previous fluorescence histochemical methods. Preterminal axons are routinely observed and, as Lindvall and Björklund (1974b) have shown, it is possible to identify the preterminal axons of dopamine neurons as distinct from noradrenaline neurons. Dopamine axons are thin and, in the preterminal segments, contain no varicosities. At sites of termination they collateralize into dense plexuses of fine intertwined axons with small varicosities as in the neostriatum (Fig. 5B), or into dense pericellular arrays with larger varicosities as in the septum (Color Plate C of the frontispiece). In contrast to this, preterminal noradrenaline axons, particularly of locus coeruleus origin, are thin but have fusiform varicosities. In areas of terminal distribution, noradrenaline axons collateralize into a plexus of fine axons with regularly spaced varicosities. In general, the varicosities of noradrenaline axons of locus coeruleus origin are smaller than those arising from other brainstem cell groups. It is not possible to make any statement at this time about the appearance of adrenaline axons. The varicosities observed in GA-method material occurring in a suspected terminal area are interpreted, as are those in Falck–Hillarp material, as terminals of the axons which would contain vesicles and make synaptic contacts. Certainly, the high level of fluorescence in the varicosities implies a high neurotransmitter content.

The disadvantages of the GA method are that it does not demonstrate 5-hydroxytryptamine, the sectioning is quite difficult and only a relatively few sections can be obtained from a small piece of tissue in a day, and the method is not applicable to large blocks of tissue as these cannot be accommodated in the Vibratome. To some extent this latter problem has been obviated by some recent adaptations of the technique for cryostat sections. (Battenberg & Bloom, 1975; Bloom & Battenberg, 1976; de la Torre & Surgeon, 1976; Watson & Barchas, 1975). One of these will be described in greater detail below.

2. THE CRYOSTAT–GLYOXYLIC ACID METHOD (WATSON & BARCHAS, 1975)

The method has been described only for use with rats. Untreated rats are decapitated, and the brains are removed, frozen and placed in a cryostat at $-17°C$. Sections (10–15 μm) are cut and picked up on warm slides. The slides are then immediately placed in a GA solution (0.1 M phosphate buffer containing 0.5% magnesium chloride with 2% glyoxylic acid, pH adjusted to 5.0 with sodium hydroxide) for 12 min and then dried in warm air (45°C for 5 min) and treated with GA vapor at atmospheric pressure for 2–5 min at 100°C.

This method gives material which, at its best, approaches the quality of the Vibratome–GA method. It and the other variants referred to above have the distinct advantages of being much more rapid than the Vibratome–GA method and do not require the difficult Vibratome sectioning. Further, with larger specimen holders they should be readily adaptable to larger brains. The authors' experience, which as been limited to the present time, indicates that good material is not as often achieved with this method as with the Vibratome–GA method. This may only reflect a lack of experience, however, and the anatomist's natural inclination to keep using methods which have proven successful.

III. Conclusions

The Falck–Hillarp technique and its more recent modifications have introduced a new era into neuroanatomy. With their use, one studies neurons in a way which is not conventional in the anatomic sense. That is, neurons and their cell bodies, dendrites, and axons are identified on the basis of neurotransmitter content rather than on the basis of the location of cell bodies and the distribution of axonal projections. This has proved to be a powerful technology because the identification of neurotransmitter, *and* the organization of the neuron system, makes the tools of pharmacology as well as those of other disciplines applicable to the study of the function of the neurons. A rapidly growing literature on the function of monoamine neurons, particularly in the area of neuroendocrine regulation, attests to the usefulness of this approach. As noted above, the techniques are not without difficulty, but the authors are confident that they can be applied successfully in any laboratory where the methodology is undertaken with care.

NOTE ADDED IN PROOF: Recent advances in fluorescence histochemistry have combined the use of freeze-drying with glyoxylic acid perfusion (Lorén, Björklund, and Lindvall, *Brain Research* **117**, 1976, 313–318; Lorén, Björklund, Falck, and Lindvall, *Histochemistry* **49**, 1976, 177–192). This combination of treatments, together with high levels of magnesium ions in the perfusion solution (Lorén, Björklund, and Lindvall, *Histochemistry* **52**, 1977, 223–239) can provide sections with the superior fluorescence quality of Vibratome-glyoxylic acid material and the convenience and ease of preparation afforded by freeze-drying.

Acknowledgments

The preparation of this review was supported in part by U.S. Public Health Service Grant No. NS-12080. The authors are grateful to Bart Ziegler, who prepared many of the sections shown as well as the photographs.

References

Aghajanian, G.K., Kuhar, M.J., & Roth, R.H. Serotonin-containing neuronal perikarya and terminals: Differential effects of *p*-cholrophenylalanine. *Brain Research,* 1973, **54,** 85–101.

Amin, A.H., Crawford, T.B.B., & Gaddum, J.H. The distribution of substance P and 5-hydroxy-tryptamine in the central nervous system of the dog. *Journal of Physiology (London),* 1954, **126,** 596–618.

Andén, N.-E., Dahlström, A., Fuxe, K., Larsson, K., Olson, L., & Ungerstedt, U. Ascending monoamine neurons to the telencephalon and diencephalon. *Acta Physiologica Scandanavica,* 1966, **67,** 313–326.

Battenberg, E.L.F., & Bloom, F.E. A rapid, simple and more sensitive method for the demonstration of central catecholamine-containing neurons and axons by glyoxylic acid induced fluorescence. I. Specificity. *Psychopharmology Communications,* 1975, **1,** 3–13.

Bertler, Å., & Rosengren, E. Occurrence and distribution of dopamine in brain and other tissues. *Experientia,* 1959, **15,** 10–11.

Björklund, A., Falck, B., & Lindvall, O. Microspectrofluorometric analysis of cellular monoamines after formaldehyde or glyoxylic acid condensation. In P.B. Bradley (Ed.), *Methods in brain research.* New York: Wiley, 1975. Pp. 249–294.

Björklund, A., Falck, B., & Owman, C. Fluorescence microscopic and microspectrofluorometric techniques for the cellular localization and characterization of biogenic amines. In S.A. Berson (Ed.), *Methods of investigative and diagnostic endocrinology.* Vol. I. *The thyroid and Biogenic Amines.* Amsterdam: North-Holland Publ., 1972. Pp. 318–368. (a)

Björklund, A., Lindvall, O., & Svensson, L.-Å. Mechanisms of fluorophore formation in the histochemical glyoxylic acid method for monoamines. *Histochemie,* 1972, **32,** 113–131. (b)

Björklund, A., & Moore, R.Y. *The central catecholamine neuron.* New York: Raven Press, in press.

Bloom, F.E., & Battenberg, E.L.F. A rapid, simple, and sensitive method for the demonstration of central catecholamine-containing neurons and axons by glyoxylic acid-induced fluorescence. II. A detailed description of methodology. *Journal of Histochemistry and Cytochemistry*, 1976, **24**, 561-571.

Corrodi, H., & Jonsson, G. The formaldehyde fluorescence method for the histochemical demonstration of biogenic monoamines. *Journal of Histochemistry and Cytochemistry*, 1967, **15**, 65-78.

Dahlström, A., & Fuxe, K. Evidence for the existence of monoamine-containing neurons in the central nervous system. I. Demonstration of monoamines in the cell bodies of brain stem neurons. *Acta Physiologica Scandinavica*, 1964, **62**, Suppl. 232, 1-55.

Dahlström, A., & Fuxe, K. Evidence for the existence of monoamine neurons in the central nervous system. II. Experimentally induced changes in the intraneuronal amine levels of bulbospinal neuron systems. *Acta Physiologica Scandinavica, Suppl.*, 1965, **247**, 1-36. (b)

de la Torre, J.C., & Surgeon, J.W. Brain and tissue monoamine histofluorescence visualization in 18 minutes. *Federation Proceedings, Federation of American Societies for Experimental Biology*, 1976, **35**, 242.

Ehringer, H., & Hornykiewicz, O. Verteilung von Noradrenalin und Dopamin (3-hydroxytyramin) in Gehirn des Menschen und ihr Verhalten bei Erkrankungen des Extrapyramidalen Systems. *Klinische Wochenschrift*, 1960, **38**, 1236-1239.

Eränkö, O. The practical histochemical demonstration of catecholamines by formaldehyde-induced fluorescence. *Journal of the Royal Microscopical Society*, 1967, **87**, 259-276.

Falck, B. Observations on the possibilities of the cellular localization of monoamines by a fluorescence method. *Acta Physiologica Scandinavica*, 1962, **56**, Suppl. 197, 1-25.

Falck, B., Hillarp, N.-A., Thieme, G., & Torp, A. Fluorescence of catecholamines and related compounds condensed with formaldehyde. *Journal of Histochemistry and Cytochemistry*, 1962, **10**, 348-354.

Falck, B., & Owman, C. A detailed methodological description of the fluorescence method for the cellular demonstration of biogenic monoamines. *Acta Universitatis Lundensis, Sectio 2*, 1965, **7**, 1-23.

Falck, B., & Torp, A. A new evidence for the localization of noradrenaline in adrenergic nerve terminals. *Medicina Experimentalis*, 1961, **5**, 169-172.

Fuxe, K. Evidence for the existence of monoamine neurons in the central nervous system. IV. Distribution of monoamine nerve terminals in the central nervous system. *Acta Physiologica Scandinavica*, 1965, **247**, 39-85.

Fuxe, K., Hökfelt, T., Jonsson, G., & Ungerstedt, U. Fluorescence microscopy in neuroanatomy. In W.J.H. Nauta, & S.O.E. Ebbesson (Eds.), *Contemporary research methods in neuroanatomy*. Berlin & New York: Springer-Verlag, 1970. Pp. 275-314.

Fuxe, K., Hökfelt, T., & Ungerstedt, U. Localization of indolealkylamines in CNS. *Advances in Pharmacology*, 1968, **6A**, 235-251.

Fuxe, K., & Jonsson, G. A modification of the histochemical fluorescence method for the improved localization of 5-hydroxytryptamine. *Histochemie*, 1967, **11**, 161-166.

Hamburger, B.T., Malmfors, T., & Sachs, C. Standardization of paraformaldehyde and of certain procedures for the histochemical demonstration of catecholamines. *Journal of Histochemistry and Cytochemistry*, 1965, **13**, 147-150.

Hökfelt, T.G.M., & Ljungdahl, A. Modification of the Falck-Hillarp formaldehyde fluorescence method using the Vibratome: Simple, rapid, sensitive localization of catecholamines in sections of unfixed or formalin fixed brain tissue. *Histochemie*, 1972, **29**, 324-339. (a)

Hökfelt, T.G.M., & Ljungdahl, A.S. Histochemical determination of neurotransmitter distribution. *Research Publications, Association for Research in Nervous and Mental Disease*, 1972, **50**, 1-24. (b)

Kuhar, M.J., Aghajanian, G.K., & Roth, R.H. Tryptophan hydroxylase activity and synaptosomal uptake of serotonin in discrete brain regions after midbrain raphe lesions: Correlations with serotonin levels and histochemical fluorescence. *Brain Research*, 1972, **44**, 165-176.

Lindvall, O., & Björklund, A. The glyoxylic acid fluorescence histochemical method: A detailed account of the methodology for the visualization of central catecholamine neurons. *Histochemistry*, 1974, **39**, 97–127. (a)

Lindvall, O., & Björklund, A. The organization of the ascending catecholamine neuron systems in the rat brain. *Acta Physiologica Scandinavica, Suppl.*, 1974, **412**, 1–48. (b)

Lindvall, O., Björklund, A., Hokfelt, T., & Ljungdahl, Å. Application of the glyoxylic acid method to Vibratome sections for improved visualization of central catecholamine neurons. *Histochemie*, 1973, **35**, 31–38.

Lindvall, O., Björklund, A., & Svensson, L.-Å. Fluorophore formation from catecholamines and related compounds in the glyoxylic acid fluorescence histochemical method. *Histochemistry*, 1974, **39**, 197–227.

Lorén, I., Björklund, A., Falck, B., & Lindvall, O. An improved histofluorescence procedure for freeze-dried paraffin-embedded tissue based on combined formaldehyde-glyoxylic acid perfusion with high magnesium content and acid pH. *Journal of Histochemistry and Cytochemistry*, in press.

Olson, L., & Ungerstedt, U. Monoamine fluorescence in CNS smears: Sensitive and rapid visualization of nerve terminals without freeze-drying. *Brain Research*, 1970, **17**, 343–347. (a)

Olson, L., & Ungerstedt, U. A simple high capacity freeze-drier for histochemical use. *Histochemie*, 1970, **22**, 8–19. (b)

Twarog, B.M., & Page, I.H. Serotonin content of some mammalian tissues and urine. *American Journal of Physiology*, 1953, **175**, 157–161.

Ungerstedt, U. Stereotaxic mapping of the monoamine pathways in the rat brain. *Acta Physiologica Scandinavica*, 1971, **82**, Suppl. 367, 1–48.

Vogt, M. The concentration of sympathin in different parts of the central nervous system under normal conditions and after the administration of drugs. *Journal of Physiology (London)*, 1954, **123**, 451–481.

von Euler, U.S. Specific sympathomimetic ergone in adrenergic nerve fibers (sympathin) and its relations to adrenaline and noradrenaline. *Acta Physiologica Scandinavica*, 1946, **12**, 73–97.

Watson, S.J., & Barchas, J.D. Histofluorescence in the unperfused CNS by cryostat and glyoxylic acid: A preliminary report. *Psychopharmacological Communications*, 1975, **1**(5), 523–531.

Chapter 6

Single-Cell Staining Techniques

Charles D. Tweedle

Departments of Biomechanics and Zoology
Michigan State University
East Lansing, Michigan

141

I.   Introduction

A.   Goals

The goal of this chapter is to present an explanation of current procedures for intracellular dye techniques as used in neurobiology. Emphasis will be focused on the Procion dyes and cobalt as used both by direct intracellular injection and axonal iontophoresis, as well as the relatively new procedure of intracellular injection of horseradish peroxidase (HRP). The use of autoradiographical techniques for delineating single-cell morphology has been covered elsewhere in this series (Globus, 1973). Applicability and advantages and disadvantages of the various stains and techniques will be discussed. An earlier article by Kater, Nicholson, and Davis (1973) ''Intracellular Staining in Neurobiology,'' is recommended for a more in-depth exposition of the modifications of intracellular staining that have been used in various experimental preparations.

B.   Overview

The original idea behind intracellular staining of neurons was a simple one. With the advent of glass microelectrodes (Ling & Gerard, 1949), it became possible, in effect, to insert a fine glass tube full of electrolyte solution into a nerve cell. Why not, then, just use a visible dye solution in the microelectrode and, hopefully, after recording the physiological parameters desired, eject the dye into the cell? This would not only tell where the microelectrode was, but perhaps also give some information about the overall morphology of the physiologically examined neuron. Stains used up to the late 1960's usually failed to mark the neuron beyond the site of electrode penetration (see Nicholson & Kater, 1973, for an historical review). Stretton and Kravitz (1968) were the first to report, using the fluorescent Procion dyes (mainly Procion yellow M4RS), a method for staining virtually the entire neuritic structure of an individual nerve cell by passing the dye through a microelectrode after recording with it. The dye remained intact throughout histological processing and could be observed under fluorescence microscopy. Procion yellow has been used subsequently in a large number of both invertebrate and vertebrate preparations. It proved possible, although somewhat tedious, to do ultrastructural studies on Procion yellow-filled neurons even though the dye was not electron-opaque (Kellerth, 1973; Purves & McMahan, 1972).

In an attempt to obtain an electron-opaque intracellular marker, Pitman, Tweedle, and Cohen (1972) developed a procedure whereby cobaltous chloride could be injected into cockroach motoneurons and subsequently reacted with ammonium sulfide to form a black precipitate (cobaltous sulfide). This also withstood histological processing for both light and electron microscopy. The cobalt sulfide-filled neurons were more visible than those filled with Procion yellow under conventional microscopy, and often lent themselves well to examination in cleared whole-mount preparations. Unfortunately, the treatment with ammonium sulfide led to considerable ultrastructural disruption. The simultaneous development of two more procedures for electron opaque markers followed in 1973. Gillette and Pomeranz (1973a) modified the cobalt method to include reaction of the cobalt-filled cell with peroxide followed by 3,3'-diaminobenzidine, forming a polymer that further reacted with osmium during postfixation. With this procedure, overall fixation was reported as "superior" and that of the filled cell as "fair." The deleterious ammonium sulfide reaction with cobalt was avoided in this technique. Christensen (1973) used a nonfluorescent Procion dye containing chromium (Procion brown MX5-BR) to inject lamprey giant interneurons. The chromium renders the dye electron opaque, and the injected cell processes can be studied at both the light and electron microscope levels. Procion rubine (MX-B), which contains copper, may be similarly utilized (A. Selverston & B. N. Christensen, personal communication).

Recently, buffered horseradish peroxidase (HRP) has been introduced by microelectrode injection into mammalian spinal neurons (Jankowska, Radstad, & Westmann, 1976; Snow, Rose, & Brown, 1976) or leech neurons (Muller & McMahan, 1975, 1976) and subsequently stained histochemically to reveal neuronal structure at both the light and ultrastructural levels.

In 1971, Iles and Mulloney reported that dye could be introduced into the cut end of an axon or nerve tract. These workers passed current between a saline pool containing the preparation and one containing the nerve trunk immersed in Procion yellow. The charged dye thus was carried into specified nerve trunks of interest. The cobalt axonal procedure does not require current passage; the positive ions are carried into and along the axons by some unknown means (*vide infra*).

Recently, the axonal iontophoresis technique (especially utilizing the more rapidly moving cobalt) has been widely used for determination of single-cell dendritic geometry as well as the mapping of groups of cells. The method has greatly simplified intracellular staining both of groups of axons and individual nerve cells. No specialized electrophysiological equipment or expertise is required. Of course, the axonal delivery technique does not furnish the electrophysiological information usually obtained with microelectrode injection. Therefore, the means of cell filling used will depend largely on the kind of problem addressed. As far as this author is aware, the use of axonal delivery of HRP in invertebrate preparations has not been reported although the copious work in using the technique in the vertebrate nervous system (LaVail, 1975) would certainly indicate its feasibility.

C. Applications of Intracellular Staining

The original development of the Procion staining technique was motivated by a desire to ask whether an identified, functionally defined, invertebrate nerve cell had a unique or variable geometric shape (Stretton & Kravitz, 1968). This question has been investigated since in various invertebrate preparations using both intracellularly injected Procion yellow and cobalt procedures. It has also been proved possible to utilize single-cell staining to document developmental changes in neuritic structure of identified invertebrate neurons, as well as the effects brought about by axotomy or deafferentation. Variations in neuronal structure brought about by differences in sex, age, training, environment, etc., could and probably will eventually be similarly investigated in invertebrate preparations. The relationship of electrophysiological properties to specific regions of a neuron has been studied by Procion yellow injection combined with electrical recording in both invertebrates and vertebrates. Procion yellow has also been widely used for morphological identification of functionally identified neurons, especially in the vertebrate central nervous system (CNS). After recording from a penetrated unit, the cell is filled with stain for identification. This has led to improved understanding of the cells of the retina, visual cortex, spinal cord, and cerebellum. Physiologically defined ''silent cells'' were finally identified as being glial in nature by the use of Procion yellow (Dennis & Gerschenfeld, 1969).

Dye can be transported up cut axonal trunks to fill the cell bodies of invertebrate ganglia. Therefore, the location of the somata of the axons going out a particular nerve trunk can be determined fairly simply. This aids not only in understanding the organization of the ganglion, but also in determining where to record. Intracellular staining has also been used to delineate possible sites of synaptic contact between identified neurons. A particularly spectacular means of doing this is to fill the abutting cells with different colored or types of dye and look for points of contact.

Intracellular staining using the Procion dyes, cobalt, and HRP has been combined with electron microscopy to examine ultrastructural details of synaptic connections of the stained neuron in both vertebrate and invertebrate preparations.

Finally, the axonal delivery of both cobalt and Procion dyes has been successfully used to trace projection pathways in the vertebrate CNS (e.g., Fuller & Prior, 1975; Precht, Richter, Ozawa, & Shimazu, 1974).

D. Choice of Stain

The intracellular stain to be used depends on the information desired, the experimental preparation, and the availability of equipment or technical expertise. The relatively good recording properties and low toxicity of Procion yellow-filled microelectrodes have made this a predominant dye for single-cell light microscopic studies. However, in many cells, the dye does not always fill the finer dendritic branches or fine axons revealed by intracellularly injected cobalt or horseradish peroxidase. Cobalt is more toxic to impaled neurons, and the cobalt electrode is more likely to become clogged than either HRP or Procion electrodes. The apparent low toxicity of HRP, the good recording properties of the HRP microelectrode, and the distribution of the dye into fine branches make it a valuable dye for future intracellular work.

Axonal delivery of dye is a very simple technique to obtain information concerning the structure and distribution of groups of cells, including those too fine to fill with a microelectrode. By using axonally delivered cobalt, in certain preparations, it also has been possible to obtain information on the morphology of individually identified neurons, often in cleared whole-mount preparations. Owing to its more rapid movement up axons and penetration of fine branches, cobalt has been used predominantly in this fashion. In heavily pigmented tissues, however, Procion yellow may be advantageous because of its fluorescent properties. The relative merits of the various dyes for ultrastructural studies are discussed in section V.

Intracellular injection of dye by microelectrode requires electrophysiological equipment and expertise. If neither is available, axonal delivery of dye may be the only alternative that is usable to provide the necessary information. The use of Procion yellow also requires the availability of a fluorescence microscope.

II. Microelectrode Injection

Ejection of dye from intracellular micropipettes can be accomplished either by current passage or pressure. Those using standard intracellular electrophysiological techniques require little additional equipment or expertise for dye delivery by iontophoresis. Owing to the necessarily limited scope of this chapter, a certain proficiency in intracellular techniques is assumed.

A. *Iontophoresis*

1. CURRENT PASSING

All the dyes under discussion in this article are charged and, therefore, can be injected iontophoretically into cells. The small size of most neurons prohibits intracellular injection by the pressure technique. Standard intracellular electrophysiological equipment can be used for current passage. Passing current and recording potential changes in a cell can be done simultaneously using the principle of the Wheatstone bridge (Araki & Otani, 1955). The microelectrode and a large resistor (e.g., 10 Ω) form one arm, and a pair of resistors, the ratio of which is variable, form the other. The bridge is balanced so that the stimulating

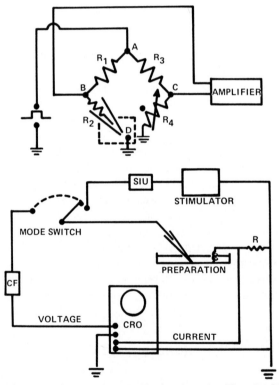

FIG. 1. *Upper:* Wheatstone bridge circuit used with microelectrodes. The ratio of the resistors $R_3 : R_4$ is varied by variable resistor R_4 until the signal put between A and ground (D) does not change the potential between points B and C. The bridge is thus balanced so that the stimulating current does not cause a steady deflection in the recorded potential. *Lower:* Circuit in which one can change the microelectrode from a recording to a stimulating mode, where current may be passed for iontophoresis of dye. CF, Cathode follower; SIU, stimulus isolation unit; CRO, oscilloscope; R, dropping resistor.

current does not cause a steady deflection in the recorded potential (Fig. 1). Another circuit that has been used for current passage is also shown in Fig. 1. Here the microelectrode can be switched from a recording to a stimulating role.

Llinás (1973) uses a relay between the recording amplifier and micropipette. Activation of this relay disconnects the amplifier and connects a dry cell battery capable of delivering current to the microelectrode. After the cell has been penetrated and its electrical activity recorded, the relay is activated for dye passage. This apparatus is used for a short current passage of up to 10^{-6} A, and is sufficient to fill a number of different types of vertebrate neurons.

2. STAIN PREPARATION AND THE FILLING OF MICROELECTRODES

Recently Procion dyes have become more readily available to biologists (e.g., Wilson Diagnostics, Inc.; Polysciences). Cobaltous chloride or nitrate and sodium or ammonium sulfide can be obtained from standard chemical suppliers. Horseradish peroxidase may be obtained from Sigma in either the type II or the more purified type VI variety. Both Procion yellow M4RAN and M4RS have been used; published reports do not suggest the superiority of either. The Procion dyes are usually made up in a 4–6% (w/v) aqueous solution. Mixing the dye solution in 0.15 M KCl instead of distilled water will reduce the electrode resistance, which is often high in electrodes filled with Procion (Barrett, 1973). Lower electrode resistance can also be achieved by the use of a supersaturated solution of Procion yellow (10% w/v) made by gently heating the dye solution and allowing it to cool slowly. This higher dye concentration is more likely to lead to electrode clogging, however, especially at cool temperatures. Beveling the electrode is another useful way of reducing resistance (*vide infra*). The Procion dyes always should be fresh preparations as hydrolysis is fairly rapid.

Cobaltous chloride for use in microelectrodes is made up in solutions that range from 10 mM to 3 M. The lower concentrations are less likely to cause electrode blockage but the higher ones reduce electrode noise. Theoretically, other metallic chlorides (e.g., nickel and copper) with insoluble sulfides should also work for intracellular staining (Pitman *et al.*, 1972). Cobalt nitrate, which is more soluble in water than cobaltous chloride, has been reported to provide better recording properties for extracellular microelectrodes (Kien, 1976) and may prove to be advantageous for intracellular microelectrodes, as well.

Jankowska *et al.* (1976) filled microelectrodes with 15–20% Sigma, type II horseradish peroxidase in 0.2 M NaOH (pH 8.5). Using the more purified Sigma type VI peroxidase, Snow *et al.* (1976) used 4% HRP in 0.05 M Tris HCl buffer which contained 0.2 M KCl per liter (pH 8.6).

Dye solutions for microelectrodes should be put through a Millipore filter (0.22 μm) to remove larger particles that might clog the tips. Electrode glass should be cleaned in hot (80°C) sulfuric acid and rinsed in hot distilled water before use.

It is often a problem to fill the small tips of the microelectrodes without breaking them or introducing air bubbles. Larger-tipped pipettes can be filled directly through the shaft with a syringe. Storage for a few hours in a hydrated atmosphere usually assures that the tip will fill with stain. Pipettes with tips less than 1 μm may be filled first with alcohol by gently boiling at reduced pressure. A few days before they are to be used, these pipettes can be transferred to distilled water and then to the filtered stain solution and filled by simple diffusion. The "fiber-fill" technique of Tasaki, Tsukahara, Ito, Wayner, and Yu (1968) is widely used for filling Procion yellow microelectrodes. In this procedure, a few strands of fiberglass are inserted into the capillary glass before pulling. The fiberglass fibers carry the stain solution down to the electrode tip. Unfortunately, in our work with fiber-filled cobalt microelectrodes, we found that they were more likely to clog than electrodes filled by other means. Recently, glass capillary tubing containing an inner glass fiber has become commercially available. Microelectrodes made from this can be filled easily and quickly by simply inserting the shank into the appropriate solution. The current-carrying and dye-passing properties of the filled microelectrode should be determined before the stage of cell impalement.

3. BEVELED ELECTRODES

In recent years, the use of beveled electrodes for intracellular recording has grown and various beveling techniques published (Barrett & Graubard, 1970; Barrett & Whitlock, 1973; Brown & Flaming, 1974; Kripke & Ogden, 1974). This procedure produces both a sharper electrode tip and a larger orifice for dye passage. The electrodes are reported to cause less damage during cell penetration and are possessed of better recording properties than conventional electrodes. Beveled electrodes may be particularly useful in studies on smaller neurons or processes normally difficult to penetrate.

4. INJECTION PROCEDURE

After impaling the nerve cell with a dye-filled electrode, the physiological parameters of interest may be recorded before dye injection. Depending on the dye used, recording with the electrode may present various degrees of difficulty. Procion electrodes have rather high resistance. Beveling has been used as one means of lowering resistance of Procion yellow electrodes to obtain more accurate recording of transient voltage responses. Fiberglass-filled Procion yellow microelectrodes also have lower resistance and current-carrying ability. The tendency of Procion yellow electrodes to plug may also be lowered if the Procion solution is made up in a 0.15 M KCl (Kater et al., 1973).

For most invertebrate cells, the dye is usually passed by applying square wave (cathodal) hyperpolarizing current pulses of $1–5 \times 10^{-8}$ A, 500-msec duration at 1 Hz for 5–60 min. With small vertebrate cells (e.g., retina), success has been obtained with constant current $(2-3 \times 10^{-9}$ A) for 1–30 min. The lower current prevents cell damage and lessens the chance of losing the cell. At the other extreme, cathodal currents of $1-10 \times 10^{-6}$ A have been used to pass a single large pulse of dye for a few seconds (Kaneko and Kater, 1973; Llinás, 1973). Although the electrode quickly blocks and/or the cell is lost, the neuron usually contains an adequate amount of stain for recognition of major branches. The large amount of current probably damages the cell, but this technique is useful for cells that can only be held for a short period of time.

Cobalt-filled electrodes tend to block more easily than Procion-filled electrodes, especially with higher concentrations of cobalt. Depolarizing anodal current of $1-5 \times 10^{-8}$ A, 0.5 sec in duration at 1 Hz for 30–60 min was found to be satisfactory for cockroach motoneurons. Cobalt-filled electrodes also show considerable rectification in current-passing capability, especially with higher cobalt concentrations. If electrode polarization decreases current passage, voltage may have to be raised to maintain the desired current.

Horseradish peroxidase electrodes seem to have good recording and current-passing properties when the HRP is dissolved (2–20%) in a slightly basic ionic solution. HRP is ejected by anodal (depolarizing) current $(5-40 \times 10^{-9}$ A for 5–25 min) or by pressure injection.

B. Pressure Injection

Remler, Selverston, and Kennedy (1968) first used pressure injection of Procion yellow into giant fibers of crayfish. For similar large fibers or motoneurons >100 μm, the large microelectrode tips necessary for pressure may be used without disrupting the neuron. When successful, the results obtained are quite pleasing. The large amount of dye delivered can produce a cell that is well filled. There is also an advantage in that noncharged compounds may be passed by pressure. On the negative side, the large electrode tip may damage the cell and cause dye leakage; the cell may rupture from overfilling and allow dye leakage; or fine processes of the neuron may be "blown out." One probably wouldn't choose to use pressure injection of dye in an ultrastructural study. The large electrode tips also results in micropipettes which are often not very useful for intracellular recording. A combination of first recording electrophysiological parameters of interest with a conventional microelectrode and then pressure filling the cell with a second dye-filled microelectrode has proven successful (e.g., Selverston, 1973).

PROCEDURE

Remler et al. (1968) apparently just used a micrometer syringe connected by polyethylene tubing to the dye-filled microelectrode held in a microelectrode holder to pass Procion yellow. Pitman et al. (1972) used a similar device to deliver intracellular cobalt. Kater *et al.* (1973) have described a more elaborate but probably more controllable device with the pressure source being compressed air. It requires: (1) a tank of compressed air; (2) a regulator, 0–100 lb; (3) one lever action toggle switch and associated high-pressure tubing (Clippard Instruments, Cincinnati, Ohio); and (4) a pressure injection microelectrode holder (e.g., Model MPH-1, W. P. Instruments, Hamden, Connecticut).

Microelectrode glass is cleaned and the chosen dye prepared as described above. A conventional microelectrode is then pulled. Beveling may then be used to increase the tip aperture or the tip may alternatively be broken by bumping it against a glass slide. Beveling usually offers the advantage of a sharper electrode tip. The microelectrode is then filled with the filtered stain solution by a fine guage hypodermic needle inserted into the back of the pipette. The pipette can then be attached to the pressure apparatus and put into a holder. Pressure may then be passed to move the dye into the electrode tip and also to determine the rate of stain delivery out of the tip. Observation of the outflow of stain is best done under liquid and viewed through a dissecting microscope. Experience will indicate the proper rate of dye ejection for a given preparation.

Some of the neurons or giant fibers to be injected by pressure are so large and identifiable that penetration can be determined visually. With other neurons, this is often not possible. In these cases, the microelectrode must also have the potential for recording or for stimulating to identify the neural unit and/or to determine that the cell has been penetrated.

Once the neural unit is impaled, pressure is produced to eject dye. This can be either by gradually engaging the micrometer syringe or, in the other setup, by passing steady air pressure of 1–70 psi. Davis (1970) found that dye ejected at a rate sufficient to expand the cell body by 50% in 5 sec, repeated for ½ to 1 hr, was adequate for filling crustacean motoneurons. The amount of pressure used, site of injection, and size of the injected neuron will determine the length of time required for dye passage. Pressure injection may also be useful for getting a short, but large "blast" of dye into a neuron that cannot be held for any length of time. The goal is to fill the cell adequately while avoiding dye leakage or bursting of the cell.

It is advisable to watch the cell filling under a dissecting microscope in order to constantly balance the amount of pressure delivered against the swelling of the cell and migration of the dye.

III. Axonal Iontophoresis: Procedure

Intracellular staining of groups of nerve cells by axonal delivery of dye is becoming an increasingly popular procedure for certain neuroanatomical problems. The equipment required is minimal and technical heroics unnecessary. The cut end of a nerve trunk is simply separated out and exposed to a bath of dye isolated from the bulk of the preparation by a suitable insulating seal. The stain molecules then pass up the nerve trunk for several hours. This may or may not require the passage of current, depending on the dye used. The dye apparently travels in efferent and afferent fibers at the cut end, allowing investigation of both cell bodies of origin and terminal projections of the nerve tract. Somata positions thus have been mapped in invertebrate and vertebrate preparations, and pathway projections mapped in the vertebrate central nervous system. Axonal delivery of cobalt has also been utilized in the visualization of the dendritic structure of identified invertebrate neurons.

The technique for delivery of dye up nerve trunks can be modified to the preparation. One must have a section of nerve sufficiently long to place in dye solution and be able to isolate it effectively from the bulk of the preparation by an insulating seal. For invertebrate ganglia, the apparatus shown in Fig. 2A is widely used. The chamber consists simply of two adjacent holes drilled in a plastic block. The chambers are connected by a narrow slit, and the nerve trunk of interest is placed into one chamber and the ganglion in the other. The slit, through which the nerve trunk extends, is then filled with petroleum jelly or silicone stopcock grease, taking care that the seal is complete. The chamber containing the nerve trunk is then filled with dye (e.g., 4% Procion yellow, or 5–15% $CoCl_2$, in water or saline). The chamber containing the ganglion is filled

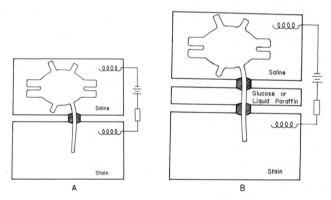

FIG. 2. (A) A simple arrangement for delivery of dye up cut axons. Two adjacent chambers drilled in a lucite block are connected by a narrow slit. The nerve trunk of interest is then draped through this slit and sealed with petroleum jelly. One chamber holding the bulk of the preparation is filled with saline or tissue culture medium; the other, containing the nerve ending, is filled with dye. If current passage is needed, chlorided silver or platinum electrodes may then be put in place and current of the appropriate polarity passed. (B) A variation on the arrangement above for use where extracellular migration of dye is a problem. A third chamber or trough containing either glucose or liquid paraffin is interposed between those containing the saline and dye. The conductivity of the extracellular space is reduced and current flow is preferentially through the nerve trunk.

with saline or tissue culture medium (Tyrer & Altman, 1974). For Procion dyes, electrodes are then put into place, one into the saline and one into the dye solution. After sealing the chambers with an overlying coverslip to retain moisture, cathodal current up to 1×10^{-6} A is passed, usually for several hours at 4°–10°C. For cobalt, no current passage is needed. The preparation is simply left for an appropriate time (usually several hours) either at room temperature or in the refrigerator. It has been reported by Strausfeld and Obermeyer (1976) that the addition of bovine serum albumin (BSA) to the cobalt solution (13 mg BSA/ 100 ml cobalt solution) enhances the staining of fine nerve fibers and also decreases the time necessary for backfilling in the insect CNS.

For axonal iontophoresis of Procion, the current source need only be capable of passing 10^{-6} A D.C. across approximately 10 MΩ resistance. A standard 12-V dry cell battery or a constant current source will work. With the battery, current can be regulated by a fixed and a variable resistor placed in series in the circuit. Assuming constant electrode resistance, and with a knowledge of the battery voltage, the current can be regulated by altering the series resistance of the circuit. Since most of the resistance of the system is in the current-limiting resistors and not the electrodes or preparation, current flow is usually fairly constant. Alterations in the electrode resistance can occur due to electrolytic polarization of the metal electrodes in salt solution when constant current is passed. The use of large-diameter chlorided platinum or silver wire helps circumvent this problem (see Silver, 1958, for technique).

The use of a commercially obtained constant current source is advantageous in that it adjusts current flow to alterations in resistance during the course of the iontophoresis. Current flow may be monitored by inserting a microammeter in the circuit.

In experiments using either cobalt or Procion, prolonged dye passage can lead to migration of dye in the extracellular space inside the nerve sheath. This can be partially circumvented by placing a trough of slightly hypotonic glucose solution or liquid paraffin between the stain and saline pools (see Fig. 2B). The intervening solution removes the electrolytes from the section of nerve passing through it. This increases the extracellular resistance and thus greatly cuts down on extracellular stain migration. Such an arrangement is especially helpful in molluscan or crustacean preparations where axons are often more loosely packed inside the nerve sheath.

The dye moves most rapidly up larger diameter axons (see Fig. 3). This characteristic is useful if one is interested in preferentially staining the soma and dendrites of giant fibers (e.g., Mendenhall & Murphey, 1974). Different axons seal each time the nerve is cut, and as dye will only travel up open axons, one can obtain variable results from dye delivery up a given nerve trunk. This can be useful or hindering depending on the goal desired. Wine, Mittenthal, and Kennedy (1974) intentionally limited the number of axons stained by damaging most

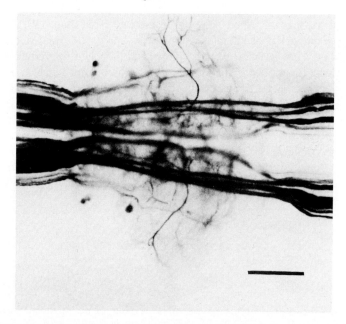

FIG. 3.   Photograph of a cleared cockroach metathoracic ganglion seen from the dorsal surface. Cobalt was introduced by axonal delivery through the ventral nerve cord between the third and fourth abdominal ganglia. Primarily the giant fibers and their collaterals were stained. The small cell bodies are probably interneurons. Calibration bar: 0.25 mm. Modified from Pitman *et al.* (1973). By courtesy of the authors and permission of Springer-Verlag.

of the axons in a root. If one is interested in staining as many as possible of the axons in a nerve trunk, dipping the end of the nerve trunk into distilled water for a few minutes before putting it into dye solution is advantageous.

Various other setups for dye delivery have been initiated. In one technique, the nerve trunk is pulled into a cobalt-filled polyethylene suction electrode attached to a hypodermic syringe. The electrode can then be sealed and the tip placed with the preparation in the saline solution. In an intact animal, one can also build a "well" constructed of vaseline around the end of a nerve branch, and then fill the well with stain. If needed, one electrode can be put into contact with the stain solution and one with the rest of the animal for current passage.

A number of workers (Fuller & Prior, 1975; Pearl & Anderson, 1975; Prior & Fuller, 1973) have found that the perfusion of the brain with cold saline before cobalt delivery does not affect the passage of the dye up nerve trunks. Even more intriguing are a few reports that axonal delivery of cobalt works perfectly well on fixed vertebrate tissue (Fuller & Prior, 1975; Pearl & Anderson, 1975). Current passage has no effect on the rate of cobalt migration in these vertebrate preparations. Even without current, small ($< 2 \mu$m) fibers carried dye about 1 mm/hr.

Other workers, however, have not had success with fixed tissue (W. Armstrong & C. A. Mason, personal communication; the author's own experience) and thus it remains an open question whether movement of cobalt up vertebrate axons requires living tissue. If not, then it is puzzling why the maintenance of invertebrate preparations in good physiological condition has been reported to provide better cell findings (Tyrer & Altman, 1974). In any case, with present techniques, a distance limitation of 2–3 cm exists for cobalt movement into the CNS.

Results from the above experiments, using axonally delivered cobalt, have corroborated findings of earlier silver degeneration studies on various vertebrate pathways. Some advantages of the technique are its simplicity, the staining of both efferent and afferent connections, and the ability to view the stained pathways in whole-mount preparations.

IV. Tissue Processing for Light Microscopy

A. *Procion Yellow*

After intracellular injection of one or a number of cells, the preparation is usually incubated in saline in the cold to allow further dye diffusion. The length of this incubation time depends largely on the size of the injected neuron. Gillette and Pomeranz (1973b) left injected *Aplysia* preparations up to 20 hours before fixation, while Kaneko (1973) left injected retinal cells only for a few minutes. Jankowska and Lindström (1973) perfused cats with cold Ringer's solution after injection of motor neurons with Procion yellow and left the spinal cords at 4°C for 16–32 hr before fixation. They reported that this increased the staining of axons. This report raises the question of how the Procion dye is transported, since unoxygenated, saline-perfused tissue presumably cannot be carrying on much axonal transport. Perhaps diffusion or osmotic pressure is playing the major role.

After incubation, the tissue is fixed. Stretton and Kravitz (1973) have found that the particular fixative used and its pH greatly influence background fluorescence levels. A glutaraldehyde/formaldehyde mixture at pH 4.0 was found to give low background fluorescence, as did formaldehyde alone at pH 7.4. Bouin's or Susa fixatives have also been used with good results. Complete dehydration is necessary to obtain a clear whole-mount preparation. The clearing agents of choice for viewing of whole-mount preparations seem to be methyl benzoate or beechwood creosote. Prolonged storage in either of these compounds can be deleterious, however, as methyl benzoate makes the tissue brittle and creosote turns it dark. Much information about overall neuronal geometry can be obtained from the whole-mount preparation. If the desired information is available at this level, the problems of serial reconstruction can be circumvented.

If sectioning of the tissue is necessary, it can be frozen or embedded in a suitable medium. A variety of embedding media has been used and includes paraffin, celloidin, and various plastics. Celloidin must be completely removed from the tissue before viewing under the fluorescence microscope. Epon plastic was used originally but Maraglas seems to have a smaller amount of the auto-fluorescence that might mask weak fluorescence from fine processes of filled cells. Paraffin offers the advantage of allowing one to cut ribbons of serial sections. The cut sections are mounted on slides and coverslipped. Nonfluorescent mounting media include Lustrex (Monsanto Chemical Company), Entellan (Merck), and Fluoromount (Gurr Company, London).

Fluorescence microscopy for use with Procion yellow has been described in detail by Van Orden (1973). A high-intensity white light source can be used, with Schott BG-12 and BG-38 activation filters and a barrier filter that passes light above 500 nm (e.g., Schott OGI or Leitz K-510). Prolonged exposure of preparations at their excitation maxima can lead to photodecomposition of Procion yellow.

Three-dimensional reconstruction of neurons filled with Procion yellow or other dyes can be made from serial sections. Stretton and Kravitz (1968) made photographic enlargements of phase-contrast images of Procion yellow material onto semitransparent paper. The positions of dye-containing profiles were then marked, and the photographs oriented relative to each other. Other workers have used similar arrangements by tracing color micrographs of the sections onto transparent plastic sheets to form the reconstruction (e.g., Davis, 1970). Selverston (1973) recommends the use of color slides projected and traced onto clear plastic sheets and separated to scale with spacers. The use of computers to create and analyze such reconstructions has also been implemented at both the light (Reddy, Davis, Ohlander, & Bihary, 1973; Selverston, 1973) and ultrastructural (Levinthal, Macagno, & Tountas, 1974) levels to provide quantification of neuronal geometry. In the use of such techniques, however, one should be aware of the large alteration in neuronal structure that may have taken place through histological processing (Barrett, 1973).

B. Procion Brown

For light microscopy, routine histological procedures using frozen or embedded material can be carried out for Procion brown. The dye is not fluorescent, but is dark enough to be seen in sectioned material. A light counterstain may be used on sections, if desired.

C. Cobalt

After intracellularly injecting a neuron with cobalt, the preparation is usually allowed to incubate in saline for a short while before reaction with sulfide. A good rule of thumb is to allow at least 30 min from the beginning of the injection to the sulfide treatment. Prolonged periods of incubation result in the disappearance of the cobalt from the previously filled neuron, and this makes it difficult to obtain multiple fills using intracellularly injected cobalt. The preparation is then usually reacted with ammonium sulfide in saline (1:40) for 10–15 min, rinsed in saline, and fixed. With sulfide precipitation, the filled neuron turns black. Sodium sulfide, which can be adjusted in saline to a physiological pH, is gentler on the tissue than ammonium sulfide and is thus preferable. Recently, Székely (1976) also used buffers saturated with hydrogen sulfide and obtained good results.

With axonal iontophoresis of cobalt, the above procedure is followed with a couple of modifications. The preparation is rinsed in saline for several minutes before sulfide precipitation to remove interstitial cobalt. After axonal infusion of cobalt into large pieces of neural tissue (e.g., the vertebrate brain) the sulfide precipitation time may have to be extended up to 60 min.

Tissue containing cobalt sulfide-filled neurons can be processed by normal histological techniques. As the precipitate is slightly soluble in acid solutions, prolonged storage in the alcohols (usually slightly acid) may diminish the density of staining. Since a large part of the value of cobalt preparations is often the information available in whole mounts, the proper choice of clearing agents is important (see Fig. 4). Methyl benzoate, methyl salicylate, and creosote have been found to be suitable; however, prolonged storage in these solutions is deleterious. Slabs of brain material a few millimeters in thickness or 30- to 100-μm fixed frozen sections can be cut and cleared to give information about pathway projections, etc. Unfortunately, cobalt-stained material tends to fade after a few days, so photographs or data should be obtained as soon as possible.

A more permanent preparation can be obtained with the Timm's Silver Intensification Technique. Cobalt sulfide is dense enough to be visible under the light microscope in sectioned material when it occurs in high concentration or in relatively large processes. The smallest branches of a neuron, however, often cannot be distinguished with certainty from the background. In 1974, Tyrer and Bell used a modification of Timm's silver sulfide technique to enhance cobalt-filled neuronal profiles in locust ganglia. This treatment brought out previously unresolvable branches and defined large structures more clearly (see Fig. 5).

A stock solution of the Timm silver developer can be made up and stored in the refrigerator for up to 2 weeks. This stock solution contains 30 gm gum acacia (powdered), 0.43 gm citric acid, 0.17 gm hydroquinone, and 10 gm sucrose dissolved in 100 ml of hot distilled water.

Silver nitrate solution is added to the stock solution immediately before use. For 10 μm paraffin sections the final developer contains 1 part of 1% $AgNO_3$ to 9 parts of stock solution. Dewaxed sections are incubated in this medium for 10–20 min in the dark at 60°C. (Slides can be removed from the solution, rinsed, and

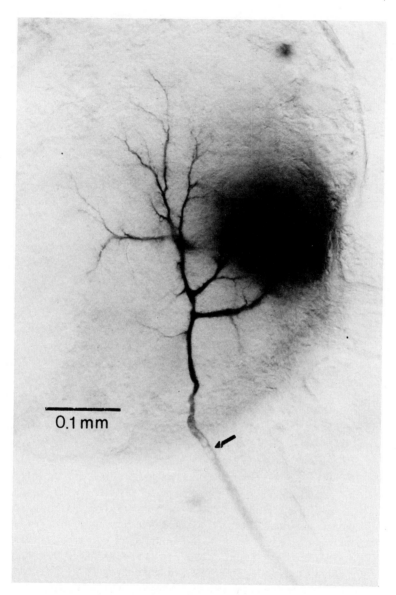

FIG. 4. Photograph of a cleared whole-mount preparation of an identified motoneuron in the cockroach metathoracic ganglion stained by intracellular injection of cobalt chloride. Focus is on the major branches of the dendritic tree. The cell body and initial process are out of focus. The axon may be seen leaving the ganglion (arrow). Modified from Pitman *et al*. (1973). By courtesy of the authors and permission of Springer-Verlag.

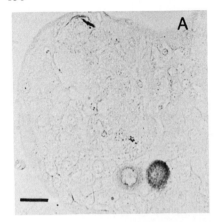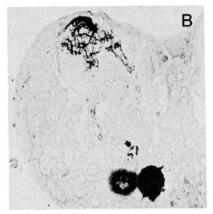

FIG. 5. The Timm's silver enhancement technique. (A) Transverse section through the locust metathoracic ganglion. A group of four motorneurons was filled with cobalt by axonal delivery. Two cell bodies and a few larger branches may be seen to be stained. (B) The same section after treatment with silver developer. The processes seen in (A) are now much more visible and others, originally undetectable, become obvious. Calibration bar: 50 μm. Modified from Tyrer and Bell (1974). By the courtesy of the authors and permission from *Brain Research*.

checked for intensification every 5–10 min.) The cobalt profiles should stain black against a brown background. If the silver solution begins to precipitate before the intensification is done, the slides may be rinsed and put into fresh solution. When the intensification is satisfactory, the slides are rinsed in warm water, dehydrated, and mounted. The brown background serves as a counterstain.

Thick, plastic sections require 1 part of 5% $AgNO_3$ to 9 parts stock solution; incubation time and temperature are as above. It may be necessary to counterstain the sections with toluidine blue.

Thin sections for electron microscopy may also be stained with the procedure, but as osmium and copper grids also react with the silver solution, sections should be floated in the Timm's solution and removed before the less reactive osmium starts to precipitate the silver (Székely and Kosaras, 1976). Only many sections and experience will delineate the timing involved for individual cases.

Whole small invertebrate brains or ganglion can be intensified by a modification of the Timm's technique (Strausfeld and Obermayer, 1976). Cleared brains are dehydrated to 30% alcohol and then brought through ascending concentrations of gum arabic to 25% gum arabic in 30% alcohol at 40°C. They are then incubated for 1 hr in 25% alcoholic gum arabic containing 1.6% citric acid and 1% hydroquinone. Sections are then incubated in the dark in the same mixture, but adding 0.5 gm of silver nitrate / 100 ml. They should then turn tobacco brown. Impregnated tissue is then washed in alcoholic gum arabic, and then in 30% alcohol, dehydrated, embedded, and sectioned (see also Section VIII).

Mason and Lincoln (1976) have used axonal delivery of cobalt sulfide combined with intensification to delineate the retino-hypothalamic projection of the rat. For 40-μm frozen sections, they used 1 part of 5% AgNO$_3$ and 9 parts of Timm's stock solution for 15–20 min at 60°C in the dark. They were able to trace 0.5- to 2-μm fibers from the optic nerve to the suprachiasmatic and arcuate nuclei. This was further than they were able to trace the fibers with the cobalt and no enhancement (Mason, 1975). It was also found that, even without cobalt delivery, certain neurosecretory cells contained misleading clumps of black dots that, upon further investigation, turned out to be silver reaction with lysosomes.

D. Horseradish Peroxidase

A more detailed protocol for this histochemical technique is given in Chapter 13 by LaVail, this volume. Briefly, after iontophoretic injection, the HRP is allowed to diffuse from 30 min to several hours. The animal is perfused with saline which is followed by a mixture of a low concentration of paraformaldehyde (1–4%) and glutaraldehyde (0.5–2%), then fixed for 1–24 hr. Following a rinse in buffer, whole mounts of ganglion or frozen section of CNS tissue are prepared. Sections of the injected and fixed material must be thin enough to react with the histochemical reagents, yet be thick enough to demonstrate overall neuronal geometry. For vertebrate tissue, frozen sections may be taken or thick sections cut with a tissue chopper. Invertebrate ganglia are often small enough not to require sectioning (e.g., Muller & McMahan, 1976). Sections or whole mounts are processed according to the method of Graham and Karnovsky (1966). Care should be taken in the handling of the reputedly carcinogenic 3,3'-diaminobenzidine (DAB) solution.

Muller and McMahan (1976) found that the use of benzidine dihydrochloride in place of 3,3'-diaminobenzidine tetrahydrochloride produced a reaction product that was more satisfactory for visualization of filled leech neurons in whole-mount preparations and light microscopy. Overreaction of the benzidine and H$_2$O$_2$ with HRP resulted in the formation of crystals and distortion, however. This did not happen with diaminobenzidine. Diaminobenzidine formed a more electron dense product and was, therefore, used for ultrastructural studies.

E. Photography

Many intracellularly stained preparations are studied in whole mounts. It is sometimes difficult to photograph these preparations owing to the great depth of field involved. Pitman, Tweedle, and Cohen (1973) and Mulloney (1973) have outlined their methods both for pbotography and the production of stereo pairs. For transferring information from whole-mount preparations a Wild drawing tube or *camera lucida* is also useful.

V. Ultrastructural Use of Intracellular Dyes

A. *Procion Yellow*

It is possible to examine Procion yellow-filled processes at the ultrastructural level even though the dye is not electron opaque. Purves and McMahan (1972) were the first to do this when examining the synaptic endings on an identified leech motoneuron. Later, Kellerth (1973) did a similar study on injected cat spinal motoneurons. The cells were injected with a 6% Procion yellow dye through fairly high-resistance (200–400 MΩ) microelectrodes; current ranged from 1×10^{-8} to 5×10^{-8} A. The low current was an aid in avoiding cellular disruption. Electrodes with lower resistance often caused cellular distortion, while those with higher resistance did not deliver enough dye. Dye was passed until the cell was deeply yellow. This was followed by a few hours diffusion time, sufficient for staining large dendritic branches nearest the site of injection. The Procion dye did not appear to diffuse out of the cell. Aldehyde fixation was used and was followed by a buffer rinse and osmication. Purves and McMahan found that placing tissue for short periods of time in the osmium (e.g., 3 min) was necessary to avoid loss of fluorescence and yet provide tissue preservation. This was no problem with the small leech ganglion but could be a hindrance if larger blocks of tissue were being used.

Purves and McMahan used three methods for locating fluorescent profiles seen with the light microscope for study with the electron microscope.

1. Thick (3-μm) sections were cut until a fluorescent profile was found. Then adjacent thin sections were cut and compared to the thick section for positive determination of dye-injected processes.

2. The same thin section was examined in both fluorescence and electron microscopes. 1000-Å thin sections were transferred to glycerol-coated slides and examined and photographed under the fluorescent microscope. The section was then floated off, put on a grid, stained, and examined under the electron microscope.

3. A thick section was cut, put on a slide, examined and phtographed, mounted on a dummy block with epoxy, and then thin sections of it were taken.

Both of the above-mentioned studies reported some changes in fine structure, especially in the cell body, where injection occurred. Most features of the cell were preserved, however. Distortion of the endoplasmic reticulum and mitochondria was seen, along with many small vacuoles in the cytoplasm. In a leech motoneuron, no indication was found that the injection procedure significantly altered presynaptic profiles. In a Procion yellow-filled leech sensory cell, fluorescent profiles with large numbers of normal-looking vesicles could also be seen. Thus, both pre- and postsynaptic structures seemed fairly well preserved after dye injection.

In the soma of the cat motoneuron, however, the ultrastructure of the apposing boutons was altered; they could be found deeply invaginated in the stained motoneuron. Signs of glial phagocytosis of boutons were also seen. This discrepancy between the two studies probably is due to the fact that the synaptic structures examined in the leech were out in the neuropil and not so near the site of intracellular injection.

B. Procion Brown

In 1973, Christensen reported the development of a chromium-containing electron-opaque Procion dye, Procion brown (MX5-BR). Although it does not fluoresce, the dye has a reddish color and is readily identified in the light microscope. Under the electron microscope, the dye appears as a dense, unstructured precipitate filling the cytoplasm. Christensen used short (1–2 min) osmication times, sufficient for tissue preservation in his preparation while still allowing visualization of the cell processes in the block. Presumably, longer osmication times would be possible if desired.

The stained processes were positively recognized by reembedding 5-μm thick sections and examining thin sections. The considerable electron density of the stained processes, however, leaves little doubt about their identity. The dye was contained within the plasmalemma, and cellular integrity seemed fairly normal. In Procion brown-stained lamprey giant interneurons, presynaptic and postsynaptic membrane thickenings were seen and did not appear to be altered by the technique (Christensen, 1973). Christensen (1976), in fact, was able later to use the dye to identify specific synaptic contacts in the lamprey and to examine the ultrastructural differences brought about at the contacts by nerve stimulation. On the other hand, Christensen and Ebner (1975) reported that a number of Procion brown-injected vertebrate cortical cells had presynaptic terminals invaginated into the postsynaptic cell membrane, and also found evidence of glial phagocytosis of boutons. This is very similar to the findings presented by Kellerth (1973) for Procion yellow-injected motoneurons (*vide supra*). The preservation of synaptic morphology in Procion-injected cells thus seems best when cell injection is at a site removed from the synapses examined, e.g., the neuropil of invertebrates following dye injection into the soma. This may limit the use of the Procion dyes for ultrastructural analyses of synaptic structure in vertebrate neurons (at least following intrasomal injection). A. Selverston (personal communication) has indicated success with the use of a copper-containing Procion dye, Procion rubine (MX-B), for ultrastructural studies.

C. Cobalt

Intracellularly injected cobaltous chloride subsequently reacted with ammonium sulfide in saline before fixation produces cobalt sulfide, which is electron dense (Pitman et al., 1972, 1973). The ammonium sulfide solution, however, is very basic and harmful to ultrastructural integrity. It is also, unfortunately, virtually impossible to buffer. Sodium sulfide, or buffers saturated with hydrogen sulfide, both of which can be adjusted to a physiological pH, may be used instead. Owing to the better preservation of tissue, these sulfide solutions should probably be used even for light microscope studies. Following precipitation of the cobalt sulfide, the tissue is rinsed, fixed, and run through normal electron micrographic procedures. If one is dealing with small pieces of tissue and a short osmication time is possible, visualization of the stained processes in the block face is possible. This aids in orientation, trimming, etc. Longer periods of osmication are not detrimental, however.

The injected processes appear brown in thick sections. Adjacent thick and thin sections are helpful for identification of the stained profiles. Thick (5 μm) and thin plastic sections can be treated with Timm's silver enhancement technique (Tyrer & Bell, 1974) (see Fig. 5).

The cobalt sulfide reaction product disrupts the intracellular contents although apposing neuropil and synaptic structures appear unaffected. The distribution of cobalt in the injected (or back-filled) cell is quite variable and ranges from very dense (in fine nerve endings) to diffuse in larger branches. In well-filled cells, it is relatively easy to trace processes but it is difficult in not-so-well-filled cells. Advantages of cobalt are that (at lease in insects) it fills fine processes not shown by Procion yellow (e.g., Burrows, 1973; Tyrer & Altman, 1974), and it readily and rapidly diffuses for long distances up axon trunks, and thus does not require the use of microelectrodes.

Gillette and Pomeranz (1973a) modified the cobalt technique for ultrastructural studies. The intracellular cobalt will catalyze an osmium binding polymer from 3,3'-diaminobenzidine. The procedure is as follows (as slightly modified by Atwood and Pomeranz, 1974).

1. After delivery of cobalt, the preparation is rinsed in saline to remove interstitial $CoCl_2$ and exposed to 0.03% H_2O_2 in saline for 5–10 min.

2. It is then put into 0.5 mg/ml 3,3'-diaminobenzidine tetrahydrochloride (DAB) in saline adjusted to pH 7.4–7.8 with NaOH. At this point, filled neurons

turn dark blue owing to the formation of a polymer and can be examined if desired. The preparation is left in this solution for several minutes.

3. After rinsing in saline, aldehyde fixation is carried out followed by osmication and embedding.

In thick sections, the larger branches of the filled cell appear yellow-orange under the light microscope and a dark-field condenser. In the electron microscope, the large major axons of the injected cell are distinguished by the accumulation of the darkly staining polymer at the inside of the neuron membrane. This gives the cell processes a distinctly contrasting outline. Mitochondria are also often full of reaction product and dark granules are spread throughout the cytoplasm. Fine processes are often completely filled with the reaction product.

This method provides a useful marker for localizing presynaptic structures impinging on the injected neuron. Synaptic endings onto back-filled fast flexor motoneurons of the crayfish thus were studied by Atwood and Pomeranz (1974). The ultrastructure of the filled cells was described as "fair" by Gillette and Pomeranz (1973a) in their original paper. The author's unpublished observations using their technique on cobalt-filled insect interneurons indicates that this is a correct appraisal. Organelles of the injected cell are recognizable but abnormal. The usefulness of this technique for staining presynaptic nerve endings has yet to be determined.

For use in the vertebrate nervous system, a tissue chopper may be necessary to render the material thin enough to be reacted with the DAB solution. If it is found consistently that cobalt can be transported along the axons of fixed tissue (Fuller & Prior, 1975; Pearl & Anderson, 1975) an interesting variation might be to fix the tissue, transport the cobalt down the axons of choice, slice the tissue, and then carry it through the rest of the histochemical procedure.

D. Horseradish Peroxidase

A new and very promising light and ultrastructural marker is the intracellular injection of horseradish peroxidase, which has already been shown to provide good results in the leech (Muller & McMahan, 1975, 1976) and in vertebrate spinal neurons (Jankowska *et al.*, 1976; Snow *et al.*, 1976). The general procedure used in these studies is as follows.

1. Snow *et al.* (1976): Fill microelectrodes with 4% HRP (type VI, Sigma) in 0.05 M Tris-HCl buffer, which contains 0.2 mole of KCl per liter (pH = 8.6). Jankowska *et al.* (1976): Fill microelectrodes with 10–20% HRP (Sigma, type II) in 0.2 M NaOH (pH = 8.5). Muller and McMahan (1976): Fill microelectrodes with 2% HRP (Sigma, type VI) in 0.1 M KCl

2. Inject the cells by current passage (or pressure could be used in larger cells)

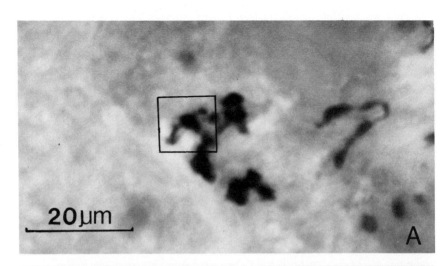

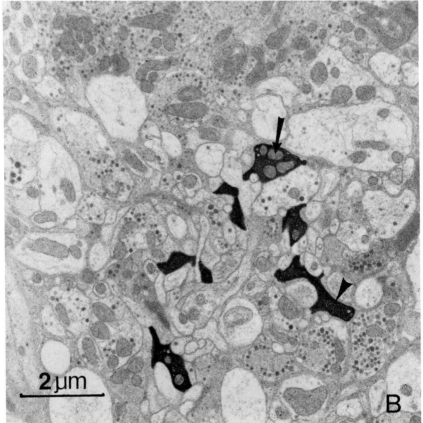

3. Allow the HRP to diffuse (30 min to several hours)
4. Perfuse the animal with saline and follow by using a fixative
5. After rinsing in buffer, section the tissue with a tissue chopper (e.g., Sorval TC-2) at 120 μm; 30- to 50-μm frozen sections may be taken; small invertebrate ganglia do not require sectioning.
6. Incubate for peroxidase activity (Graham & Karnovsky, 1966)
7. Process for light and electron microscopy

Alternate thick and thin sections or thin sections taken from reembedded thick sections may be examined.

Very thin axon collaterals (not revealed by previous studies using Procion) were stained in vertebrate spinal neurons. Some vertebrate cells appeared quite damaged by the injection, while others seemed well preserved. In well-preserved cells viewed under the electron microscope, fine granular deposits were found in the soma, dendrites, and axon. Cell organelles appeared normal. Possible disruptive changes in the nerve endings onto the HRP-filled vertebrate cells comparable to those reported after staining with Procion yellow (Kellerth, 1973) or Procion brown (Christensen & Ebner, 1975) have not yet been systematically examined. Both pre- and postsynaptic structures in HRP-filled leech cells were well preserved (Muller & McMahan, 1976).

The advantages of the technique are the mobility of the dye, good ultrastructural preservation of the filled cell, and easy identification of the electron dense deposits. Also, the enzyme-filled micropipettes have acceptable recording properties and are relatively nontoxic to the impaled cell.

E. Advantages and Disadvantages of the Various Intracellular Markers for Ultrastructural Analysis

Procion yellow is technically rather tedious to use in ultrastructural analysis as it is not electron dense and requires the examination of alternate or reembedded sections. Short osmication times are also required as osmication quenches fluorescence. The advantages of the dye seem to be its low toxicity and the fact that the ultrastructure of the injected cell is fairly well preserved, especially in sites away from the point of injection. Thus, it may be possible to study pre- and postsynaptic structures of filled cells. However, horseradish peroxidase offers all these same advantages without the same amount of labor.

FIG. 6. (A) Phase micrograph of 3-μm-thick plastic section showing sensory nerve processes in the leech which have been stained by pressure injection of horseradish peroxidase into the soma. (B) Electron micrograph from the region of the section above indicated by the box. The considerable electron density of the reaction product may be seen. It is still possible to recognize cell organelles such as mitochondria (arrow) and synaptic vesicles (arrowhead). Courtesy of K. J. Muller and U. J. McMahan.

Although more toxic to the injected cell than Procion yellow, Procion brown is much easier to work with, as it is electron opaque. However, it does not appear to be as mobile as cobalt or HRP and, thus, is less likely to fill fine brances well in large neurons.

The use of microelectrode-delivered cobalt as an ultrastructural dye is hindered somewhat by a tendency of the micropipette to block during dye passage. This could lead to incomplete cell filling, especially when using small tip diameter microelectrodes. Subsequent ammonium sulfide treatment before fixation has, in the past, led to mediocre fixation. However, with the use of pH-adjusted sodium sulfide, one can obtain fair overall fixation (unpublished observations). The cobalt sulfide within filled cells obliterates most cell structure, giving the stained processes an appearance similar to that of Golgi-stained material at the ultrastructural level.

The reaction of the intracellular cobalt with DAB lending to the formation of an osmiophillic polymer (Gillette & Pomeranz, 1973a) leaves most of the intracellular structure at least recognizable. This latter technique, like the procedure with HRP, involves the use of 3,3'-diaminobenzidine, which is reportedly carcinogenic and must be handled very carefully. Advantages of the cobalt procedures are that the ion is readily transported along axons by back-filling and will fill very fine nerve endings. It is also possible, at least in vertebrates, to deliver the cobalt in saline-perfused or fixed tissue. Fixation of the tissue before cobalt delivery and reaction with DAB or sulfide may lead to an improved ultrastructural appearance of the filled cells. These experiments have yet to be done, however.

Horseradish peroxidase is a relatively new intracellular dye for single cells and thus only a scarce body of literature exists concerning it. It appears to have great promise. Electrode recording properties are good, the dye is not too toxic, the product is electron dense, and it migrates better in vertebrate neurons than the Procion dyes. Intracellular ultrastructure is also not greatly disrupted (see Fig. 6). It may prove possible to use the stain delivered up axonal trunks by retrograde transport in invertebrate preparations as has been done in the vertebrate CNS (e.g., LaVail, 1975).

VI.   Possible Limitations of Intracellular Staining

A.   *Physiological Effects of Intracellular Dyes*

Procion yellow has been reported to have very little deleterious physiological effect when iontophoretically injected into the neurons of invertebrate preparations. In vertebrate CNS neurons, there are reports that injected Procion yellow leads to prolongation and even blockage of action potentials (Barrett, 1973; Barrett & Graubard, 1970; Van Keulen, 1971; Zieglgänsberger & Reiter, 1974) as well as a reduction in resting potential (Jankowska & Lindström, 1973). Barrett (1973) feels that these effects are probably due to reaction of the Procion dye with neuronal proteins as no such behavior was observed with just the passage of similar amounts of current from standard electodes filled with 1 M potassium acetate. In an ultrastructural study of similar Procion yellow-filled motoneurons, Kellerth (1973) reported the occurrence of boutons deeply invaginated into the soma as well as glial phagocytosis of boutons. However, these morphological changes probably are not rapid enough to explain the physiological abberrations reported. They presumably took place during the postinjection incubation period. They do indicate that the injection procedure induces trauma to the cell, which might be especially noticeable, physiologically, in vertebrate prepations owing to the large synaptic input to the somal region. Osmotic effects brought about by the injected Procion (or any dye) may be an important factor.

Procion brown (MX5-BR), useful because of its electron density, appears to be fairly toxic (Christensen, 1976). In lamprey giant interneurons, membrane potential was lost during the 10- to 30-min dye injection period. Synaptic morphology of the injected cells at the ultrastructural level seemed normal in the lamprey, although in Procion brown-filled cortical cells Christensen and Ebner (1975) reported a similar invagination of presynaptic terminals and glial reaction as reported above.

Cobalt-filled microelectrodes have been used to record normal action potentials from insect preparations (Pitman et al., 1972). However, there are other reports of significant alterations in electrophysiological characteristics of neurons impaled with electrodes filled with cobaltous chloride, even when the electrode is just serving a monitoring function (Kater et al., 1973). Resting membrane potential and action potentials were reportedly affected. This is presumably from leakage of the ions out of the electrode and indicates that a backing current might be necessary where possible. Thus, for certain physiological applications, the use of cobalt microelectrodes may be contraindicated. Lux and Schubert (1969) reported that intracellular injection of cobalt into cat spinal motoneurons produced prolonged changes in the equilibrium potential of IPSP's. The interaction of injected cobalt with intracellular calcium might be expected to lead to a variety of such effects.

During horseradish peroxidase injection from microelectrodes, the response of the injected cell deteriorates somewhat. Snow et al. (1976) report that the somadendritic spike of spinocervical neurons disappears in the first minute after penetration. However, postsynaptic potentials could still be recorded. Jankowska et al. (1976) also noticed changes in postsynaptic potentials, although these recovered greatly in several cells after injection. Muller and McMahan (1976) found that successful HRP filling of leech cells by pressure injection was accompanied by a gradual decrease in cell input resistance and lengthening of the duration of the action potential. Injection of the enzyme only transiently affected the frequency and amplitude of spontaneous synaptic potentials. Experience with this technique is limited due to its recent appearance, so information is scarce. However, the dye appears to be only mildly toxic.

B. Dye Passage across Gap Junctions

In 1969, Payton, Bennett, and Pappas reported that Procion yellow crossed the crayfish septate axon. This finding has been substantiated at electrotonic junctions in the retina (Kaneko, 1973) and in *Aplysia* neurons (Gillette & Pomeranz, 1973b). Indeed, the passage of Procion yellow between cells may be a useful way of locating gap junctions. Procion yellow movement between cells was seen in cricket giant interneurons across a presumed electrotonic junction (Murphey, 1973). Zieglgänsberger and Reiter (1974) reported Procion yellow transfer between cat spinal neurons, apparently through dendrodendritic contacts. Tritiated glycine also passes between cells at these contacts, although gap junctions have yet to be demonstrated (Globus, Lux, & Schubert, 1968; Kreutzberg & Schubert, 1975). Whether or not Procion brown crosses electrotonic junctions has not been specifically reported, although Christensen (1976) presented a photograph of a mixed chemical and electrotonic junction without apparent dye passage.

In their original paper, Pitman *et al.* (1972) reported that cobalt did not cross the crayfish septate junction. However, in a more detailed study of the matter, Politoff, Pappas, and Bennett (1974) found that with slow injection and low junctional resistance, cobalt could cross at this junction. If larger amounts of cobalt were injected rapidly, junctional resistance increased and no cobalt precipitate was seen across the septum. Apparently, a sufficiently high concentration of cobalt uncouples the junction and prevents dye transfer. This may explain the result of O'Shea and Williams (1974) who observed no cobalt movement between electrically coupled locust visual interneurons when one was filled through a microelectrode. Similarly, Smith, Sunderland, Leake, and Walker (1975) found no cobalt transfer between coupled Retzius cells in the leech. Strausfeld and Obermayer (1976) have reported recently that prolonged axonal delivery of low (2–4%) cobalt concentrations at 4°C led to the transynaptic staining of "functionally contiguous" assemblies of interneurons in the insect brain. Whether gap junctions are involved is not clear.

It was originally reported that horseradish peroxidase did not cross the crayfish septate junction (Reese, Bennett, & Feder, 1971) but later work indicated that, following fixation, some of the enzyme was mobile and could then move across the septum (Bennett, Feder, Reese, & Stewart, 1973). In the leech, Carbonetto and Muller (1977) found no movement of HRP across S cell gap junctions, however.

As can be seen, one must be cautious in interpreting neuronal geometry as a result of intracellular staining. The neuronal profiles stained may be of more than one neuron.

VII. Prospectus

Intracellular nerve staining is less than 10 years old. Already the "older" techniques have been modified and parameters for their more successful use determined. Owing to the large number of scientists now using intracellular dyes, the next decade is likely to produce an even greater output of theoretical and technical knowledge. Structure–function analysis at the single-cell level is rapidly moving out of the realm of theory and into that of the microelectrode and the computer.

VIII. Addendum

A new fluorescent, non–electron-opaque Procion dye, Procion Lucifer yellow CHL, has been introduced by Nicholson, Kater, and Stewart (1977), who report it to be more visible under fluorescence microscopy than Procion Yellow. This attribute should increase the percentage of recognizably stained cells and filled processes obtained with a given amount of injected dye. In their study, Nicholson *et al.* (1977) injected a 3% aqueous solution of the dye into cerebellar elements with negative current for 0.5–10 min. Four hours later the material was fixed and processed histologically and sections were viewed under the fluorescent microscope with a Schott B 12 excitation filter with a 470 nm cutoff barrier filter. The migration of the dye in cerebellar glial cells was notably less than in neurons, which were often stained to the level of dendritic spines. The greater fluorescence of the dye may make it particularly useful for marking CNS neurons, which cannot be held for long with a microelectrode.

Coggshall (1977) has described the back-filling of *Drosophila* neurons by pressure injection of horseradish peroxidase into the dorsal longitudinal flight muscles which they innervate. Whether the mechanism of uptake of HRP by the intramuscular nerves was by pinocytosis or diffusion into damaged axons is unknown. This technique may be useful for anatomically delineating the nerve cells which innervate specific muscles in other preparations as well. Chapters 10 and 13 may be consulted for recent modifications of HRP methodology.

An improved method for silver intensification of cobalt-filled neurons in insect whole-mount ganglion preparations has recently been published (Bacon & Altman, 1977). This procedure involves presoaking the ganglion in a modified Timm's developer base solution at 60°C for 1 hr to allow the tissue to equilibrate with the solution before the addition of silver nitrate. The modified developer base solution consists of 3 gm of gum acacia, 0.8 gm of citric acid, 0.17 gm of hydroquinone, and 10 gm of sucrose in 100 ml of distilled water. The ingredients should be added to the water at 60°C and stirred continuously until dissolved. The pH should be 2.6 and can be adjusted with citric acid. After presoaking, the preparation is transferred to a mixture of 10 parts of the above developer base solution and 1 part of 1% silver nitrate at 60°C and kept in the dark. This developer solution should be replaced with freshly made developer every 30 min. The development should be checked every 10 min and stopped if the ganglion sheath starts to darken. Incubation time is normally 40–120 min for insect material. Termination of the development is by transferral into warm distilled water followed by distilled water at room temperature. Dehydration and clearing may then be carried out. Darkening of the preparation sometimes occurred in material fixed in aldehydes, due to aldehyde reaction with silver. Fixation in alcoholic Bouin's was recommended for insect material.

A number of reports have appeared (Rehbein, Kalmring, & Romer, 1974; Hoy & Casaday, 1976; Kien, 1976; Rehbein, 1976) where complete single-cell staining of invertebrate neurons has been achieved with *extracellular* recording microelectrodes filled with either cobalt chloride (2–$3\ M$) or cobalt nitrate (30%). In his study of locust visual interneurons, Kien (1976) recommended filling the extracellular microelectrodes with a 30% cobalt nitrate solution (tip resistance, 15–25 MΩ). The electrodes were then stored overnight at 40°C. The use of cobalt nitrate and the heat treatment were reported to dramatically reduce electrode noise when compared to electrodes filled with cobalt chloride. After recording from fibers in the neuropil, cobalt was ejected using 0.5-sec, 1-Hz positive pulses of about 20 nA for 1 hr, after which the preparation was removed and reacted with ammonium sulfide. Staining with extracellular electrodes is successful only when the tip of the recording electrode is in the immediate vicinity of the neuron under study. When single units are recorded, only single neurons are usually found to have taken up the cobalt (Rehbein et al., 1974; Kien, 1976). An advantage of extracellular recording and staining is that it permits structural analysis of neurons with processes of small diameter. However, one does have to satisfy certain criteria to claim that the stained neuron is the same one that was recorded (see Rehbein et al., 1974; Kien, 1976). How the cobalt ions get inside the cell from an extracellular electrode remains unknown, although it seems likely to be through damage to the neuronal membrane. Interestingly, current passage is not required to obtain filled cells (Rehbein et al., 1974).

Acknowledgments

The author should like to thank Drs. K. Muller and U. J. McMahan for allowing me to see their manuscript before publication, Drs. R. M. Pitman, I. Korr, E. Kuntz, J. Burnett, and G. Hatton for critically reading the manuscript, and Mr. Norman St. Pierre for drawing Fig. 2.

References

Araki, T., & Otani, T. Response of single motoneurons to direct stimulation in toad's spinal cord. *Journal of Neurophysiology,* 1955, **18,** 472–485.

Atwood, H. L., & Pomeranz, B. Crustacean motor neuron connections traced by back filling for electron microscopy. *Journal of Cell Biology*, 1974, **63**, 329-334.

Bacon, J. P., & Altman, J. S. A silver intensification method for cobalt-filled neurones in wholemount preparations. *Brain Research*, 1977, **138**, 359-363.

Barrett, J. N. Determination of neuronal membrane properties using intracellular staining techniques. In S. B. Kater & C. Nicholson (Eds.), *Intracellular staining in neurobiology*. Berlin & New York: Springer-Verlag, 1973. Pp. 281-300.

Barrett, J. N., & Graubard, K. Fluorescent staining of cat motoneurons *in vivo* with bevelled micropipettes. *Brain Research*, 1970, **18**, 565-568.

Barrett, J. N., & Whitlock, D. G. Technique for bevelling glass microelectrodes. In S. B. Kater & C. Nicholson (Eds.), *Intracellular staining in neurobiology*. Berlin & New York: Springer-Verlag, 1973, Pp. 297-299.

Bennett, M. V. L., Feder, N., Reese, T. S., & Stewart, W. Movement during fixation of peroxidases injected into the crayfish septate axon. *Journal of General Physiology*, 1973, **61**, 254-262.

Brown, K. T., & Flaming, P. G. Bevelling of fine micropipettes by a rapid precision method. *Science*, 1974, **185**, 693-695.

Burrows, M. The morphology of an elevator and a depressor motoneuron of the hindwing of a locust. *Journal of Comparative Physiology*, 1973, **83**, 165-178.

Carbonetto, S., & Muller, J. J. A regenerating neurone in the leech can form an electrical synapse on its severed segment. *Nature (London)*, 1977, **267**, 450-451.

Christensen, B. N. Procion brown: An intracellular dye for light and electron microscopy. *Science*, 1973, **182**, 1255-1256.

Christensen, B. N. Morphology correlates of synaptic transmission in lamprey spinal cord. *Journal of Neurophysiology*, 1976, **39**, 197-212.

Christensen, B. N., & Ebner, F. F. The ultrastructure of physiologically studied cortical cells. *Neuroscience Abstracts*, 1975, **1** (200), 127.

Coggshall, J. C. Neurons associated with the dorsal longitudinal flight muscles of *Drosophila melanogaster*. *Journal of Comparative Neurology*, 1977, **177**, 707-720.

Davis, W. J. Motoneuron morphology and synaptic contacts: Determination by intracellular dye injection. *Science*, 1970, **168**, 1358-1360.

Dennis, M. V., & Gerschenfeld, H. Some physiological properties of identified mammalian glial cells. *Journal of Physiology (London)*, 1969, **203**, 211-222.

Fuller, P. M., & Prior, D. J. Cobalt iontophoresis techniques for tracing afferent and efferent connections in the vertebrate CNS. *Brain Research*, 1975, **88**, 211-220.

Gillette, R., & Pomeranz, B. Neuron geometry and circuitry via the electron microscope: Intracellular staining with osmiophillic polymer. *Science* 1973a, **182**, 1256-1258.

Gillette, R., & Pomeranz, B. A study of neuron morphology in *Aplysia californica* using Procion yellow dye. *Comparative and Biochemical Physiology, A*, 1973b, **44**, 1257-1259.

Globus, A. Iontophoretic injection technique. In R. F. Thompson & M. M. Patterson (Eds.), *Bioelectric recording techniques. Part A. Cellular processes and brain potentials*. New York: Academic Press, 1973. Pp. 24-38.

Globus, A., Lux, H. D., & Schubert, P. Somatodendritic spread of intracellularly injected glycine in cat spinal motoneurons. *Brain Research*, 1968, **11**, 440-445.

Graham, R. C., & Karnovsky, M. J. The early stages of absorptions of injected horseradish peroxidase in the proximal tubules of mouse kidney. *Journal of Histochemistry*, 1966, **14**, 291-302.

Hoy, R. R., & Casaday, G. Physiological and anatomical properties of cricket auditory interneurons. *Neuroscience Abstracts*, 1976, **2**, 486.

Iles, J. F., & Mulloney, B. Procion yellow staining of cockroach motor neurons without the use of microelectrodes. *Brain Research*, 1971, **30**, 397-400.

Jankowska, E., & Lindström, S. Procion yellow staining of functionally identified interneurons in the spinal cord of the cat. In S. B. Kater & C. Nicholson (Eds.), *Intracellular staining in neurobiology*. Berlin & New York: Springer-Verlag, 1973. Pp. 199–210.

Jankowska, E., Radstad, J., & Westmann, J. Intracellular application of horseradish peroxidase and its light and electron microscopical appearance in spino-cervical tract cells. *Brain Research*, 1976, **105**, 557–562.

Kaneko, A. Morphological identification of single cells in the fish retina by intracellular dye injection. In S. B. Kater & C. Nicholson (Eds.), *Intracellular staining in neurobiology*. Berlin & New York: Springer-Verlag, 1973. Pp. 157–179.

Kaneko, C. R. S., & Kater, S. B. Intracellular staining techniques in gastropod molluscs. In S. B. Kater & C. Nicholson (Eds.), *Intracellular staining in neurobiology*. Berlin & New York: Springer-Verlag, 1973. Pp. 151–156.

Kater, S., Nicholson, C., & Davis, W. J. Guide to intracellular staining techniques. In S. B. Kater & C. Nicholson (Eds.), *Intracellular staining in neurobiology*. Berlin & New York: Springer-Verlag, 1973. Pp. 307–325.

Kellerth, J. Intracellular staining of cat motoneurons with Procion yellow for ultrastructural studies. *Brain Research*, 1973, **50**, 415–418.

Kien, J. A preliminary report on cobalt sulphide staining of locust visual interneurons through extracellular electrodes. *Brain Research*, 1976, **109**, 158–164.

Kreutzberg, G. W., & Schubert, P. The cellular dynamics of intraneuronal transport. In M. W. Cown & M. Cuénod (Eds.), *The use of axonal transport for studies of neuronal connectivity*. New York: American Elsevier, 1975. Pp. 83–112.

Kripke, B. R., & Ogden, T. E. A technique for bevelling fine micropipettes. *Electroencephalography and Clinical Neurophysiology*, 1974, **36**, 323–326.

LaVail, J. H. The retrograde transport method. *Federation Proceedings, Federation of American Societies for Experimental Biology*, 1975, **34**, 1618–1624.

Levinthal, C., Macagno, E., & Tountas, C. Computer-aided reconstruction from serial sections. *Federation Proceedings, Federation of American Societies for Experimental Biology*, 1974, **33**, 2336–2340.

Ling, G., & Gerard, R. W. The normal membrane potential of frog sartorius muscles. *Journal of Cellular and Comparative Physiology*, 1949, **34**, 383–396.

Llinás, R. Procion yellow and cobalt as tools for the study of structure-function relationships in vertebrate central nervous systems. In S. B. Kater & C. Nicholson (Eds.), *Intracellular staining in neurobiology*. Berlin & New York: Springer-Verlag, 1973. Pp. 211–226.

Lux, H. D., & Schubert, P. Post-synaptic inhibition. Intracellular effects of various ions in spinal motoneurons. *Science*, 1969, **166**, 625–626.

Mason, C. A. Delineation of the rat visual system by the axonal iontophoresis-cobalt sulfide precipitation technique. *Brain Research*, 1975, **85**, 287–293.

Mason, C. A., & Lincoln, D. W. Visualization of the retina-hypothalamic projection in the rat by cobalt precipitation. *Cell and Tissue Research*, 1976, **168**, 117–131.

Mendenhall, B., & Murphey, R. K. The morphology of cricket giant interneurons. *Journal of Neurobiology*, 1974, **5**, 565–580.

Muller, K. J., & McMahan, U. J. The arrangement and structure of synapses formed by specific sensory and motor neurons in segmental ganglia of the leech. *Anatomical Record*, 1975, **181**, 432.

Muller, K. J., & McMahan, U. J. The shapes of sensory and motor neurons and the distribution of their synapses in ganglia of the leech: A study using intracellular injection of horseradish peroxidase. *Proceedings of the Royal Society of London, Series B*, 1976, **194**, 481–499.

Mulloney, B. Microelectrode injection, axonal iontophoresis, and the structure of neurons. In S. B. Kater & C. Nicholson (Eds.), *Intracellular staining in neurobiology*. Berlin & New York: Springer-Verlag, 1973. Pp. 99–114.

Murphey, R. K. Characterization of an insect neuron which cannot be visualized *in situ*. In S. B. Kater & C. Nicholson (Eds.), *Intracellular staining in neurobiology*. Berlin & New York: Springer-Verlag, 1973. Pp. 135–150.

Nicholson, C., & Kater, S. B. The development of intracellular staining. In S. B. Kater & C. Nicholson (Eds.), *Intracellular staining in neurobiology*. Berlin & New York: Springer-Verlag, 1973. Pp. 1–19.

Nicholson, C., Kater, S. S., & Stewart, W. Neuronal and glial elements visualized with new intracellular stain. *Neuroscience Abstacts,* 1977, **3,** 178.

O'Shea, M., & Williams, J. L. D. The anatomy and output connection of a locust visual interneuron; the lobular giant movement (LGMD) neurone. *Journal of Comparative Physiology,* 1974, **91,** 257–266.

Payton, B. W., Bennett, M. V. L., & Pappas, G. D. Permeability and structures of junctional membranes at an electrotonic junction. *Science,* 1969, **166,** 1641–1643.

Pearl, G. S., & Anderson, K. V. Use of cobalt impregnation as a method of identifying mammalian neural pathways. *Physiology and Behavior,* 1975, **15,** 619–622.

Pitman, R. M., Tweedle, C. D., & Cohen, M. J. Branching of central neurons: Intracellular cobalt injection for light and electron microscopy. *Science* 1972, **176,** 412–414.

Pitman, R. M., Tweedle, C. D., & Cohen, M. J. The form of nerve cells: Determination by cobalt impregnation. In S. B. Kater & C. Nicholson (Eds.), *Intracellular staining in neurobiology*. Berlin & New York: Springer-Verlag, 1973. Pp. 83–98.

Politoff, A., Pappas, G. D., & Bennett, M. V. L. Cobalt ions cross an electrotonic synapse if cytoplasmic concentration is low. *Brain Research,* 1974, **76,** 343–346.

Precht, W., Richter, A., Ozawa, S., & Shimazu, H. Intracellular study of frog's vestibular neurons in relationship to the labyrinth and spinal cord. *Experimental Brain Research,* 1974, **19,** 377–393.

Prior, D. J., & Fuller, P. M. The use of cobalt iontophoresis technique for identification of the mesencephalic trigeminal nucleus. *Brain Research,* 1973, **64,** 472–475.

Purves, D., & McMahan, U. J. The distribution of synapses on a physiologically identified motor neuron in the central nervous system of the leech: An electron microscope study after the injection of the fluorescent dye Procion yellow. *Journal of Cell Biology,* 1972, **55,** 205–220.

Purves, D., & McMahan, U. J. Procion yellow as a marker for electron microscopic examination of functionally identified nerve cells. In S. B. Kater & C. Nicholson, (Eds.), *Intracellular staining in neurobiology*. Berlin & New York: Springer-Verlag, 1973. Pp. 71–82.

Reddy, D. R., Davis, W. J., Ohlander, R. B., & Bihary, D. J. Computer analysis of neuronal structure. In S. B. Kater & C. Nicholson (Eds.), *Intracellular staining in neurobiology*. Berlin & New York: Springer-Verlag, 1973, Pp. 227–254.

Reese, T. S., Bennett, M. V. L., & Feder, N. Cell-to-cell movement of peroxidases injected into the septate axon crayfish. *Anatomical Record,* 1971, **169,** 409.

Rehbein, H. G., Kalmring, K., & Romer, H. Structure and function of acoustic neurons in the thoracic ventral nerve cord of *Locusta migratoria* (Acrididae). *Journal of Comparative Physiology,* 1974, **95,** 263–280.

Rehbein, H. G. Auditory neurons in the ventral cord of the locust: Morphological and functional properties. *Journal of Comparative Physiology,* 1976, **110,** 233–250.

Remler, M. P., Selverston, A. I., & Kennedy, D. Lateral giant fibers of crayfish: Location of somata by dye injection. *Science,* 1968, **162,** 281–283.

Selverston, A. The use of intracellular dye injections in the study of small neural networks. In S. B. Kater & C. Nicholson, (Eds.), *Intracellular staining in neurobiology*. Berlin & New York: Springer-Verlag, 1973. Pp. 255–280.

Silver, I. A. Other electrodes. In P. E. K. Donaldson (Ed.), *Electronic apparatus for biological research*. London: Butterworth, 1958. Pp. 568–584.

Smith, P. A., Sunderland, A. J., Leake, L. D., & Walker, R. J. Cobalt staining and electrophysiological studies of Retzius cells in the leech, *Hirudo medicinalis*. *Comparative Biochemistry and Physiology, A*, 1975, **51**, 655-661.

Snow, P. J., Rose, P. K., & Brown, A. G. Tracing axons and axon collaterals of spinal neurons using intracellular injections of horseradish peroxidase. *Science*, 1976, **191**, 310-311.

Strausfeld, N. J., & Obermayer, M. Resolution of intraneuronal and transynaptic migration of cobalt in insect visual and central nervous systems. *Journal of Comparative Physiology*, 1976, **110**, 1-12.

Stretton, A. O. W., & Kravitz, E. A. Neuronal geometry: Determination with a technique of intracellular dye injection. *Science*, 1968, **162**, 132-134.

Stretton, A. O. W., & Kravitz, E. A. Intracellular dye injection: The selection of Procion yellow and its applications in preliminary studies of neuronal geometry in the lobster nervous system. In S. B. Kater & C. Nicholson, (Eds.), *Intracellular staining in neurobiology*. Berlin & New York: Springer-Verlag, 1973. Pp. 21-40.

Székely, G. The morphology of motoneurons and dorsal root fibers in the frog's spinal cord. *Brain Research*, 1976, **103**, 275-290.

Székely, G., & Kosaras, B. Dendro-dendritic contacts between frog motoneurons shown with the cobalt labelling technique. *Brain Research*, 1976, **108**, 194-198.

Tasaki, I., Tsukahara, Y., Ito, S., Wayner, M. J., & Yu, W. Y. A single, direct, and rapid method for filling microelectrodes. *Physiology and Behavior*, 1968, **3**, 1009-1010.

Tweedle, C. D. Intracellular dye techniques for neurobiology. *Federation Proceedings, Federation of American Societies for Experimental Biology*, 1975, **34**, 1616-1617.

Tyrer, N. M., & Altman, J. S. Motor and sensory flight neurons in a locust demonstration using cobalt chloride. *Journal of Comparative Neurology*, 1974, **157**, 117-138.

Tyrer, N. M., & Bell, E. M. The intensification of cobalt-filled neurone profile using a modification of Timm's sulphide-silver method. *Brain Research*, 1974, **73**, 151-155.

Van Keulen, L. C. M. Morphology of Renshaw cells. *Pfluegers Archiv*, 1971, **328**, 235-236.

Van Orden, L. S. Principles of fluorescence microscopy and photomicrography with applications to Procion yellow. In S. B. Kater & C. Nicholson (Eds.), *Intracellular staining in neurobiology*. Berlin & New York: Springer-Verlag, 1973. Pp. 61-70.

Wine, J. J., Mittenthal, J. E., & Kennedy, D. The structure of tonic flexor motoneurons in crayfish abdominal ganglia. *Journal of Comparative Physiology*, 1974, **93**, 315-335.

Zieglgänsberger, W., & Reiter, C. Interneuronal movement of Procion yellow in cat spinal neurones. *Experimental Brain Research*, 1974, **20**, 527-530.

Chapter 7

Electron Microscopy and the Study of the Ultrastructure of the Central Nervous System

Jerald J. Bernstein and Mary E. Bernstein

Departments of Neuroscience and Ophthalmology
University of Florida College of Medicine
Gainesville, Florida

I. Introduction

The electron microscope has revealed the wealth of fine structure of the nervous system which now is taken for granted. The impact and scope of information

175

from the ultrastructure of the nervous system are incalculable. This information has been gathered in a relatively short span of time since the ability to do reliable research on central nervous system ultrastructure was made possible only by the use of perfusion fixation (Palay, McGee-Russell, Gordon, & Grillo, 1962) which was developed well after the advent of the electron microscope. Cell ultrastructure has confirmed not only histological (light microscopic) and biochemical hypotheses but has revealed new and often puzzling aspects of cell structure and function (Peters, Palay, and Webster, 1970).

Although the study of ultrastructure as revealed by transmission electron microscopy is an exciting area of neuroanatomical and neuroscientific research, the process is long, often tedious, requires special skills; surprisingly, to yield data, 85% of the time is spent in specimen preparation, 10% at the electron microscope, and 5% in photography. The great interest in this field of neuroanatomical research has led to rapid advances in and improved methodology for specimen preparation and the availability of automatic instruments for specimen preparation and sectioning. In addition, the contributions of polymer science have been invaluable through the synthesizing of new resins for specimen embedding.

This chapter briefly surveys the most commonly utilized methodologies for the study of central nervous system ultrastructure. Included are some of the more recent techniques which yield data on the morphology and function of the cellular elements of the central nervous system.

II.   Specimen Preparation

The techniques for specimen preparation can be divided into several categories. These include fixation, dehydration, infiltration and embedding, specimen sectioning, and staining.

Fixation of the tissue is an extremely critical process in the study of ultrastructure. The preferred method of fixation is perfusion, in which the blood supply is replaced with fixative (Palay *et al.*, 1962). Perfusion fixation rapidly places the fixative in the tissue and minimizes the effects of anoxia and autolysis. Many variations of the perfusion technique for rapid *in situ* fixation have been developed. These techniques control the pH and osmolarity of the fixing solution to render the fixative physiologically balanced and to control for swelling or shrinkage of nervous system elements. The dehydration of the tissue removes the water from the specimen and replaces tissue water with solvents that are compatible with the embedding resin with a minimal loss of ultrastructural detail. The infiltration and embedding of the tissue places the specimen in a matrix suitable for sectioning and allows for stability of the specimen under the electron beam. The sectioning of the embedded specimen is a difficult process. Thin sections of tissue of approximately 80-nm thick give an excellent image in the microscope but are difficult to reproduce reliably. The only stains available to date are heavy metals or electron opaque reaction products which produce differential contrast under the electron beam. The various procedures will now be dealt with in further detail.

A. *Fixation*

The optimum fixation procedure will preserve the extra- and intracellular organization of the tissue. Preservation of the central nervous system without changes due to autolysis or anoxia necessitates a fixation process in which the fixative reaches a particular nervous system element rapidly and in high concentration. A permanent fixation must be achieved which will not only maintain structures as they exist *in vivo,* but will also stabilize these structures. If these criteria are met, the loss of cellular components will be minimized during the dehydration and embedding procedures. The fixative should react with the cellular chemical matrix and provide stable bonds to prevent the distortion of ultrastructure. The fixatives which best meet these criteria include glutaraldehyde, glutaraldehyde–formaldehyde mixtures and osmium tetroxide with various buffering systems and additives to control for pH and osmolarity.

Glutaraldehyde, a 5-carbon dialdehyde, has been found to give the best preservation of any aldehydes when used alone. Concentrations of glutaraldehydes from 1.5 to 6.0% (Karlsson & Schultz, 1965; Sabatini, Bensch, & Barnett, 1963) have been recommended for rat central nervous system. The rate of glutaraldehyde penetration into cells is more rapid than that of osmium tetroxide but less than that of formaldehyde. In addition, the rate of penetration of glutaraldehyde at room temperature is almost double that which it is in the cold (4°C) (Chambers, Bowling, & Grimley, 1968). In spite of this fact, chilled glutaraldehyde fixatives and/or chilled animals are often used during fixation to compensate for anoxic and autolytic effects. Glutaraldehyde increases both inter- and intramolecular cross-bonding of proteins. A high percentage of the tissue glycogen is also preserved, which is important for cytological identification of tissue components. However, reaction with lipids is limited, which is a drawback since myelin is not well preserved.

The best fixation procedure for general work utilizes a glutaraldehyde or a glutaraldehyde–formaldehyde perfusate as the primary fixative followed by a secondary fixation with osmium tetroxide (double fixation). This method combines the good points of both fixatives, since aldehydes, while yielding stabilized proteins fail to preserve lipids, whereas osmium gives excellent lipid preservation while imparting increased stability to the membranes. The preservation of myelin is closely related to the preservation of lipid since its main components are cerebrosides and sphingomyelin. The nuclear chromatin, as well as general cytoplasmic details, are better preserved by double fixation than by the use of either fixative alone.

Osmium tetroxide alone is an excellent fixative. However its poor penetration into cells and high cost make double fixation with postosmication by immersion the most commonly used procedure. Osmium tetroxide is extremely volatile and poisonous (its fumes being injurious to nose and eyes) and should be used in a hood. It is a strong oxidizing agent and is reduced in the presence of organic matter and on exposure to light. As a result, a hydrated dioxide is formed which does not act as a fixative. Therefore, increasing reduction of the osmic acid solution decreases the effectiveness of preservation. After perfusion with glutaraldehyde, the tissue retains its susceptibility to change. Therefore, not only should the wash be buffered to preserve structure, but the osmium tetroxide fixative should be buffered as well (Karlsson & Schultz, 1965).

The use of primary fixative composed of a combination of glutaraldehyde and formaldehyde is preferred by many authors. A variety of proportions of glutaraldehyde to formaldehyde have been used, ranging from 4% formaldehyde–0.5% glutaraldehyde to 1% formaldehyde–1% glutaraldehyde. Other variations include use of solutions of varying concentrations serially introduced (Peters, 1970). When glutaraldehyde–formaldehyde fixatives are used, formaldehyde fixative is prepared immediately before use from paraformaldehyde powder to ensure formaldehyde purity. The formaldehyde is prepared by mixing paraformaldehyde powder with distilled water at 60°C to which $1 N$ NaOH is added (with stirring) until the solution clears. Either $0.12 M$ phosphate buffer (containing sucrose and $1.5–2\mu g$ of calcium chloride/ml of fixative solution) or $0.1 M$ cacodylate buffer (containing 5 μg calcium chloride/ml of fixative) may be utilized.

The quality of fixation is affected by the pH as well as the composition of the buffer, the osmolarity, and perhaps tonicity of the perfusate. Other factors are the concentration and temperature of the perfusate, and the duration and pressure of the perfusion. The optimal pH for the best overall fixation and the maximum binding of glutaraldehyde to proteins however, depends on the area of nervous system examined and age and species of the animal. Most authors interested in mammalian nervous system use a pH of 7.2–7.4. Slightly hypertonic buffered fixatives have been found to yield good results when fixation is with aldehydes. Buffer systems which, in addition to the buffer, contain either sucrose, dextran, or cacodylate or ionic substances such as sodium, potassium, and other ions are used to bring the fixative to a compatible osmolarity and tonicity.

Cacodylate is an arsenic derivative often used as a buffer. The cacodylate buffer has the capacity to absorb more calcium chloride than phosphate buffers as it does not produce insoluble calcium phosphate with an excess of Ca^{2+} ions (Peters, 1970). Since Ca^{2+} ions are excellent membrane stabilizers, cacodylate may yield better tissue preservation as more Ca^{2+} ions may be maintained in the fixative solution.

The fixation procedure used in the authors' laboratory yields reliable fixation of thoracic and lumbar spinal cord. The animal is deeply anesthetized, the chest cavity opened, pericardium incised, and the heart exposed. Heparin, to prevent clotting of the blood, and 1.0% sodium nitrite (a vasodilator) can be injected prior to perfusion either intravenously (i.v.) or directly into the left ventricle. The fixative is passed through a cannula into the aorta or more commonly into the left ventricle. Depending on the method of perfusion a return is provided by severing the right auricle, inferior vena cava, or removing the apex of the heart in aortic perfusions. The perfusate is introduced through syringes, drip bottles, peristaltic pumps, or similar methods. The first solution perfused is a buffered wash solution without fixative (amount varies with size of animal) to remove the blood from the circulatory system, immediately followed by the fixative solution. Perfusion pressure may be modified by suspending drip bottles containing solutions at a desired height above the animal to obtain a proper flow rate. A medical i.v. drip system is desirable to eliminate bubbles from the system, particularly when switching from buffered wash to fixative solution, and to enable one to monitor the solution flow rate during perfusion. A better alternative to the gravitation method is the peristaltic pump which gives a known pulsatile flow rate. In the author's laboratory, a fixative at a flow rate of 20 ml/min for 15 min is followed by 45 min of perfusion at 10 ml/min. Some investigators perfuse for a shorter time, but leave the nervous system *in situ* for at least an hour or for as long as overnight (the animal is refrigerated) before dissecting out the tissue to be studied. Such procedures often result in better preservation of tissue as protein binding by the aldehyde is maximized. Some investigators utilize artificial respiration during the perfusion process to reduce the effects of anoxia. On the other hand, increased carbon dioxide tension of blood can produce a dilation of cerebral blood vessels, so that some anoxia might be an advantage. This advantage must be weighed in lieu of anoxic alterations of ultrastructure. The mechanical trauma of introducing the cannula for perfusion may cause a reflex constriction of cerebral blood vessels which could result in a poor perfusion.

The buffer system used in the author's laboratory is a modified phosphate buffer containing sucrose for osmolar balance and $CaCl_2$ for membrane stabilization.

Phosphate Buffer Stock Solution (pH 7.4)

5.0 $NaH_2PO_4 \cdot 4H_2O$	101.28 ml
5.0% NaOH	23.12 ml
H_2O (glass distilled)	26.50 ml
Sucrose	12.00 gm
0.2% $CaCl_2$	2.00 ml

Buffer Wash
    Phosphate buffer stock solution   1 part
    H_2O (glass distilled)   1 part

5% Glutaraldehyde Fixative
    Phosphate buffer stock solution   1 part
    10% Glutaraldehyde   1 part

1% Osmium Tetroxide Fixative
    Phosphate buffer stock solution   1 part
    2% Osmium tetroxide   1 part

After the 1-hr perfusion, 1-mm-thick tissue blocks are immersed overnight in buffered glutaraldehyde fixative. The tissue is washed in two changes of buffer and immersed in 1% buffered osmium tetroxide and gently agitated (2 hr). The tissue is then rapidly dehydrated.

Dehydration
    70% ETOH   3 min
    80% ETOH   3 min
    95% ETOH   5 min
    100% ETOH   30 min (three changes)
    Propylene oxide   20 min (two changes)

Infiltration
    1:1 Propylene oxide to resin   2 hr (mild agitation)
    1:3 Propylene oxide to resin   2 hr (mild agitation)

Embedding
    Complete resin   Overnight (mild agitation)
    Orient tissue and polymerize resin

B. Embedding Media

Embedding media are selected to give good sectioning characteristics (i.e., homogeneity, hardness, plasticity, and elasticity) and stability under the electron beam. The embedding medium should have uniform polymerization, minimal extraction of cellular constituents, easy penetration into tissue, and little change in volume during sectioning. Epoxy resins are the embedding media in most common use today. These resins polymerize with very little shrinkage (1–2%) and are degraded very slowly by the electron beam. Among the epoxy resins, Araldite, Epon, and combinations of the two are the most often used.

Araldite is a glycerol-based aromatic resin that shows little volume shrinkage (1–2%) after polymerization, but is viscous (3000–6000 Hz at 23°C). The resin appears to be less grainy than other embedding resins with high resolution microscopy. The following is a good basic Araldite resin mixture.

Araldite 502	27 ml
DDSA [Dodecenyl succinic anhydride (hardener)]	23 ml
DMP-30 [2,4,6-Tri(dimethylaminomethyl) phenol (accelerator)]	0.75–10.0 ml (Luft, 1961)

The araldite–DDSA mixture may be stored at 4°C for up to 6 months, the accelerator being added immediately before use to avoid premature polymerization.

Epon 812 appears to be the most widely used embedding medium. It is a glycerol-based aliphatic epoxy resin of relatively low viscosity. Epon 812 polymerizes uniformly at a low temperature (60°–80°C) with the addition of acid anhydrides and amine accelerator. Commonly used acid anhydrides are dodecenyl succinic anhydride (DDSA) and nadic methyl anhydride (NMA). The amine accelerator is 2,4,6-tri(dimethylaminomethyl)phenol (DMP-30). The use of two anhydride polymerizing agents (DDSA and NMA) gives a wide range of hardness to suit varying tissues and preferences. Factors influencing the sectioning quality of epon blocks are the anhydride–epoxy ratio, and the temperature and the duration of polymerization. The anhydride–epoxy ratio is easily varied according to instructions provided by companies distributing the media. The following mixture has proven successful in the authors' laboratory:

Mixture A
    Epon 812 5 ml
    DDSA   8 ml

Mixture B
    Epon 812 8 ml
    NMA   7 ml

Final embedding mixture (to be varied according to hardness desired) is as follows:

Mixture A	13 ml
Mixture B.	15 ml
DMP-30	16 drops (Luft, 1961)

The method for dehydration and embedding utilized is a graded series of ethanols. The resin is infiltrated into the tissue in propylene oxide as most resins are not readily miscible in alcohols. An alternative would be to dehydrate in a graded series of acetones followed directly by infiltration with the resin. The tissue is infiltrated in a 1:1 mixture, then with a 1:3 mixture, of propylene oxide to complete resin for 2 hr each with agitation. Infiltration and embedding are completed with the resin mixture alone for an additional 20 hr with mild agitation. A three-step polymerization of the resin is usually used: 35°C overnight, 45°C the following day, and 60°C for the second night. Polymerization continues at room temperature for several days (Luft, 1961) so tissue should not be sectioned immediately after resin polymerization. Other resins are available which are used to help solve the particular problems of the investigator (water-based resins, etc.).

In the final infiltration process the tissue is embedded in a mold which will fit in the ultramicrotome. Tissue may be embedded in plastic capsules or may be flat embedded. Flat embedding of the tissue allows for more latitude in tissue orientation in the ultramicrotome.

C. Sectioning

The embedded tissue is trimmed into the shape of a truncated pyramid with sloping sides short enough to provide good support. The block face should be a trapezoid with parallel top and bottom edges. It should be as small as possible yet include the material desired as the larger the block face, the greater the difficulties encountered in sectioning.

The specimen is placed in the ultramicrotome, and the glass or diamond knife edge aligned with the lower edge of the trimmed tissue block face. The fluid level is adjusted in the knife boat so the miniscus wets the knife edge (sections are floated off the knife edge). The object is to produce and float sections of 60 to 80 nm (silver to gray refraction) on the liquid in the boat mounted on the knife. Section thickness is judged by interference color: purple, about 200 to 150 nm; gold, 150 nm–90 nm; silver, 90 nm–to about 60 nm; thinner sections are gray. Sectioning difficulties are usually due to dull knives, debris on the knife edge, or poor trimming of the block. However, sectioning difficulties can be due to the specimen. Insufficient polymerization of resin may cause bonded, noncleanable debris on the diamond or glass knife edge which results in striations, streaks, and block compression. Soft tissue blocks can result from poorly mixed polymers, or the uptake of water from the atmosphere by the block.

In addition to the condition of the knife and specimen, mechanical factors such as speed of sectioning influence the amount of compression and the uniformity of thickness of sections. External vibration of the ultramicrotome causes chatter, streaks, striations, compression, and lack of thickness uniformity. Placing the microtome on a clamping block is helpful in reducing vibration. Shifting air currents in the room can differentially change the temperature of the block face and result in poor sections.

After the sections are floating on the fluid in the boat they are picked up on grids that can be placed in the electron microscope. Orientation, comparisons, and identification of structures may be accomplished through the study of relatively thick sections (0.5–1.0 μm) by light microscopy. Alteration of thick and thin sections can be an invaluable aid to orientation of the tissue for the study of the ultrastructure.

D. Staining

The ultrastructural image is produced by relative electron opacity resulting in differential contrast. Heavy metal ions not only increase contrast and resolution but increase the resistance of sections to damage caused by the electron beam. Certain stains also provide a means of obtaining cytochemical information about cell components.

Some fixatives act as stains as well as fixation agents. For instance, osmium tetroxide acts as a stain and a fixative for unsaturated lipids and binds with lipoprotein membranes, resulting in stained membranous structures, ribosomes, Golgi complexes, and multivesicular bodies. Potassium permanganate reacts with a variety of protein groups and acts as a protein stain. It is a poor fixative but a valuable stain for certain ultrastructural features.

The sectioned tissue on the grid is usually stained for additional contrast with uranium in the form of uranyl acetate and lead in the form of lead citrate. In general, multiple staining enhances the density of all cell components beyond the degree that can be obtained with any stain alone. Uranium and lead consistently yield good differential contrast when used with a variety of embedding media. Grids may be stained by floating them tissue side down on drops of the staining solution in a covered petri dish or by a variety of ingenious techniques which vary in different laboratories.

Staining with lead citrate is done in an alkaline solution (ca. pH 12.0). Exposure to air must be minimized because in all solutions of lead salts, a lead carbonate precipitate is formed on exposure to carbon dioxide. Lead citrate solution may be prepared in the laboratory or is commercially available. Tissue is stained with 0.1–0.4% lead citrate for 2 to 15 min depending on the material.

Prolonged treatment with uranyl acetate results in the staining of most proteins, including nucleoproteins. Material may be stained with uranyl acetate in block before or during the dehydration process, or it may be stained after the tissue has been embedded, prepared as thin sections, and placed on grids. Only freshly prepared solutions should be used, as a precipitate forms with time. Uranyl acetate dissolves slowly in water, but its solubility increases with temperature and in alcoholic solutions, so both alcoholic and aqueous staining of sections on grids with uranyl acetate have been used. Staining in a 6% solution of uranyl acetate in 1% acetic acid for 10 min at room temperature (Westrum & Black, 1971), staining in a 3% aqueous solution of uranyl acetate for 30 min at 50°C (Venable & Coggeshall, 1965), or staining for 10 min at room temperature in 2% uranyl acetate in 50% methanol or ethanol all yield good results.

After staining sectioned material on grids with uranyl acetate, the grids are washed in glass distilled water, stained in lead citrate, washed again in glass-distilled water, and dried on filter paper. The material is now ready to be examined with the electron microscope.

III. Additional Selective Methods for Electron Microscopy

The electron microscope is extensively utilized as a tool in experimental research in which more than a description of fine structure is desired. Impressive methods have been developed which enable one to use histochemical and immunological methods for the identification of cell components. Histochemical techniques vary from the identification, location, and comparative quantification of various enzymes (enzyme histochemistry) to the use of differential staining to elucidate anatomical structure as well as to attempt to relate this structure to a specific chemical constituency. Electron microscopic autoradiography has been used to study cellular metabolism and proliferation, neuronal protoplasmic transport, as well as whole neurons in anterograde tracer studies. These studies have increased our knowledge of the internal components and cellular mechanisms within neurons and neuroglia.

A wide range of enzymes may be studied histochemically with the electron microscope (Hayat, 1973). Tissue prepared for histochemical studies requires minimum loss and shifting of biochemical activity. Since the localization of an enzyme can best be determined in relation to preserved cellular organelles, adequate preservation is required. Therefore, in histochemical enzyme studies the usual fixative is glutaraldehyde, which yields minimal loss of biochemical activity while maintaining adequate preservation.

Distribution patterns of dense material in the presynaptic element have been revealed by specialized staining techniques. Zinc oxide–osmium tetroxide (ZIO) and ethanolic phosphotungstic acid (EPTA) have been used to reveal synaptic morpbology (Akert, Kawana, & Sandri, 1971; Bloom, 1972; Pfenniger, 1971). The ZIO method was found to impregnate both cholinergic (Akert & Sandri, 1968) and aminergic nerve endings (Pellegrino de Iraldi & Suburro, 1970). The formation of developing synaptic contacts in rat has been studied with EPTA, which selectively stains sites of synaptic junctions and facilitates the analysis of synaptogenesis. In addition, alterations in the size and shape of dense projections may be used for a quantitative estimate of synaptogenesis as well as a maturation index (Bloom, 1972).

Identified neuronal populations have been studied by a combination of Golgi techniques and electron microscopy. By this method of marking, groups of isolated cells are studied for their cytoarchitectural and ultrastructural relationships (Blackstad, 1975). Other methods make use of retrograde transport by tracing the movement of a protein within neurons from nerve terminals to perikaryon. This technique has led to the development of neuroanatomical tracing techniques using horseradish peroxidase (HRP) (LaVail & LaVail, 1974; LaVail, Chapter 13 in this volume). HRP results in reaction product deposition in perikaryon and is useful in the study of cytoarchitectonics. In addition, anterograde transport of incorporated radioactive markers have been studied by autoradiography (see Chapter 9 by Hendrikson in this volume). The autoradiographic technique increases the ability to study a wide range of problems and quantify results (Saltpeter & Bachmann, 1972).

Specific and individual cell markers which will identify individual neurons are most desirous in attempts to correlate electrophysiological studies with fine structure, to yield, in particular, a knowledge of synaptic complement to a neurophysiologically studied neuron (see Chapter 6 by Tweedle in this volume). Procion brown which is electron-opaque (Christensen, 1973; Tweedle, 1975) and cobalt (Gillette & Pomeranz, 1973) can be delivered by iontophoresis directly from an intracellular recording micropipette. Procion brown and cobalt have been used to locate a physiologically studied neuron with some success but the ideal neuronal marker remains an elusive but much sought after goal since these techniques are difficult to make compatable with good ultrastructural preservation.

Immunohistochemical localization of some transmitter synthesizing enzymes in the central nervous system offers the promise of identifying specific neuronal systems because of the characteristic localization of synthesizing enzymes in a particular class of neurons. Antibodies have been obtained and immunohistochemical localization studies carried out for tyrosine hydroxylase (Pickel, Tong, & Reis, 1974), choline acetyltransferase (Hattori, Singh, McGeer, & McGeer, 1976) and other enzymes in catecholamine systems, tryptophan hydroxylase (Reis, Pickel, Shikimi, & Joh, 1975) in serotonin systems; glutamic acid decarboxylase in γ-aminobutyric acid systems (McLaughlin, Barber, Saito, Roberts, & Wu, 1975); and choline acetyltransferase for acetylcholine systems (McGeer, McGeer, Singh, & Chase, 1974). Since specialized staining methods, as well as the use of immunochemistry methods, frequently do not yield adequate ultrastructural detail, these techniques are frequently coupled with traditionally prepared material.

IV. Ultrastructure: The Interpretation of the Electron Microscopic Image

A. Neuron Cell Body

Now that the tissue has been prepared and the electron microscopic image is seen on the phosphorescent screen of the microscope, another skill is required, i.e., interpretation of the electron microscopic image. Different regions of the central nervous system present a different appearance of neurons and neuropil to the electron microscopist. However, basic components are ubiquitous throughout the central nervous system.

Basically, neurons from different regions of the central nervous system have similar organelles but vary in size and shape. The nucleus of the neuron may be round or ovoid and may or may not have membranes which infold into the karyoplasm (Fig. 1). The nucleus (N) is enclosed by a double-layered nuclear envelope perforated by 70-nm-diameter pores (np) which are bridged by fine septa. The inner nuclear membrane is smooth with a thin adherent fibrous lamina. The fibrous lamina is most pronounced near and under nuclear pores (np). The outer nuclear membrane is irregular and ruffled in appearance, is considered to be continuous with the endoplasmic reticulum of the cytoplasm, and may contain ribosomes (r) irregularly dispersed over its surface.

The nuclear organelles include nuclear chromatim (C) consisting of fine strands (*ca* 20 nm in diameter) dispersed throughout the nucleus or in small aggregates to give a slightly variegated appearance. The nucleolus (n) is a dense spherical mass composed of dense granules and fine filaments which are closely packed. A nucleolar satellite (ns) containing the heterochromatic X chromosome and consisting of dense coiled fibrils is found in a paranucleolar position in females.

Within the cytoplasm, the most distinctive ultrastructural features are membranes. These take various forms. The most conspicuous of these is the tubular

---

Fig. 1.   (a) The nuclear membrane of a neuron in the ventral horn of the adult rat spinal cord contains nuclear pores (np) 70 nm in diameter which are spanned by septa. The double nuclear membrane is composed of an inner smooth membrane beneath which appears a fibrous lamella (↑) (especially prominent under the nuclear pore) and an outer undulating membrane. Within the cytoplasm, granular or rough endoplasmic reticulum (R), polysomes (p), and free ribosomes (r) are prominent as well as smooth endoplasmic reticulum (S). Magnification: × 55,000; uranyl acetate, lead citrate stain. (b) This small neuron in the female adult rat spinal cord contains finely dispersed chromatin (C) within the nucleus (N). The nucleolus (n) is composed of a mass of coarse fibrillar material intertwined with finer filamentous material. A nucleolar satellite (ns) is present containing X chromosomal material. Within the cytoplasm, mitochondria (M) may be branched or unbranched. Rough endoplasmic (R) reticulum is present, often stacked, and interspersed with ribosomes to form the characteristic light microscopic organelle, the Nissl body (Ni). Golgi apparatus (G) is present in a perinuclear position. Lysosomes (L) are scattered throughout the cytoplasm of the perikaryon. Synaptic complexes (SY) abut the plasma membrane of the perikaryon as well as processes of astrocytes (As). Magnification: × 6,000; uranyl acetate, lead citrate stain.

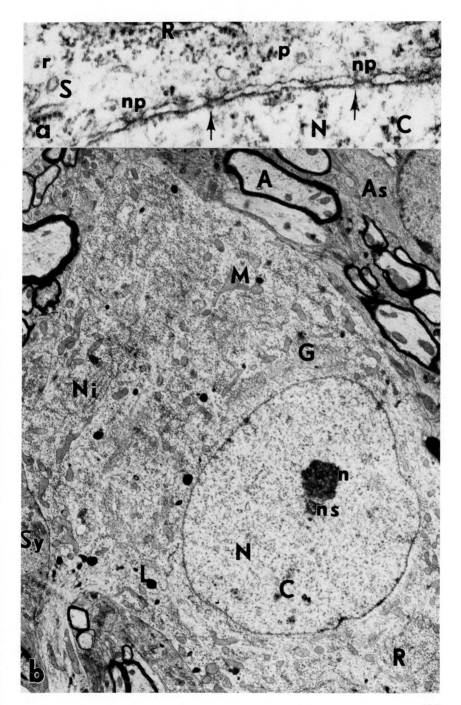

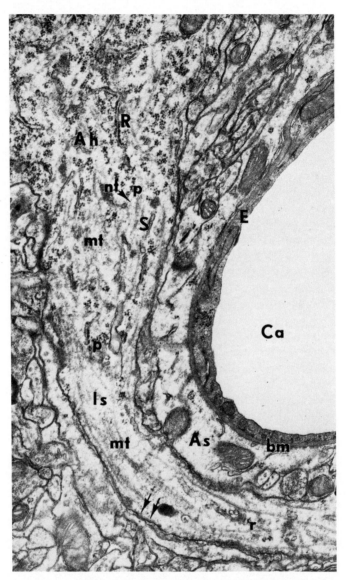

FIG. 2.   The axon hillock (Ah) and initial segment of the axon (Is) of this neuron in the adult rat cortex contains a decreased density of neuronal cytoplasmic organelles [granular (R) and smooth (S) endoplasmic reticulum, polysomes (p), microtubules (mt), neurofilaments (nf)]. Microtubules are fasciculated in these regions of the neuron. The initial segment has a fibrous lamellae (↑↑) subjacent to the cytoplasmic membrane as well as ribosomes (r) and polysomes. The lumen of a capillary (Ca) is observed surrounded by an endothelial cell (E) which forms the capillary. A basement membrane (bm) surrounds the endothelial cell, separating the capillary from the central nervous system by a pericapillary space. Astrocytic (As) processes form neuroglial endfeet which surround the capillary. Magnification: × 42,000; uranyl acetate, lead citrate stain.

network which passes three-dimensionally throughout the cytoplasm. These tubular membranes form into various organelles by different conformational aspects and associated structures. The tubular network termed endoplasmic reticulum (ER) can have areas that have smooth membranes [smooth or agranular endoplasmic reticulum (S)] or can have associated ribosomes [granular or rough endoplasmic reticulum (R)]. The endoplasmic reticulum extends from the outer nuclear envelope and may traverse to the plasma membrane (Figs. 1, 2, 3, 6, and 8).

The free ribosome (r) is a globular element approximately 15 nm in diameter. These individual elements may form interesting arrays in the form of S's, spirals, or other complex aggregates termed polysomes (p). Free ribosomes and polysomes interspersed between stacked granular endoplasmic reticulum (which can also have agranular portions) form the ultrastructural basis of the most prominent light microscopic neuronal organelle, tbe Nissl body (Ni).

The Golgi apparatus (G) is composed of short stacks of cisterns of expanded smooth endoplasmic reticulum in various three-dimensional arrays (Figs. 1, 6, and 8). This organelle usually has associated vesicles many of which are derived from the apparatus. In addition to smooth walled, clear vesicles, there are vesicles with striae which resemble spines projecting radially from their outer surface (coated vesicles), whereas other vesicles (60–80 nm) enclose a dense granule. Many vesicles derived from the Golgi complex fall into the classification of organelles known as lysosomes (L) (Figs. 1, 6, and 8). Lysosomes are from 0.2 to 0.5 μm in diameter, are single membrane bound, and contain finely granular dense material, or may have a heterogeneous content including membranes, vacuoles, and droplets. Lipofuscin droplets, which can be classified as lysosomes, are large, lobulated organelles with vacuoles and membranous lamellae in a granular matrix. Multivesicular bodies are found throughout the cytoplasm of the neuron, including the dendrites and axon terminals (Figs. 10 and 12). These single-membrane-bound spherical bodies contain a varying number of smaller vesicles within a matrix which may appear either clear or dark and is often considered to be a type of lysosome.

Mitochondria (M) are randomly dispersed throughout the neuronal cytoplasm, including dendrites and axons. They vary in size from 0.1 to 0.5 μm in diameter and 1.0 μm (or more) in length, having been known to reach a length of 20 μm in axons. They may be simple or branched (Fig. 1). Each mitochondrion is bound by a double membrane. The inner membrane is highly folded to form fingerlike projections called cristae mitochondrials which may be longitudinally and/or transversely oriented (Figs. 3, and 6).

Microtubules or neurotubules are found throughout neuronal cytoplasm (Fig. 3) occurring in dendrites, axons, and perikaryon. Microtubules (mt) are long tubular organelles 20–26 nm in diameter with or without dense granular cores about 6

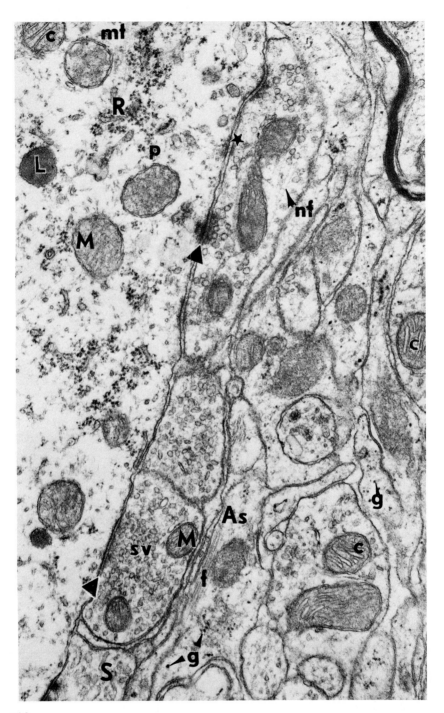

nm in diameter. Neurofilaments (nf), about 8–10 nm in diameter, are also present throughout perikaryal, dendritic, and axonal cytoplasm (Fig. 3).

B. Dendrites

Large dendrites contain the same organelles as the perikaryon: elements of granular (R) and smooth (S) endoplasmic reticulum, ribosomes (r), mitochondria (M), multivesicular bodies (mb), Golgi apparatus (G), microtubules (mt), neurofilaments (nf), vesicles (v), and lysosomes (L) can be present in the first segment of dendrites after the perikaryon (Fig. 4). Golgi apparatus is only found in large dendrites near the perikaryon. Ribosomes and granular endoplasmic reticulum (R) become less prominent distally in dendrites and are rarely observed in small dendrites. Mitochondria usually are oriented longitudinally and tend to lengthen with decreased diameter of dendrite.

In general, microtubules predominate in number over neurofilaments in dendrites. The opposite is usually the case in axons. Although neurofilaments predominate in large axons, the microtubules predominate in large dendrites, a ratio of 1:1 is found in smaller axons and dendrites. Therefore, it becomes difficult to distinguish between small dendrites and small axons. Many authors solve this problem by using the term neurite when positive identification of a small axon or dendrite is difficult.

Dendrites can have spines (sp) emanating from their surface (Figs. 5 and 6). Dendritic spines are most common after bifurcation of the initial dendrite and can increase in number exponentially from the initial portion of the dendrite to the distal portion of the dendrite. Spines vary in morphology from a shaft topped with an ovoid bulb, to stalked, sessile, twisted, straight, simple, or branched configurations. Some dendritic spines of cerebral cortical neurons contain a spine

---

FIG. 3. These synaptic complexes on a motoneuron in the ventral horn of an adult rat spinal cord demonstrate presynaptic boutons which contain synaptic vesicles (sv), neurofilaments (nf), smooth endoplasmic reticular (S), and mitochondria (M). Synaptic vesicles may be seen which are spherical or flattened and may be clear or contain a granular core. The synaptic cleft (▲) appears rigid, may contain finely granular cleft material, and is subtended by a pre- and a postsynaptic membrane with varying degrees of specialization. Synaptic vesicles are seen in association with the presynaptic membrane and the synaptic cleft. This combination of cellular specializations is often referred to as the active synaptic zone. Desmosomal junctions (*) are frequently associated with synaptic complexes. Desmosomes are specializations which are thought to be structural and have bilateral and equal distribution of fine fibrous material of the associated membranes. The cytoplasm of the neuron contains granular or rough endoplasmic reticulum (R) and polysomes (p). Double membrane-bound mitochondria (M) are present containing cristae mitochondriales (c). Single membrane bound lysosomes (L) containing dark granular material, as well as microtubules (mt) and neurofilaments are present. Within the neuropil, processes of astrocytes (As) are a prominent feature and often contain glycogen particles (g) and neuroglial fibrils (f). Magnification: × 33,000; uranyl acetate, lead citrate stain.

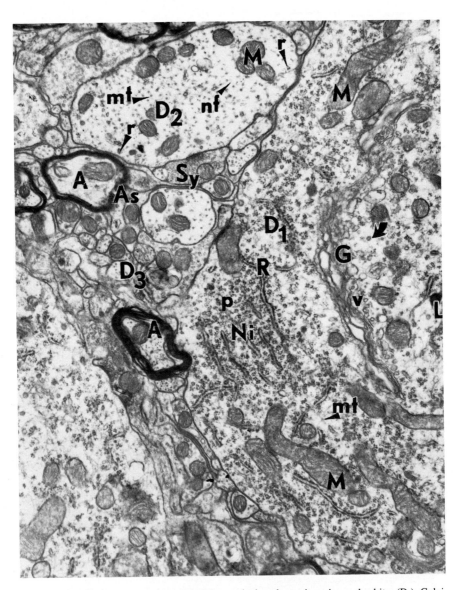

FIG. 4.   The neuropil in ventral horn of adult rat spinal cord contains primary dendrites (D_1), Golgi apparatus (G), stacked granular endoplasmic reticulum (R), and polysomes (p) forming the light microscopic organelle the Nissl body (Ni), as well as mitochondria (M), lysosomes (L), and microtubules (mt). Clear vesicles (v) and coated vesicles (⟲) are seen near a Golgi apparatus (G). Smaller dendrites (D_2) contain mitochondria (M), microtubules (mt), a few neurofilaments (nf), as well as smooth endoplasmic reticulum and ribosomes (r). Very small dendrites (D_3) are difficult to distinguish from axons (A) when the axons are unmyelinated. Synaptic complexes (Sy) on dendrites, small axons, dendrites, and astrocytic processes (As) compose the remainder of the neuropil. Magnification: × 16,500; uranyl acetate, lead citrate stain.

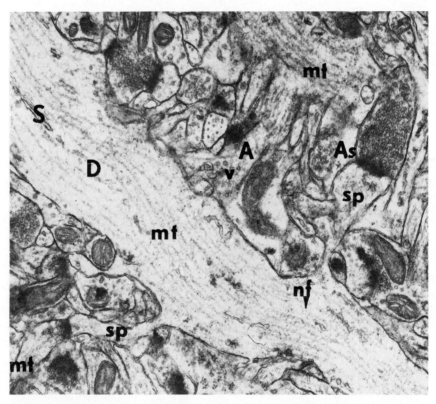

FIG. 5.   This dendrite (D) in longitudinal section in the cortex of an adult rat contains microtubules (mt) and neurofilaments (nf) as well as smooth endoplasmic reticulum (S). Dendritic spines (sp) contain a fine flocculent material. The dendritic spines are synaptically complexed with other neuronal processes. Small unmyelinated axons (A) in longitudinal section often contain vesicles (v) as well as microtubules and neurofilaments. Astrocytic processes (As) often contain glycogen particles. Axons (A) and dendrites (D) compose the remainder of the neuropil. Magnification: × 29,000; uranyl acetate, lead citrate stain.

apparatus (sa) consisting of two or three flattened cisternae alternating with thin laminae of dense material (Fig. 6b). Mitochondria are rarely found in spines. In addition, microtubules and neurofilaments have not been found to extend into dendritic spines in cortex. The spines contain a fine fibrillar or flocculent material which aids in their identification in the ultrastructural morass of the neuropil.

C.   Axons

The axon arises at the axon hillock (Ah) of the perikaryon or from an initial portion of a dendrite (Fig. 2). The cytoplasm of the axon hillock is distinguished

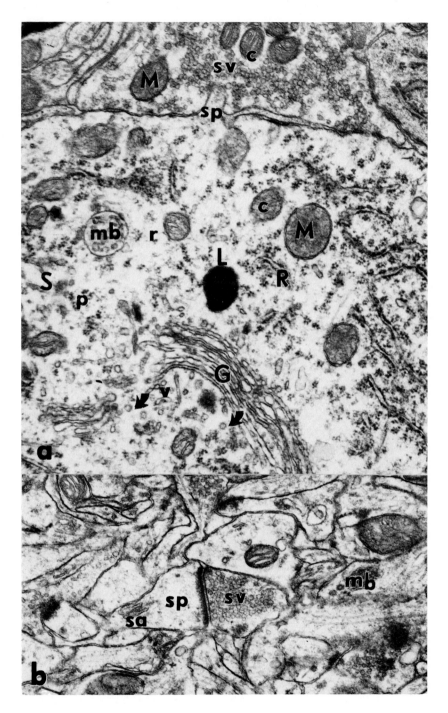

from other portions of the perikaryon by microtubular orientation (in fascicles parallel to the length of the axon) and decreased frequency of cytoplasmic organelles. The initial portion of the axon is unmyelinated and is termed the initial segment (Is). In the initial segment there are longitudinally oriented mitochondria, smooth endoplasmic reticulum, neurofilaments, and fascicles of microtubules. A layer of dense material lies beneath the plasma membrane in the initial segment. The dense material is interrupted at sites of synaptic contact. This dense layer is similar in appearance to the dense layer found under the membranes of central nodes of Ranvier. Beyond the initial segment, axons contain longitudinally oriented mitochondria, interspersed neurofilaments and microtubules, smooth endoplasmic reticulum, vesicles, and multivesicular bodies.

D. Myelin

The myelinated axon consists of an axis cylinder wrapped by concentric laminar membrane specializations which constitute myelin (My) (Fig. 7). Oligodendrocytes (O) are generally considered to be the only neuroglial cell to form the myelin sheath of central nervous system axons, however other neuroglia have been found to ensheath axons following injury. The cytoplasmic membrane of the myelinating cell is a three-layered unit membrane of two dense lines bisected by a thin intraperiod line separated from each dense line by a clear zone. Each layer is approximately 2.5 nm thick. The intraperiod line is formed by apposition of the outer portion of the central neuroglial cytoplasmic membrane while the dense line is formed by apposition of the inner cytoplasmic surfaces of the plasma membrane of the neuroglial cell. These morphological events alternate in a cross section of central myelin. Fused membranes are wrapped spirally for several turns. Internally the membranes are separate, forming an internal mesaxon (im) at the initiation of myelin laminae (Fig. 7b). Externally, in central myelin, neuroglial cytoplasm forming the myelin sheath does not form a com-

---

FIG. 6.   a. A simple spine (sp) on a neuron soma in the ventral horn of adult rat is postsynaptic to a presynaptic bouton containing spherical synaptic vesicles (sv). The neuronal cytoplasm contains double membrane-bound mitochondria (M) with cristae mitochondriales (c), single membrane-bound lysosomes (L), granular or rough endoplasmic reticulum (R), smooth endoplasmic reticulum (S), free ribosomes (r), and polysomes (p). Golgi apparatus (G) consisting of stacked cisternae of smooth endoplasmic reticulum is associated with numerous clear (v) and coated vesicles (⟲). The coated vesicle contains numerous striae radiating out from the vesicle wall. Magnification: × 24,000; uranyl acetate, lead citrate stain. b. This synaptic complex on the dendritic spine (sp) of a neuron in the cortex of an adult rat contains spherical synaptic vesicles (sv) and has asymmetric synaptic membranes (postsynaptic membrane has a heightened density). The postsynaptic dendritic spine contains a synaptic apparatus (sa) of cisternae with associated fine granular material. Magnification: × 38,000; uranyl acetate, lead citrate stain.

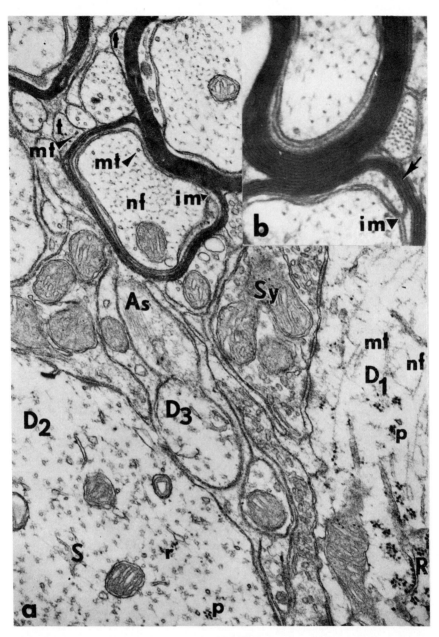

FIG. 7. (a) These myelinated axons in the ventral horn of the spinal cord in an adult rat have an external oligodendrocytic tongue (t) of cytoplasm typical of myelinated axons in the central nervous system. The oligodendrocytic cytoplasmic tongue contains microtubules (mt) in a granular background. Note the internal mesaxon (im) which is the initation of the myelin sheath. The neuronal

plete layer around the entire myelin sheath, but is confined to a narrow ridge extending the length of the internode. Ultrastructurally this process is observed as a small external tongue (t) on the myelin sheath of cytoplasm (Fig. 7a). More than one neuroglial cell can myelinate in a single internodal segment of the sheath, and one neuroglial cell usually myelinates one to six different axis cylinders.

Central nodes of Ranvier represent intervals between adjacent segments of internodal myelin (Fig. 9). Central nodes may occur at regular intervals along the length of myelinated axons but nodal frequency and length are largely unknown. As the myelin sheath nears a node, the lamellae begin to terminate serially, innermost lamallae first. The dense line of the sheath opens and is occupied by cytoplasm, creating a series of cytoplasmic pockets in longitudinal section. This area is known as the paranode (pn). Thus, the myelin sheath decreases in diameter as the node is approached. At the node, the axon often appears to be surrounded by an extracellular space which is larger than normal. A dense layer lies beneath and about 10 nm from the axonal membrane for the length of the node. The node may be an area for initiation of axonal collaterals and pre- or postsynaptic specializations for axoaxonic synapses. When the node is presynaptic the axon bulges into a protuberance which contains synaptic vesicles, mitochondria and other organelles associated with presynaptic boutons.

E. Synapses

Sites of neuronal synaptic contact which show interface membrane specializations (not a prerequisite) and asymmetrical distribution of synaptic vesicles are referred to as chemical synapses (Sy) (Figs. 3, 5, and 6). Axonal synaptic bulbs which form the predominant presynaptic elements can contain synaptic vesicles (sv), vacuoles, mitochondria (M), smooth endoplasmic reticulum (S), multivesicular bodies (mb), and microtubules (mt). The synaptic complexes are named by combining the name of the presynaptic member with the name of the postsynaptic member (axodendritic, dendrodendritic, axosomatic, axoaxonic, etc.). Presynaptic boutons or terminals contact the postsynaptic structure at the

---

portion of this myelinated axon contains neurofilaments (nf) and occasional microtubules. Dendrites of varying diameter are also present in the neuropil. A large dendrite (D₁), close to the neuronal perikaryon, contains rough endoplasmic reticulum, polysomes (p), neurofilaments (nf), and microtubules. More distal dendrites (D₂) contain polysomes (p) and free ribosomes (r), as well as smooth endoplasmic reticulum (S), neurofilaments (nf), and microtubules (mt). However, smaller dendrites (D₃) rarely contain ribosomes. The remaining neuropil is replete with synaptic complexes (Sy), astrocytic processes (As) which usually contain glycogen, as well as small axons and dendrites. Magnification: × 43,500; uranyl acetate, lead citrate stain. (b) Myelin sheath demonstrated dense lines (↑) derived from an oligodendrocyte. An internal mesaxon (im) is formed at the initiation of the myelin sheath. Magnification: × 160,000; uranyl acetate, lead citrate stain.

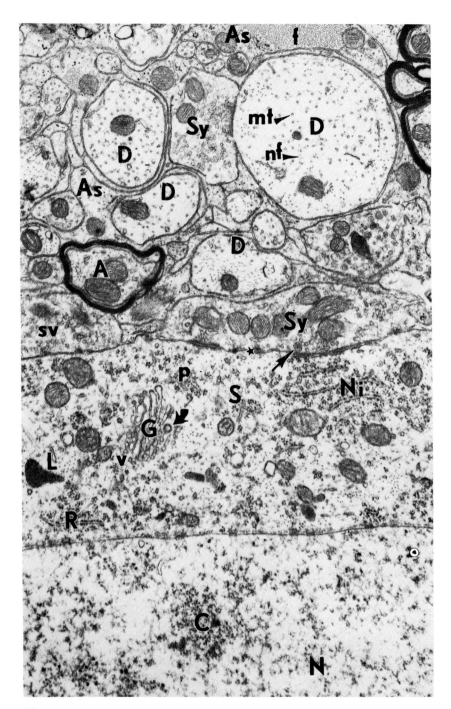

site of a distinctive junctional area with a rigid synaptic cleft (ca. 25 nm). Pre- and postsynaptic membranes may be asymmetrical (Gray's type 1) (Fig. 6b), with a dense postsynaptic membrane and a less dense presynaptic membrane or symmetrical (Gray's type 2) with little but approximately equal densities on both pre- and postsynaptic membranes. Between asymmetrical membranes, passing into the synaptic cleft, there is usually junctional dense material in an ordered array. Synaptic vesicles have been classified into several morphological types: clear round vesicles (40 nm in diameter); clear, flattened vesicles (ca. 20×60 nm); small and large granular core vesicles (ca. 45–100 nm diameter); and neurosecretory granulated vesicles (ca. 120 nm), etc.

Classification of synapses have been based on pre- and postsynaptic membrane specialization, differences in characteristics of synaptic vesicles, and presynaptic bouton profiles. For example, in the rat cerebellar cortex, six classes of synapses have been identified (Chan-Palay, 1975), four classes in carotid body (McDonald & Mitchell, 1975), and six classes in the rat and cat ventral horn of the spinal cord (Bernstein & Bernstein, 1976). This is not to imply that the descriptive morphology of these classes are the same or other classifications of synapses are not possible. Therefore, it is always necessary to operationally define classification of synapses.

Many chemical presynaptic terminals represent the termination of an axon or axon collateral. However, there are often presynaptic buoutons that are derived from a preterminal segment of axon. These presynaptic boutons have been termed terminals of passage. Although there are obvious examples of terminals of passage their frequency remains largely unknown.

Although most synapses in the mammalian central nervous system are chemical, electrotonic synapses have been demonstrated in lower vertebrates and mammals. Electrotonic synapses have been observed in rat lateral vestibular nucleus (Sotelo & Palay, 1971), mesencephalic trigeminal nucleus (Hinrichsen & Larramendi, 1970), and bipolar ganglion cell junction in primate retina (Dowling & Boycott, 1966). By morphological criteria these synapses can be one membrane specialization among several chemical presynaptic specializations occuring at the same terminal (Sotelo & Palay, 1971). Electrotonic junctions do not

---

FIG. 8.   The nucleus (N) of this motoneuron in the ventral horn of an adult rat spinal cord contains dispersed chromatin (C). The neuronal cytoplasm contains Golgi apparatus (G) with associated clear (v) and coated vesicles (◌), lysosomes (L), granular endoplasmic reticulum (R), smooth endoplasmic reticulum (S), ribosomes (r), polysomes (p), microtubules (mt), neurofilaments (nf), and the light microscopic organelle conglomerate the Nissl body (Ni). Axosomatic synapses (Sy) are present which contain spherical synaptic vesicles (sv) and occasional granular or dense cored vesicles (↑). A desmosome (*) is present on the synaptic membranes. Within the neuropil, dendrites (D) of varying diameter are observed with microtubules (mt), neurofilaments (nf), and varying organelles according to proximity to the perikaryon. Astrocytes (As) are observed. Their processes often contain glycogen and many neuroglial fibrils (f). Magnification: × 19,500; uranyl acetate and lead citrate stain.

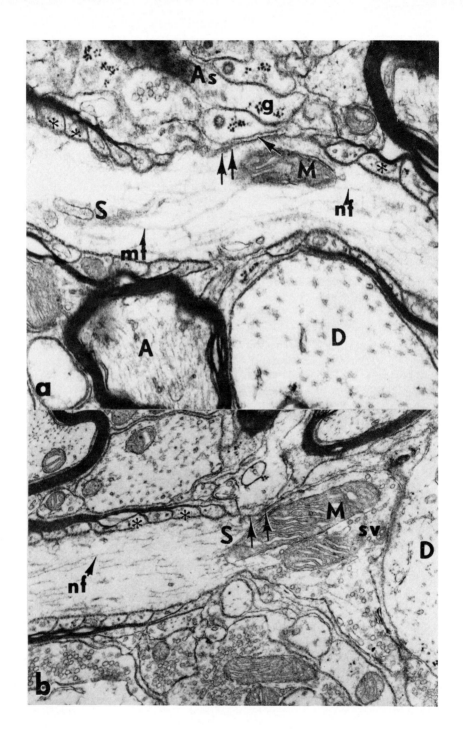

have the 25 nm synaptic cleft of chemical synapses. The synaptic membranes are closed apposed, with a space of approximately 2 nm between neuronal membranes, even though they often appear to be a fused pentalaminate structure owing to plane of section. Very thin sections oblique through the junction reveal ladderlike striations across the site of the gap about 10 nm apart, which apparently represent a hexagonal network with polyagonally arranged channels continuous with the extracellular space.

Interneuronal junctions other than active zones of synaptic junctions are common. Desmosomes (Figs. 3 and 8) have a rigid cleft (ca. 15–25 nm) with symmetrically dense membranes and symmetrically dispersed fibrous material extending into the cytoplasm on both sides of the junction. Desmosomes may be interspersed with the active zones of the synapse, but are apparently not involved in synaptic transmission. They are considered, at this time, to be adhesive and structural in nature and may occur between neurons and between neuroglia.

F. Neuroglia

Neuroglial cells classified in light microscopy as macroglia are frequently seen throughout the central nervous system. These consist of astrocytes and oligodendrocytes. Cytoplasm of astrocytes (As) intermingles with axons and dendrites to compose the major portion of the neuropil (Fig. 10). Oligodendrocytes (O), also part of the neuropil, are found throughout white and gray matter of the central nervous system but have few cytoplasmic processes (Figs. 11 and 12).

Astrocytes have irregularly shaped nuclei which may be deeply cleft with a karyoplasm of fairly even density and a fairly even distribution of chromatin. Chromatin (c) clumping may be found near the nuclear membrane. The cyto-

---

FIG. 9.   (a) Myelin lamellae terminate serially, innermost lamella first, at a central nervous system node of Ranvier within the spinal cord of an adult rat. There are characteristic cytoplasmic pockets (*) formed in the oligodendrocytic lamellae in this area known as the paranode. At the node, a layer of dense material lies beneath the axonal membrane (↑↑) for the length of the node. The extracellular space around a central nervous system node of Ranvier is often increased over normal extracellular space (↑). Astrocytic processes (As) with glycogen particles (g) overlie the node. Dendrites (D) and myelinated axons (A) are also present in the neuropil. The axon contains mitochondria (M), neurofilaments (nf), microtubules (mt), and smooth endoplasmic reticulum (S). Magnification: × 35,000; uranyl acetate, lead citrate stain. b. Synaptic complexes can be formed by terminals of passage or the final portion of an axon. This terminal axon segment in the ventral horn of the rat shows that the myelin lamellae terminate serially and form cytoplasmic pockets (*) of oligodendrocytic cytoplasm as in central nodes of Ranvier. The axonal membrane between the termination of myelin and the expansion of the axon into the presynaptic terminal has a dense matrix (↑↑) beneath the membrane as seen at the axonal initial segment and node. Smooth endoplasmic reticulum (S), microtubules (mt), and neurofilaments (nf) continue into the presynaptic terminal. Other organelles in the terminal are mitochondria (M) and synaptic vesicles (sv). Magnification: × 24,000; uranyl acetate, lead citrate stain.

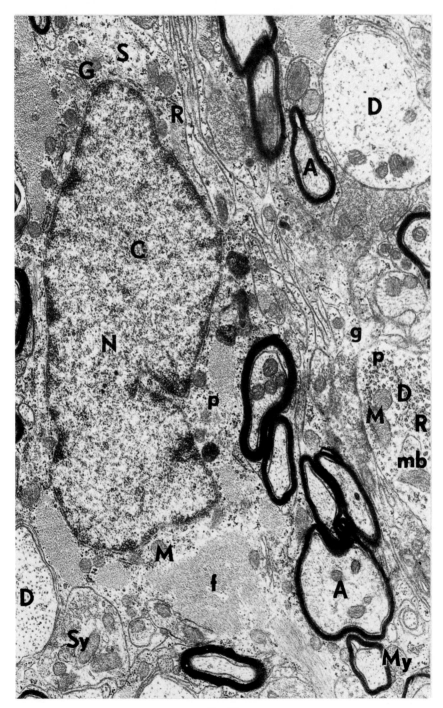

Fig. 10

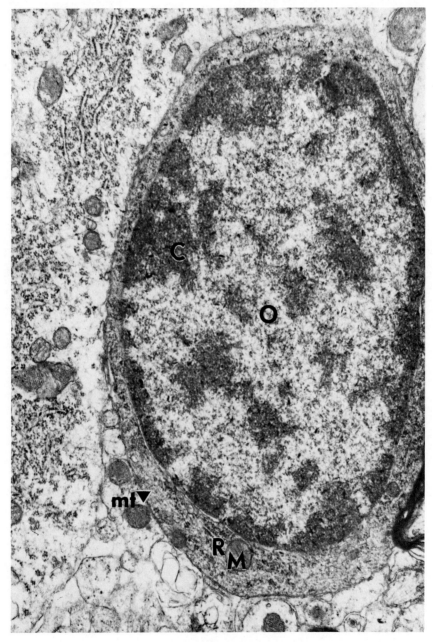

Fig. 11

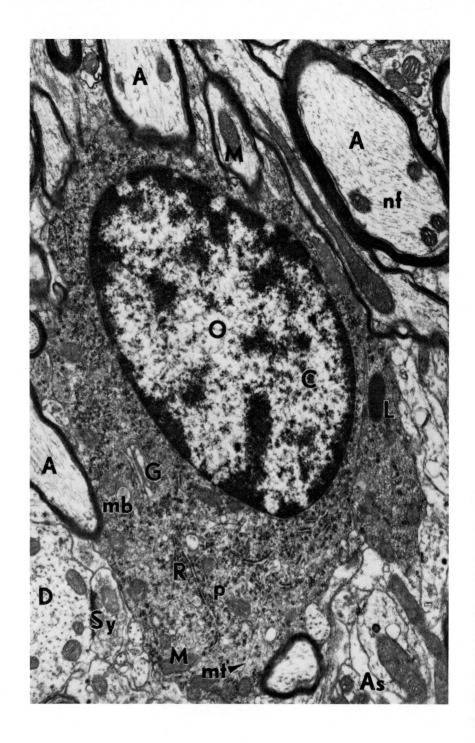

Fig. 12

plasm contains the usual organelles but there are only small amounts of granular endoplasmic reticulum (R), few free ribosomes (r), and limited Golgi apparatus (G). In fibrous astrocytes numerous neuroglial fibrils (8–9 nm in diameter) (f) occur throughout the cytoplasm extending in parallel arrays into astrocytic processes. Glycogen granules (g) are observed in astrocytic processes and are useful in the identification of small astrocytic processes in the neuropil of mammals.

Astrocytes form neuroglial endfeet around capillaries, separating neurons and their processes (except in a few specialized areas) with a complete layer of astrocytic cytoplasm from the basal lamina surrounding the endothelial cells of capillaries (Fig. 2). In addition, astrocytic endfeet form a layer (glial limitans) separating neuronal elements from the basal lamina subjacent to the pia mater of the meninges.

In contrast to astrocytes, oligodendrocytes (O) have dark staining cytoplasm with few processes and a rounded or irregular excentric nucleus containing heavily clumped chromatin subjacent to the nuclear membrane. The cytoplasm contains smooth and granular endoplasmic reticulum, numerous free ribosomes, Golgi apparatus, mitochondria, microtubules, and few fibrils or glycogen granules. The cytoplasm of oligodendrocytes appears darker than other cyto-

---

FIG. 10.   The nucleus (N) of this fibrous astrocyte (As) in the adult rat spinal cord contains finely dispersed chromatin (C). Neuroglial fibrils (f) are prominent in the cell body and processes. Mitochondria (M), sparse Golgi apparatus (G), granular (R) and smooth (S) endoplasmic reticulum, as well as ribosomes and polysomes (p) compose the astrocytic cytoplasmic organelles. Fibrils (f) and glycogen particles (g) may be observed in astrocytic processes. Myelinated (My) axons (A) usually contain more neurofilaments than microtubules in cross section. Conversely, dendrites in cross section usually contain more microtubules and fewer neurofilaments. Larger dendrites (D) in the proximity of the perikaryon can contain granular or rough endoplasmic reticulum, free ribosomes, polysomes, mitochondria (M) and multivesicular bodies (mb). Magnification: × 14,000; uranyl acetate, lead citrate stain.

FIG. 11.   The oligodendrocyte (O) not only myelinates nerve fibers but in gray matter is often in a satellite or perineuronal position to a neuron perikaryon (ventral horn of adult rat). The nucleus of the oligodendrocyte contains aggregates of dense chromatin (C) subjacent to the nuclear membrane and throughout the nucleoplasm. Oligodendrocytic cytoplasm contains granular endoplasmic reticulum (R), mitochondria (M), microtubules (mt), and Golgi apparatus. The cytoplasm of oligodendrocytes appears dense due to a background of fine, deeply staining granular particles. Magnification: × 21,000; uranyl acetate, lead citrate stain.

FIG. 12.   This oligodendrocyte in the adult rat spinal cord contains the organelles and characteristics typical of this neuroglial cell: rough endoplasmic reticulum (R), polysomes (p), mitochondria (M), Golgi apparatus (G), multivesicular bodies (mb), lysosomes, and microtubules (mt) in a dense granular cytoplasmic background. Chromatin (C) aggregates are observed in the nucleus. Myelinated axons (A) are observed in the neuropil which contain mitochondria and neurofilaments (nf). A dendrite (D) is seen with an axodendritic synapse (Sy). Astrocytic processes (As) are abundant. The remainder of the neuropil is comprised of astrocytic cell processes, small axons, and dendrites. Magnification: × 17,000; uranyl acetate, lead citrate stain.

plasm in the neuropil. This is due to the presence of numerous small granules which occupy the space between organelles. Microtubules occur in oligodendrocytic processes, (Fig. 7) which leads to difficulty in distinguishing small oligodendrocytic processes from small neuronal processes.

Another cell, encountered in normal neuropil is the microglial cell. This cell type is difficult to identify in normal material. However, the microglial cell population increases dramatically following injury. After injury the nucleus of the microglial cell is elongated or oval with dense chromatin clumps subjacent to the nuclear membrane. The cytoplasm is moderately dense and contains some microtubules and Golgi apparatus near initiation of cytoplasmic processes. Numerous small ovoid mitochondria occur throughout the cytoplasm. In the dense processes, microtubules are oriented longitudinally; lysosomes, and occasionally larger membrane-bound vacuoles containing undefinable structures can be observed. Again, microglia are difficult to identify and usually are not stressed in normal descriptive ultrastructural anatomy.

This brief description of the methodology utilized in electron microscopy and the interpretation of the ultrastructural characteristics of the central nervous system was intended as an introduction to this fascinating aspect of neuroanatomy. The literature utilizing this technique or modifications of the techniques described can only be considered to be enormous even by modest standards. Further study of specified areas are required to appreciate the value of electron microscopy. It is hoped that this initial peek under the ultrastructural rug will be the impetus to undertake research with this fascinating neuroanatomical tool.

Acknowledgments

This work was supported in part by the National Institutes of Health, (NINCDS) Grant No. NS-06164.

References

Akert, K., Kawana, E., & Sandri, C. ZIO-positive and ZIO-negative vesicles in nerve terminals. *Progress in Brain Research,* 1971, **34,** 305–317.

Akert, K., & Sandri, C. An electronmicroscopic study of zinc iodide-osmium impregnation of neurons. I. Staining of synaptic vesicles at cholinergic junctions. *Brain Research,* 1968, **7,** 286–295.

Bernstein, J.J., & Bernstein, M.E. Ventral horn synaptology in the rat. *Journal of Neurocytology,* 1976, **5,** 109–123.

Blackstad, T.W. Golgi preparations for electron microscopy: Controlled reduction of the silver chromate by ultraviolet illumination. *Golgi Centennial Symposium Proceedings,* 1975, 123– 132.

Bloom, F.W. The formation of synaptic junctions in developing rat brain. In G.D. Pappas & D.P. Purpura (Eds.), *Structure and function of synapses*. New York: Raven Press, 1972. Pp. 101–120.

Chambers, R.W., Bowling, M.C., & Grimley, P.M. Glutaraldehyde fixation in routine histopathology. *Archives of Pathology*, 1968, **85**, 18.

Chan-Palay, V. Fine structure of labelled axons in the cerebellar cortex and nuclei of rodents and primates after intraventricular infusions with tritiated serotonin. *Anatomy and Embryology*, 1975, **148**, 235–265.

Christensen, B.N. Procion brown; an intracellular dye for light and electron microscopy. *Science*, 1973, **182**, 1255–1256.

Dowling, J.E., & Boycott, B.B. Organization of the primate retina: Electron microscopy. *Proceedings of the Royal Society of London, Series B*, 1966, **166**, 80–111.

Gillette, R., & Pomeranz, B. Neuron geometry and circuitry via the electron microscope: Intracellular staining with an osmiophilic polymer. *Science*, 1973, **182**, 1256–1257.

Hattori, T., Singh, V.K., McGeer, E.G., & McGeer, P.L. Immunohistochemical localization of choline acetyltransferase containing neostriatal neurons and their relationship with dopaminergic synapses. *Brain Research*, 1976, **102**, 164–173.

Hayat, M.A. *Electron microscopy of enzymes*. Vol. 1. *Principles and methods*. Princeton, New Jersey: Van Nostrand-Reinhold, 1973.

Hinrichsen, C.F.L., & Larramendi, L.M.H. The trigeminal mesencephalic nucleus. II. Electron microscopy. *American Journal of Anatomy*, 1970, **127**, 303–320.

Karlsson, U., & Schultz, R.L. Fixation of the central nervous system for electron microscopy by aldehyde fixation. I. Preservation with aldehyde perfusates versus direct perfusion with osmium tetroxide with special references to membrane and the extracellular space. *Journal of Ultrastructure Research*, 1965, **12**, 160–186.

LaVail, J.H., & LaVail, M.M. The retrograde intraxonal transport of horseradish peroxidase in the chick visual system: A light and electron microscopic study. *Journal of Comparative Neurology*, 1974, **157**, 303–358.

Luft, J.H. Improvements in epoxy resin embedding methods. *Journal of Biophysical and Biochemical Cytology*, 1961, **9**, 409–414.

McDonald, D.M., & Mitchell, R.A. The innervation of glomus cells, ganglion cells and blood vessels in the rat carotid body: A quantitative ultrastructural analysis. *Journal of Neurocytology*, 1975, **4**, 177–230.

McGeer, P.L., McGeer, E.G., Singh, V.K., & Chase, W.H. Choline acetyltransferase localization in the central nervous system by immunohistochemistry. *Brain Research*, 1974, **81**, 373–379.

McLaughlin, B.J., Barber, R., Saito, K., Roberts, E., & Wu, J.Y. Immunocytochemical localization of glutamate decarboxylase in rat spinal cord. *Journal of Comparative Neurology*, 1975, **164**, 305–410.

Palay, S.L., McGee-Russel, S., Gordon, S., Grillo, M.A. Fixation of neural tissues for electron microscopy by perfusion with solutions of osmium tetroxide. *Journal of Cell Biology*, 1962, **12**, 385–410.

Pellegrino de Iraldi, A., & Suburo, A.M. Electron staining of synaptic vesicles using the Champy-Maillet technique. *Journal of Microscopy (Oxford)*, 1970, **91**, 99–103.

Peters, A. The fixation of central nervous tissue and the analysis of electron micrographs of the neuropil, with special reference to the cerebral cortex. In W.J.H. Nauta & S.O.E. Ebbesson (Eds.), *Contemporary research methods in neuroanatomy*. Berlin & New York: Springer-Verlag, 1970. Pp. 56–76.

Peters, A., Palay, S.L., and Webster, H. deF. *Fine Structure of the Nervous System*. New York: Harper and Row, 1970.

Pfenninger, K. The cytochemistry of synaptic densities. *Journal of Ultrastructure Research*, 1971, **34**, 103.

Pickel, V.M., Tong, H.J., & Reis, D.R. Ultrastructural localization of tyrosine hydroxylase in noradrenergic neurons of brain. *Proceedings of the National Academy of Sciences of the U.S.A.*, 1974, **72**, 659–663.

Reis, D.J., Pickel, V.M., Shikimi, T., & Joh, T.H. Rat brain tryptophan hydroxylase: Immunohistochemical localization by light and electron microscopy. *Transactions of the American Society of Neurochemistry*, 1975, **6**, 155.

Sabatini, D.D., Bensch, K., & Barnett, R.J. Cytochemistry and electron microscopy. The preservation of cellular ultrastructure and enzymatic activity by aldehyde fixation. *Journal of Cell Biology*, 1963, **17**, 19–58.

Salpeter, M.M., & Bachmann, L. Autoradiography. In M.A. Hayat (Ed.), *Principles and techniques of electron microscopy*. Vol. 2. *Biological applications*. Princeton, New Jersey: Van Nostrand-Reinhold, 1972. Pp. 221–278.

Sotelo, C., & Palay, S.L. Altered axons and axon terminals in the lateral vestibular nucleus of the rat. *Laboratory Investigation*, 1971, **25**, 653–671.

Tweedle, C.D. Intracellular dye techniques for neurobiology. *Federation Proceedings, Federation of American Societies for Experimental Biology*, 1975, **34**, 1616–1617.

Venable, J.H., & Coggeshall, R. A simplified lead citrate stain for use in electron microscopy. *Journal of Cell Biology*, 1965, **25**, 407.

Westrum, L.E., & Black, R.G. Fine structural aspects of the synaptic organization of the spinal trigeminal nucleus (*Pars Interpolaris*) of the cat. *Brain Research*, 1971, **25**, 265–287.

Study of Connections in the Nervous System

A. Techniques Based on Orthograde Processes (Chapters 8–11)
B. Techniques Based on Retrograde Processes (Chapters 12 and 13)
C. Electrophysiological Techniques (Chapter 14)

Chapter 8

The Study of Degenerating Nerve Fibers Using Silver-Impregnation Methods

Roland A. Giolli

Department of Anatomy
California College of Medicine
Department of Psychobiology
University of California
Irvine, California

Azarias N. Karamanlidis

Laboratory of Anatomy and Histology
Veterinary College
University of Thessaloniki
Thessaloniki, Greece

I. The Need for the Silver-Impregnation Methods

The silver-impregnation methods have been developed to allow a visualization of degenerating axons, and they have achieved the same basic goal as their predecessor the Marchi method, *namely to permit a study of nerve fiber connections within the nervous system.* The Marchi method, developed in 1886 by Vittorio Marchi, remained for 60 years the only procedure available to impregnate portions of neurons undergoing the anterograde (or Wallerian) type (see reviews of Bowsher, Brodal, & Walberg, 1960; Glees & Nauta, 1955) and the transganglionic type of degeneration (see Grant & Arvidsson, 1975; Grant, Ekvall, & Westman, 1970). Since its conception, the Marchi method has been used extensively to study the organization of the nervous system (see Beresford, 1966; Bowsher *et al.,* 1960). However, it must be pointed out that this method has not been proved to be the most effective tool for the neuroanatomical approach to the investigation of the nervous system. First, the "Marchi reaction" involves the impregnation of degenerating myelin sheaths, but not of either the unmyelinated portions of the axons or axon terminals (Adams, 1958, 1960; Bowsher *et al.,* 1960; Wolfgram & Rose, 1958). This drawback in the use of the Marchi method does not allow one to determine accurately the terminal distribution of the fibers. Second, the Marchi reaction typically involves not only the impregnation of degenerating myelin sheaths, but also that of normal myelin sheaths and possibly other tissue elements. Therefore, the presence of "Marchi artifact" is prevalent and requires a great deal of experience on the part of the investigator to distinguish from the true impregnation of degenerating myelin sheaths.

With the aforementioned information as background, it can be stated that the development of the silver-impregnation methods to demonstrate degenerating fibers came about primarily as a result of the drawbacks in the use of the Marchi method (Beresford, 1966; Bowsher *et al.,* 1960). Two basic silver methods have been developed and each can be regarded as a modification of the silver-impregnation method of Bielschowsky (1904). One of these methods, that of Glees (1946; Glees & Nauta, 1955; Marsland, Glees, & Erickson, 1954), had been developed for the purpose of selectively showing the degenerative changes that occur in the terminals of the axons, the *boutons terminaux.* This method does not give consistent results, and it is not always selective for degenerating boutons. The other method, that of Nauta and his colleagues (Fink & Heimer, 1967; Nauta, 1950, 1957; Nauta & Gygax, 1951, 1954; Nauta & Ryan, 1952), has been developed and refined for the purpose of illustrating degenerating axons and their terminal arborizations (and in some cases also the boutons) all against a relatively clear, unimpregnated background. Of these two basic silver-impregnation methods, the Nauta method has been by far the more popular, and it has become one of our most important tools for gaining a knowledge on how the nervous system is organized. However, the original Nauta methods (Nauta,

1950, 1957; Nauta & Gygax, 1951, 1954; Nauta & Ryan, 1952), as well as their numerous modifications (Albrecht & Fernström, 1959; De Olmos, 1969; De Olmos & Ingram, 1971; Desclin & Escubi, 1975; DeVito & Smith, 1959; Di Virgilio, German, Horn, Lewis, Scharf, Schwartz, Warwick, & Epstein, 1958; Eager, 1970; Ebbesson, 1970; Fink & Heimer, 1967; Giolli & Pope, 1973; Guillery, Shirra, & Webster, 1961; Johnstone & Bowsher, 1969; Kalaha-Brunst, Giolli, & Creel, 1974; Loewy, 1969; E. W. Powell & Brown, 1975; Sterling & Kuypers, 1966; Velayos & Lizarraga, 1973; White, 1960; Wiitanen, 1969), also possess certain drawbacks that can result in the misinterpretation of one's findings (see Section IV).

II. Methodology

As indicated above, the original Nauta methods (Nauta, 1950, 1957; Nauta & Gygax, 1951, 1954; Nauta & Ryan, 1952) have been modified by several investigators, and it would be impractical here to provide detailed descriptions of all of the numerous silver-impregnation techniques available. Therefore, the impregnation methods that have been used most extensively (or show the greatest potential) for studying the organization of neuronal pathways will be described. Descriptions are provided (a) of the procedure common to all the methods, and (b) of two of the original Nauta methods (Nauta, 1957; Nauta & Gygax, 1954) and four of their modifications (Eager, 1970; Fink & Heimer, 1967, procedure I; Kalaha-Brunst et al., 1974; Wiitanen, 1969)

A. Procedure Common to All Methods

1. POSTOPERATIVE SURVIVAL TIME

The optimal survival time will vary with the species studied, the fiber system investigated, and the proportion of thick to thin fibers comprising this system (for a discussion of these variables on the optimal survival time see Section IV,4). In rodents, fiber degeneration can generally be demonstrated best after survival periods of 5–7 days, whereas 7–10 days seem most favorable for the rabbit and cat, and 10–14 days most suitable for primates. When initiating a fiber-degeneration study, each of the points mentioned above should be considered, and a range in times tried in order that the optional survival time can be established.

2. FIXATION

Brain tissue should be fixed, preferably, by perfusion through the left ventricle. A 10% solution of formalin is the fixative of choice. The perfusion should be begin with isotonic saline (0.9% sodium chloride) and followed with 10% formalin. The brain is removed soon after the perfusion and stored in a volume of fixative at least three to four times the volume of the brain. The fixation time is usually 4 to 6 weeks, but longer fixation times have given excellent results. If electron microscopy studies are to be undertaken, it is appropriate to replace the formalin with a solution of paraformaldehyde–glutaraldehyde buffered with phosphates to a pH 7.2 (Karnovsky, 1965).

3. SECTIONING

In all of the methods to be described below, only frozen sections are used. When the fixation has been completed, the brain is immersed in a 30% solution of sucrose (made up with 10% formalin) until it sinks to the bottom of the container, which usually takes place within 2 to 3 days. The addition of the sucrose prevents the formation of ice crystals and improves both the sectioning and the overall quality of the sections. Usually blocks soaked in sucrose are more difficult to freeze, and in order to accelerate the freezing procedure, dry ice powder is placed on the freezing stage around the block. Sections are cut at a thickness of 25 to 50 μm, stored in 10% formalin, and if the impregnation process is not to be performed immediately they are refrigerated. Tissue sections have been found to be impregnated successfully even after as long a period of refrigeration as 16 months.

4. GENERAL REMARKS REGARDING THE PREPARATION OF
   SOLUTIONS AND THE STEPS IN THE METHODS

a. RINSING.   This is always done with distilled water. Quick rinse usually means one volume of distilled water, whereas thorough rinse means three to four changes of distilled water.

b. PHOSPHOMOLYBDIC ACID.   When combined with the potassium permanganate, it acts to suppress normal fiber impregnation. It can be replaced with phosphotungstic acid.

c. POTASSIUM PERMANGANATE.   This is an oxidizer. A 0.5% stock solution is prepared by heating the solution gently, then cooling and filtering it. This solution will last for several months. When used, it is diluted ten times with distilled water. The time the sections are treated in potassium permanganate is critical and must be established by the trial staining of test sections for 5, 10, 15 min, etc. Too long a treatment will suppress degenerating fibers, whereas too short a treatment will result in the impregnation of a great number of normal fibers. Material fixed for long periods has been found to require shorter treatment in potassium permanganate. During this step sections must be agitated in order that they will uniformly contact the solution.

In the Kahala-Brunst *et al.* method (1974) Mayer's hemalum is used as an oxidizer (hematein is the oxidant), and the solution is prepared by dissolving 1 gm of hematein in 19 ml of 100% ethanol mixing this with a liter of 5% aluminum potassium sulfate, and then adding 2 ml of glacial acetic acid.

d. HYDROQUINONE–OXALIC ACID SOLUTION. This is made up of equal parts of 1% solutions of hydroquinone and oxalic acid. Oxalic acid is stable, and it can be made in 1-liter stock solution. Hydroquinone, however, is very unstable, and must be prepared fresh, and the stock solution must be refrigerated. The hydroquinone–oxalic acid mixture acts to decolorize the tissue sections.

e. NITRATES. Both the silver and uranyl nitrate should be made fresh from stock solutions. The radioactive properties of uranyl nitrate are not regarded as hazardous to workers. However, the substance is extremely toxic and care should be taken in its handling. In the Fink–Heimer procedure I two different solutions of these nitrates are used. For solution A 1 gm of uranyl nitrate and 5 gm of silver nitrate are dissolved in 1 liter of distilled water. For solution B, 1.75 gm of uranyl nitrate and 16.26 gm of silver nitrate are dissolved in 1 liter of distilled water.

f. AMMONIACAL SILVER. This must always be freshly prepared as follows: Mix the silver nitrate solution with the ethanol. Let the mixture cool and then add the ammonium hydroxide (58%) followed by the 2.5% sodium hydroxide. Although the same chemicals are used in all the described methods, their proportions vary to a considerable degree. In the Nauta–Gygax method (1954) 20 ml of a 4.5% silver nitrate are mixed with 10 ml of 100% ethanol; 1.8 ml of ammonium hydroxide and 1.5 ml of sodium hydroxide are then added. In the Fink–Heimer procedure I (1967), 1 ml of ammonium hydroxide and 1.8 ml of sodium hydroxide are added to 30 ml of a 2.5% silver nitrate solution (note that in this case no ethanol is used). In the Wiitanen (1969) and Eager (1970) methods, 20 ml of 1.5% silver nitrate and 12 ml 95% ethanol are mixed; 2 ml of ammonium hydroxide and 1.6–1.8 ml sodium hydroxide are then added. Finally, in the Kalaha-Brunst *et al.* method (1974), 20 ml of a 5% silver nitrate and 10 ml of 100% ethanol are mixed; then 5 ml of ammonium hydroxide and 1.5 ml of sodium hydroxide are added.

g. LAIDLAW'S SOLUTION. This solution has been proposed by Chambers, Liu, and Liu (1956) to replace the ammoniacal silver of the Nauta–Gygax (1954) method. It has been considered as an important modification by some investigators (see Heimer, 1967). To prepare this solution, mix 20 ml of a 60% silver nitrate solution with 230 ml of a saturated lithium carbonate solution in a 250-ml graduated cylinder. Shake well and let the precipitate settle to the 70-ml mark.

Carefully remove the fluid portion and refill to 250 ml with distilled water. Shake again, let set to 70 ml, remove the liquid as before; repeat this washing procedure three times. After removing the liquid in the last washing, add ammonium hydroxide slowly and shake constantly until the precipitate is completely dissolved and the solution is almost clear (approximately 9.5 ml of ammonium hydroxide). The solution has a slight smell of ammonia. Then add distilled water to a total volume of 120 ml and filter. The solution is stored in a clean bottle and it can either be exposed to daylight for at least 2 weeks before use, or used immediately by adding two drops of 10% formalin to 50 ml of the solution.

h. REDUCER. This is made up by mixing 100% ethanol with distilled water; then adding 1% citric acid and 10% formalin. The citric acid must be kept in the refrigerator. In the Nauta–Gygax (1954), Nauta (1957), and the Kalaha-Brunst *et al.* (1974) methods, 400 ml of distilled water are mixed with 45 ml of 100% ethanol; 13.5 ml of citric acid and 13.5 ml of formalin are then added. In the Fink–Heimer procedure I (1967) the amount of water used is 460 ml, in the Wiitanen method (1969) it is 428 ml and in the Eager method (1970) 405 ml. The quantities of ethanol, citric acid and formalin used are the same in all six of the methods.

At the end of the treatment with the reducer, sections are light brown in color. If they are too dark a few drops of ammonia must be added to the ammoniacal silver (or to Laidlaw's solution); if they are too light, sodium hydroxide must be added. In the Fink–Heimer procedure I (1967), when the sections are too dark, a few drops of 1% citric acid are added to each 50 ml of the reducing solution.

i. SODIUM THIOSULFATE. This is similar to the "hypo" used in photography and acts as a fixative, preventing the fading of the stained sections. Also, this chemical tends to remove artifactual deposits of silver in the tissue sections.

j. MOUNTING, DEHYDRATION, AND CLEARING. At the end of the staining procedure sections are mounted from an alcoholic gelatin solution (equal parts of 1.5% aqueous gelatin and 80% ethanol; Albrecht, 1954) and blotted gently with filter paper. When dry, sections are dehydrated with 95 and 100% ethanol, then cleared with three baths of xylene. Finally, sections are enclosed in a neutral synthetic resin (e.g., Permount) and covered with a coverslip.

B. The Recommended Silver-Impregnation Methods

1. THE NAUTA–GYGAX METHOD (1954)

   1. Rinse section quickly
   2. 0.5% Phosphomolybdic acid, 25–30 min

3. 0.05% Potassium Permanganate, 5–15 min
4. Rinse quickly
5. Hydroquinone–oxalic acid, 1–2 min
6. Rinse thoroughly
7. 1.5% Silver Nitrate, 25–30 min
8. Rinse thoroughly
9. Ammoniacal silver, 1 min
10. Reducer, 1 min
11. Rinse quickly
12. % sodium thiosulfate, 1 min
13. Rinse thoroughly, mount sections, dehydrate, clear, coverslip

2. THE NAUTA METHOD (1957)

The important change in this method compared to the Nauta–Gygax method (1954) is the replacement of the ammoniacal silver (step 9) by Laidlaw's ammoniacal silver carbonate, as originally suggested by Chambers *et al.* (1956).

3. THE FINK–HEIMER PROCEDURE I (1967)

1. Rinse quickly
2. 0.05% potassium permanganate, 5–15 min
3. Rinse quickly
4. Hydroquinone–oxalic acid, 1 min
5. Rinse thoroughly
6. Uranyl nitrate–silver nitrate (Fink solution A), 30–60 min
7. Uranyl nitrate–silver nitrate (Fink solution B), 30–40 min
8. Rinse thoroughly.
9. Ammoniacal silver, 1–5 min
10. Reducer, 1–2 min
11. Rinse quickly
12. 0.5% Sodium thiosulfate, 1 min
13. Rinse quickly, mount, dehydrate, clear, coverslip

4. THE WIITANEN METHOD (1969)

This is a modification of the Fink–Heimer procedure II to be used primarily on sections of monkey brain.

1. Rinse quickly
2. 0.05% Potassium permanganate, 10 min
3. Rinse quickly
4. Hydroquinone-oxalic acid, 1 min
5. Rinse thoroughly.
6. 5% Uranyl nitrate, 10 min

7. Rinse thoroughly.
8. 0.3% Silver nitrate, 10 min
9. Ammoniacal silver, 2 min
10. Reducer, 1 min
11. Rinse quickly
12. 0.5% Sodium thiosulfate, 2 min
13. Rinse thoroughly, dehydrate, clear, coverslip

5. THE EAGER METHOD (1970)

1. Rinse
2. 2.5% Uranyl nitrate, 5 min
3. Ammoniacal silver, 3–15 min
4. Reducer, 2–5 min
5. Rinse
6. 0.5% Sodium thiosulfate, 2 min
7. Rinse, mount, dehydrate, clear, coverslip

6. THE KALAHA-BRUNST, GIOLLI, AND CREEL METHOD
    (1974)

1. Rinse
2. Mayer's hemalum, 5 min
3. Rinse thoroughly
4. 20% Uranyl nitrate, 10 min
5. Rinse thoroughly
6. 20% Silver nitrate, 10 min
7. Rinse thoroughly
8. Ammoniacal silver, 3 min
9. Reducer, 2 min
10. Rinse
11. 1% Sodium thiosulfate, 1 min
12. Rinse thoroughly
13. Acid alcohol (1 ml 37% hydrochloric acid and 100 ml 70% ethanol, 30
    sec)
14. Rinse thoroughly, mount, dehydrate, clear, coverslip

III. Appearance and Analysis of the Material

The development of the Nauta method in the 1950's added a new dimension to the study of neural organization. This method (Nauta, 1950, 1957; Nauta & Gygax, 1951, 1954; Nauta & Ryan, 1952) together with several of its modifications (Eager, 1970; Fink & Heimer, 1967; Wiitanen, 1969) have been widely used and have provided a wealth of information concerning the origins, trajectories, and terminations of fiber projections as demonstrated following the production of experimental lesions (or the presence of pathological lesions). The widespread use of the Nauta method and its modifications can be attributed to two facts: (a) the Nauta sections are usually easy to interpret, and (b) the method provides data that are reliable, reproducible, and supplemental to neurophysiological findings.

In a typical Nauta preparation the degenerating fibers, including their terminal processes, will be impregnated black against a background that is nearly devoid of impregnated normal fibers and has a yellow to tan color (details provided below). Artifactual staining is usually minimal and can for the most part be identified with ease (details given in Section IV).

Heimer (1967) designates a particular Nauta method as "suppressive" or "nonsuppressive" depending on whether an oxidizing agent is used (e.g., Nauta, 1957; Nauta & Gygax, 1954; Nauta & Ryan, 1952) or is omitted (e.g., Nauta, 1950; Nauta & Gygax, 1951). His thesis is that with use of an oxidizing agent there is a greater suppression of normal fiber impregnation. However, it is our feeling that the use of these terms should now be discontinued because (a) one can obtain just as great a suppression of normal fiber impregnation with the use of nonsuppressive methods (e.g., methods of Eager, 1970; Giolli & Pope, 1973) as compared with suppressive methods; and (b) as is discussed below, there can be just as complete an impregnation of the degenerating fibers, including their terminal processes, using suppressive methods (e.g., methods of Fink–Heimer, 1967; Kalaha–Brunst et al., 1974; Wiitanen, 1969) as compared with the nonsuppressive methods. Each of the six methods outlined under Section *II* relative to the appearance of the sections and to the data provided will be described.

A.   The Nauta – Gygax (1954) and Nauta (1957) Methods

These are two of the Nauta procedures that initiated a renewed interest in the field of experimental neuroanatomy. Each method can be relied on to yield a wealth of information concerning the organization of fiber projections and produces little artifactual staining. The following description is limited to the use of the Nauta–Gygax method (1954), but the Nauta method (1957) yields similar results.

Three examples of the appearance of the impregnated degenerating fibers in sections prepared by the Nauta–Gygax method are illustrated in Figs. 1–3. Figure 1 shows bundles of degenerating axons in passage, whereas Figs. 2 and 3 depict regions of pericellular fiber degeneration (degenerating terminal fiber arborizations). The fragmented, granular appearance shown by the degenerating fibers in these three figures is obvious even to the unexperienced observer, and because

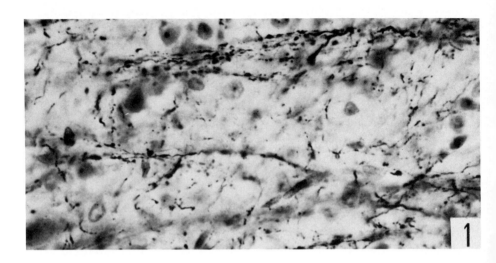

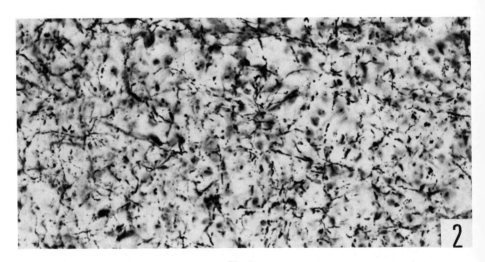

Fig. 2

FIG. 1–3.   Degenerating fibers are demonstrated in the superior colliculus 7 days after contra-lateral eve enucleation in a rabbit. Nauta–Gygax method. Figure 1 depicts bundles of degenerating axons in passage, and Figs. 2 and 3 show degenerating pericellular fibers. Magnifications: × 520 (Figs. 1 and 3). × 300 (Fig. 2).

the degenerating fibers are present in the near absence of impregnated normal fibers *and* because the background in the figures is light, the trajectory (Fig. 1) and the distribution (Figs. 2 and 3) of the degenerating fibers can be readily determined. However, it must be stressed that neither the Nauta–Gygax method (1954) nor the Nauta method (1957) can be relied on to show the precise termination of degenerating fibers, for electron microscopy studies have revealed that while these methods will impregnate degenerating pericellular fibers, they usually leave the actual endings (the boutons) unimpregnated (Heimer, 1967, 1970; Walberg, 1964, 1971, 1972).

In light and electron microscopy studies on sections prepared by the Nauta–Gygax (1954) and Nauta (1957) methods, it has been reported that the impregnation of the degenerating fibers is essentially limited to axoplasm (Guillery & Ralston, 1964; Walberg, 1964). Yet, there is electron microscopic (Eager and Barrnett, 1966) as well as chemical (Eager & Barrnett, 1966; Evans & Hamlyn, 1956; Giolli, 1965: R. D. Lund & Westrum, 1966) evidence to indicate that degenerating myelin sheaths also may be impregnated. In our opinion the contribution that degenerating myelin makes to the total silver impregnation of the degenerating fibers by these Nauta methods remains an unresolved issue.

B. The Fink–Heimer Method (1967)

Among the Nauta techniques, the Fink–Heimer method, procedures I and II, (1967) enjoys the reputation of being the most extensively used to demonstrate

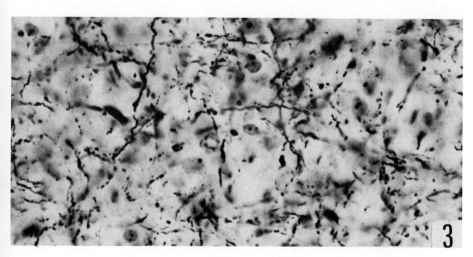

Fig. 3

degenerating nerve fibers. This is true not only because this method has been found to successfully impregnate a number of different fiber systems in a series of vertebrate species, but also because it has often proved more reliable than the other Nauta techniques.

The microscopic picture of the degenerating axons of passage and their terminal arborizations as seen in sections prepared by the Fink–Heimer method is generally comparable to that already described and illustrated for the Nauta–Gygax (1954) and Nauta (1957) methods. Where the Fink–Heimer method differ from the Nauta–Gygax (1954) and Nauta (1957) methods is that it impregnates degenerating boutons. Figures 4 and 5 depict two fields of degenerating boutons. Figure 4 shows a field of degenerating boutons in the superficial gray of the superior colliculus following an eye enucleation in a rabbit, and Fig. 5 illustrates a similar field in the dorsal lateral geniculate nucleus after the production of a

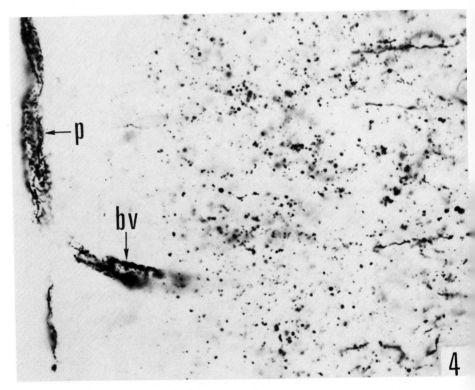

Figs. 4 and 5.   Degenerating boutons are pictured in Fink–Heimer preparations.

Fig. 4.   Within the superficial gray of the superior colliculus 7 days following contralateral eye enucleation in a rabbit. The impregnation of the pia mater (p) and a blood vessel (bv) is indicated.

lesion in the visual cortex of another rabbit. It is true that electron microscopy studies have not been conducted to determine whether the impregnated granules depicted in Figs. 4 and 5 are degenerating boutons rather than some other element of the tissue. However, it should be pointed out that the granules seen in Fig. 5 have been shown to be located in the portion of the rabbit's dorsal lateral geniculate nucleus that is expected to receive a fiber input from that sector of the visual cortex sustaining a lesion (see Giolli & Guthrie, 1971). In addition, it sbould be stressed that Giolli and Guthrie (1971) had ruled out the possibility that these impregnated granules represent the products of retrograde cell degeneration, i.e., the "fine dust" reported to represent retrograde cell degeneration present in thalamic nucleus after cortical lesions (Cragg, 1962; Guillery, 1959, 1967; T. P. S. Powell & Cowan, 1964). Moreover, similar types of granular deposits have been identified as degenerating boutons in electron microscopy studies by Heimer (1967) and Heimer and Peters (1968) on the rat's olfactory

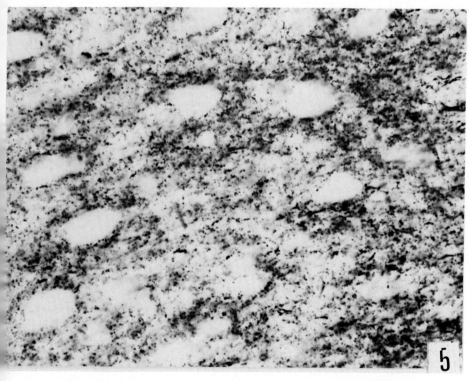

FIG. 5. Within the dorsal lateral geniculate nucleus 6 days after a lesion had been produced in the ipsilateral visual cortex of a rabbit. Magnification × 520.

system and by Walberg (1971, 1972) on the cat's cerebellum and inferior olivary nucleus.

An example of the superior results that can be obtained with the Fink–Heimer (1967) as compared with the Nauta–Gygax (1954) method is illustrated in Figs. 6 and 7. Both figures are photomicrographs of contiguous sections tbrough the same region of the superior colliculus after eye enucleation in a rabbit. It is apparent tbat there is considerably more extensive impregnation of the degenerating fibers in the Fink–Heimer preparation (Fig. 6) as compared with the Nauta–Gygax preparation (Fig. 7). Further, it is seen that degenerating boutons are impregnated in the Fink–Heimer preparation but not in the Nauta–Gygax preparation. Similar comparisons between tbe Fink–Heimer (1967) and Nauta–Gygax (1954) methods have been made through light microscopy by Heimer (1967, see

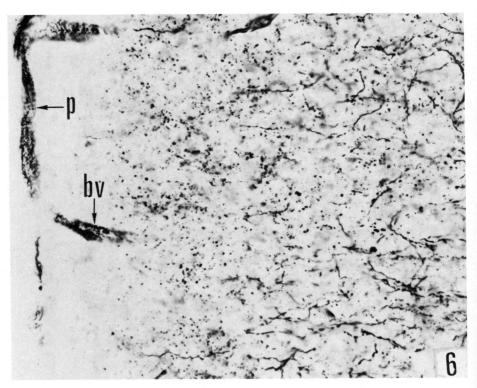

FIGS. 6 AND 7. Fiber degeneration is shown in the superficial gray of the superior colliculus 7 days after contralateral eye enucleation in a rabbit.

Figure 6 reveals the abundance of degenerating pericellular fibers and boutons in a Fink–Heimer preparation.

his Figs. 8 to 10) and electron microscopy by Heimer (1967) and Heimer and Peters (1968).

Grounds for a comparison of the effectiveness of the impregnation of degenerating fibers between the Fink–Heimer methods (1967) and the methods considered below are as yet unavailable.

C. The Wiitanen Method (1969)

This silver impregnation method was developed to fulfill a need for an effective method to bring out fiber degeneration in primates, and in the authors' view it has proven itself a valuable technique not only for the study of fiber connections in primates but also in other mammals and even in some nonmammalian vertebrates.

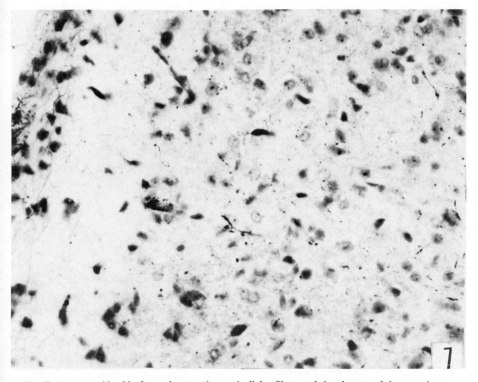

Fig. 7 shows considerably fewer degenerating pericellular fibers and the absence of degenerating boutons in a contiguous Nauta–Gygax preparation. The impregnation of the pia mater (p) and a blood vessel (bv) is indicated in Fig. 6. Magnification: × 300.

A review of the literature indicates that the Wiitanen method has been employed to successfully ascertain the course and distribution of degenerating fibers in sensory systems (visual, auditory, somatosensory and olfactory) as well as in thalamocortical, corticofugal, and corticocortical projections. This method has been utilized in studies not only on different primate species, but also on nonprimate species, *i.e.*, the cavefish, rat, hamster, cat, and tree shrew. In all cases, it has been used in combination with the Nauta–Gygax (1954), Nauta (1957), and/or Fink–Heimer (1967) methods.

Several investigators have made specific reference to the usefulness of the Wiitanen method. Thus, Wagor, Lin, and Kaas (1975) have stated that he has obtained greater success in his study of the visual corticocortical fibers of the owl monkey using this method than the Fink–Heimer method (1967). Further, Garey and Powell (1971) write that the "technique described by Wiitanen (1969) is particularly suitable for the study of terminal degeneration in the cerebral cortex for background and cellular pattern is well preserved compared with the more widely used Fink–Heimer (1967) methods." Moreover, in his research on the anatomical organization of the corticopulvinar projection of the squirrel monkey, Mathers (1972) reports that he found the Wiitanen method to give results superior to those obtained with either the Nauta–Gygax (1954) or Fink–Heimer (1967) methods. Finally, Wong–Riley (1972) states that she has obtained results with the Wiitanen method that are superior to those provided by either the Nauta–Gygax (1954) or Fink–Heimer (1967) methods.

In conclusion, the authors recommend the use of the Wiitanen silver impregnation technique for the study of fiber connections in both primates and nonprimates. Further it is recommended that this technique be used in combination with one or more of the other Nauta methods described under Section II in particular, it is strongly urged that the Fink–Heimer method be used because it has undergone electron microscopical scrutiny and has been shown to successfully impregnate degenerating boutons (Heimer, 1967; Heimer & Peters, 1968; Walberg, 1972).

D. The Eager Method (1970)

This silver impregnation method was developed to provide a simplified, yet effective procedure to selectively impregnate degenerating fibers. Eager (1970) originally used the procedure to study the course and distribution of degenerating fibers resulting from hemisection of the spinal cord in cats, from lesions of the pontine nuclei or the inferior olive in cats, and from lesions in the hippocampus in rats. Land, Eager, and Shepherd (1970) have reported that their results on the olfactory system of the rabbit are indistinguishable from those they obtained using the Fink–Heimer method (1967).

A review of the literature has revealed that the Eager method has been utilized to investigate the anatomical organization of cerebellar fiber connections, sensory systems (olfactory, visual, auditory and vestibular), and corticofugal and corticocortical projections. These studies have been conducted on the lizard, frog, chicken, rat, rabbit, cat, and monkey.

Although there is light microscopy evidence (Eager, 1970; Gregory, 1972; Land et al., 1970) to indicate that the Eager method gives results comparable to the Fink–Heimer method (1967), the electron microscopy findings of Walberg (1972) suggest that the Fink–Heimer method (1967) will heavily impregnate degenerating boutons whereas the Eager method (1970) will do so only occasionally. However, it must be recognized that Walberg's negative results are based solely upon the examination of the fiber degeneration resulting from cerebellar lesion in cats. Further electron microscopy studies are now critically needed to determine if, indeed, the Eager method is capable of impregnating degenerating boutons in other fiber systems; and this is especially true in light of the fact that sections prepared by the Eager method provide impregnation comparable to that obtained by the Fink–Heimer method (Gregory, 1972; Land et al., 1970; personal observations). The authors recommend that the Eager method be used because it is simple and, on the basis of light microscopy, provides a picture of fiber degeneration similar to that obtained with the Fink–Heimer method (1967). However, it is further recommended that the use of the Eager method be combined with that of one or more of the other Nauta methods considered in Section II.

E. The Kalaha–Brunst, Giolli, and Creel Method (1974)

As with the Eager method (1970), the method of Kalaha–Brunst et al., has been developed to provide a simplified, selective, and consistently effective technique for the impregnation of degenerating fibers. To the present, its use has been limited (Creel & Giolli, 1976; Giolli & Creel, 1974; Kalaha–Brunst et al., 1974); however, it has been successfully used to demonstrate degenerating corticocortical fibers in the squirrel monkey (Kalaha–Brunst et al., 1974), degenerating corticofugal and tectothalamic fibers in the rabbit (Kalaha–Brunst et al., 1974), and degenerating optic fibers in the rat (Creel & Giolli, 1976; Giolli & Creel, 1974). The authors would recommend, just as they have done for the Eager method (1970), that the Kahala–Brunst et al. method be used to study fiber degeneration, but this in combination with one or more of the other Nauta methods described in Section II. From our experience, based entirely upon light microscopy studies of degenerating fibers, the Kalaha–Brunst et al. method provides a picture of the fiber degeneration that is comparable to that obtained with the Fink–Heimer method.

IV.   Limitations and Artifacts of the Methods

A.   Limitations

1.   POSTOPERATIVE SURVIVAL TIME

A striking example of the manner in which the postoperative survival time can influence the impregnation of degenerating fibers, including their terminals, is provided by Schneider (1968). This investigator enucleated an eye from each of a series of hamsters; he then sacrificed the animals at intervals between 1 and 7 days and processed the brains by the Fink–Heimer method (1967). His results show that different segments along the lengths of the degenerating optic fibers will be differentially impregnated depending upon the survival time.

E. W. Powell and Schnurr (1972) have studied the optimal survival times for the Nauta–Gygax (1954) and Fink–Heimer (1967) methods subsequent to lesions of the septal region in rats, cats, and squirrel monkeys. They have reported that the optimal time varies with the species studied and the Nauta procedure used; further, they indicate that this time averages one week. In another study, Cottle and Mitchell (1966) have investigated the optimal survival time for fiber degeneration using the Nauta method (1957) after transection of the vagus nerve in cats. They have reported a survival period of 3 to 6 days to give the best results.

It is our view that the optimal postoperative survival time to be used cannot be predicted without taking into consideration factors such as the species of animals to be studied, the fiber system(s) to be analyzed, and the degree to which any fiber system is composed of course and fine fibers. It is strongly recommended that one initially utilize a wide range of postoperative survival times (e.g., in mammalian species a range of 2 days to 2 weeks).

2.   FIXATIVES AND FIXATION TIMES

The classical fixative used for the Nauta technique has been a solution of 10% formalin (see Fink and Heimer, 1967; Nauta, 1957; Nauta and Gygax, 1954). E. W. Powell and Schnurr (1972) have studied the effects of 10% formalin, buffered 10% formalin (pH 7.2), and a 4:1 solution (%) of paraformaldehyde–glutaraldehyde on the results obtained with the fiber degeneration methods. They have found that 10% formalin is superior because it provides a yellow background for the impregnated segments of degenerating fibers as compared with the paraformaldehyde–glutaraldehyde, which gives a brownish background. They have not reported any differences between buffered and unbuffered 10% formalin.

Results of an actual controlled study to determine the optimal fixation time have not been published. The usual time stated is in the range of 1 week to 6 months (Fink and Heimer, 1967; Nauta, 1957; Nauta and Gygax, 1954). In the experience of the authors, brain material that has been fixed in 10% unbuffered formalin for a period of as short as 2 weeks to as long as 16 months impregnates well with the Fink–Heimer (1967) and Giolli and Pope (1973) methods. We recommend a fixation time of 2 weeks to 2 months.

3. ADVANTAGES OF SOME NAUTA METHODS OVER OTHERS

Because of the ability of the Nauta methods to impregnate degenerating fiber systems often varies, this subject deserves special consideration. The single most important factor here is the refractiveness that the Nauta (1957) and Nauta–Gygax (1954) methods have toward providing a complete impregnation of the degenerating fibers. In this regard, Tigges and Tigges (1969) have commented that the Fink–Heimer method (1967) allows for a considerably more complete

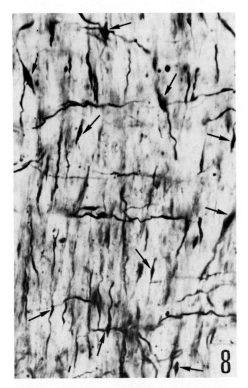

FIGS. 8-11.   Examples of the artifactual impregnation of nondegenerating fibers are illustrated in these figures.

FIG. 8.   Normal fiber impregnation in the optic tract of a rabbit; some of these fibers exhibit fusiform swelling (indicated by arrows) along their lengths. Nauta–Gygax method.

impregnation of the degenerating optic fibers than does the Nauta–Gygax technique (1954), and Carpenter and Peter (1972) have reported that they had been able to impregnate degenerating nigrostriatal fibers using the Wiitanen method (1969), whereas the Nauta–Gygax technique (1954) had yielded negative results. Other authors have found that the Fink–Heimer (1967) and/or Wiitanen (1969) methods succeed in cases in which the Nauta–Gygax method gave less than satisfactory results (Ebbesson, 1970; Heimer, 1967; Heimer, Ebner, & Nauta, 1967; J. S. Lund & Lund, 1970). The development of the Wiitanen method (1969), particularly for studies on primates, has proven to be unusually useful for tracing degenerating fibers to their nuclei of termination, where other methods either have failed or have provided less than desired results (Garey & Powell, 1971; Mathers, 1972; Wagor *et al.*, 1975; Wiitanen, 1969; Wong–Riley, 1972).

4. CALIBER OF THE NERVE FIBERS

Studies have indicated that fine fibers undergo a more rapid rate of anterograde degeneration than do thick fibers (Cottle & Mitchell, 1966; Joseph & Whitlock,

FIG. 9. Normal fiber impregnation in the inferior colliculus of a rabbit. Nauta–Gygax method.

1972; Ohmi, 1961; Weddell & Glees, 1941). By contrast, investigations by Nauta (1957), van Crevel and Verhaart (1963a, 1963b), and Guillery (1970) support the finding of Ramón y Cajal (1928) that coarse fibers degenerate more rapidly than do fine fibers. Guillery (1970) has hypothesized that the observation made by some workers that the coarser fibers degenerate at a slower rate than the finer ones may have resulted from these workers having mistaken as fine axons in passage the terminal arborizations of coarse fibers. With this question as yet unanswered, it is recommended that a range in the survival time be used to ensure an optimal impregnation of fibers of different calibers.

5.   EFFECTS OF TEMPERATURE

It has been reported that the ambient temperature can affect the rate at which the anterograde type of degeneration proceeeds (Armstrong, 1950; Ebbesson, 1970; Gamble, Goldby, & Smith, 1957; Joseph & Whitlock, 1972; Torrey, 1934). To the present, this influence on the rate of fiber degeneration, as it can be demonstrated with the silver impregnation methods, has been studied primarily

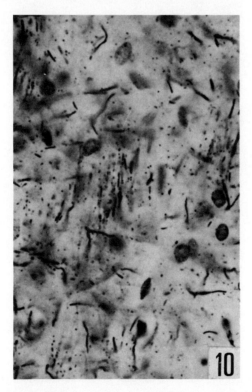

FIG. 10.   Granular deposits of silver as seen in the medial thalamus in a rabbit. Eager method.

on poikilothermic species (see Ebbesson, 1970), although Merzbacher (1903) and Gamble and Jha (1958) have reported increased rates of fiber degeneration with increased environmental temperature in the bat and rat, respectively.

B. Artifacts

In order that the investigator may be able to accurately identify the degenerating fibers and their terminal processes, it is important to gain a knowledge of the type of silver deposits that represent the impregnation of normal fibers and of nonnervous structures. Four basic types of artifactual impregnation are recog-nized.

1. A common type of misinterpretation involves mistakenly identifying as degenerating fibers the staining of segments of normal fibers (Figs. 8 and 9). Figure 8 illustrates a picture as typically seen in Nauta sections in which some

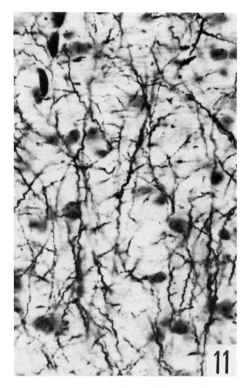

Fig. 11. Impregnation of the fibers of glial cells in the optic tract in a rabbit. Fink–Heimer method. Magnification: × 520.

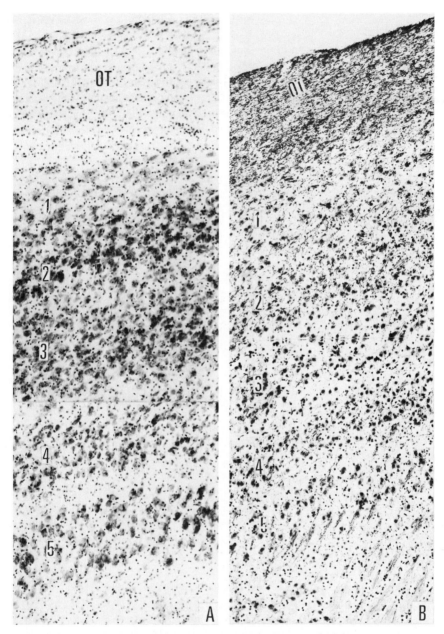

FIG. 12.   (A) The five cellular layers of the dorsal lateral geniculate nucleus of the tree shrew as seen in a Nissl preparation. (B) The pattern of the pericellular fiber degeneration 2 weeks after contralateral eye enucleation as revealed in a Nauta–Gygax preparation contiguous with (A). A comparison of (A) with (B) reveals that nearly all of the pericellular fiber degeneration is located within cellular layers 2 and 4 of the geniculate nucleus. (B) is not as long as (A) because of a greater shrinkage of the tissue in (B) as compared with (A). OT: optic tract. Magnification: × 120.

normal fibers show fusiform swelling along their lengths. The impregnation of a opulation of normal fibers usually will occur, but these fibers customarily do not show signs of physical breakdown and can be readily identified as normal (Fig. 9).

2. A second common type of misinterpretation can result when granular silver deposits are mistaken for degenerating boutons (Fig. 10). This staining can occur using any of the Nauta methods, but it is most often seen in sections prepared by the Fink–Heimer (1967), Eager (1970), and Kalaha-Brunst et al. (1974) methods.

3. A third type of artifactual impregnation involves the fibers of glial cells (Fig. 11). This type of impregnation is less common than are the other types;

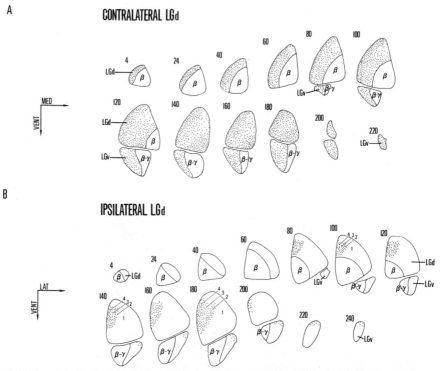

FIG. 13. The distribution of the pericellular fiber degeneration in the dorsal lateral geniculate nucleus (A) contralateral and (B) ipsilateral to the enucleated eye in a rabbit is shown in serial transverse sections oriented from the most rostral (smallest numbers) to the most caudal (largest numbers). Seven days survival time; Nauta–Gygax method. The beta sector of the dorsal lateral geniculate nuclei is labeled as β, the beta–gamma sectors of the ventral lateral geniculate nuclei as β–γ. The alpha sectors of dorsal and ventral nuclei are indicated by LGd and LGv, respectively. Note that the fiber degeneration is limited to the alpha sectors of each nucleus. [Modified slightly from Fig. 3A of Giolli and Guthrie (1969) and reproduced with the kind permission of the Wistar Institute of Anatomy and Biology, Philadelphia, Pa.].

however, it is important to recognize because it can easily be mistaken for degenerating fibers.

4. A fourth type of misinterpretation could result from identifying as degenerating fibers, impregnated reticular fibers either in the walls of blood vessels or in the pia mater (Figs. 4 and 6).

V. Interpretation of the Results

A primary step in the interpretation of the results obtained with the silver impregnation methods is to determine the distribution of the degenerating nerve fibers. This is accomplished (a) by preparing contiguous series of sections that have been stained by the Nissl and Nauta methods; and (b) by relating the pattern

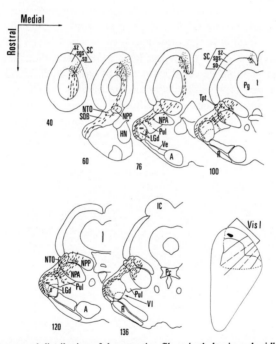

FIG. 14.   The course and distribution of degenerating fibers in thalamic and midbrain nuclei of the rabbit are illustrated after the production of a lesion in visual cortical area I (Vis 1) as this area has been demarcated by Thompson, Woolsey, and Talbot (1950). The outline of the lesion is shown on the cerebral hemisphere. The degenerating axons in passage (line segments) and degenerating pericellular fibers (dots) are drawn on serial horizontal sections numbered from dorsal to ventral (40 to 136). Six days survival; Fink–Heimer method. The only structures in this figure relevant to the subject matter of this chapter are the dorsal lateral geniculate nucleus, ventral lateral geniculate nucleus, and superior colliculus—labeled as LGd, LGv, and SC, respectively.

of the pericellular fiber degeneration, as shown on Nauta preparations, to the
cytoarchitecture as seen on adjacent Nissl preparations. The advantage of this
approach is illustrated in Fig. 12A,B, in which it is possible to see the distribution of the pericellular fiber degeneration as limited primarily to layers 2 and 4 of
the doral lateral geniculate nucleus contralateral to eye enucleation in a tree
shrew. What is done to preserve the information that is provided by this approach
is to map the pattern of the fiber degeneration in Nauta sections onto drawings
made from adjacent Nissl sections. This procedure is illustrated in Fig. 13A,B
for the pattern of the pericellular fiber degeneration to the alpha sector of the
dorsal and ventral lateral geniculate nuclei located contralateral (A) and ipsilateral (B) to the enucleated eye in a rabbit.

If it is desirable to determine the course and distribution of a total fiber
projection, the above-mentioned approach can also be used. In this case, portions
of, or entire, sections are drawn (Fig. 14) rather than individual nuclei (Fig. 13).
In addition, not only the zones of pericellular fiber degeneration (small dots in
Fig. 14) but also the course of degenerating axons in passage (line segments in
Fig. 14) can be included in the drawings.

To carry the interpretation of the results to a final step, the anatomical organization of fiber projections can be documented as has been done in Fig. 15. In this
figure, the locations of several lesions on the rabbit's visual cortical area I are
shown and the resulting zone of pericellular fiber degeneration in the ipsilateral
dorsal lateral geniculate nucleus for each lesion is depicted. In this way, data on

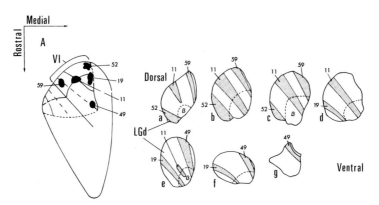

FIG. 15.   Data are provided on the organization of the projection from visual cortical area I to the
dorsal lateral geniculate nucleus of the rabbit. The field of pericellular fiber degeneration present in
tbe geniculate nucleus for each of the cortical lesions (indicated on the cerebral hemisphere) is drawn
on outlines of the nucleus as seen in the horizontal plane and labeled from dorsal to ventral (a–g). The
alpha sector of the nucleus is indicated by LGd, the beta sector by β. [Taken from Fig. 3A of Giolli
and Guthrie (1971) and reproduced with the kind permission of the Wistar Institure of Anatomy and
Biology, Philadelphia, Pa.].

the topographical organization of the projection from visual cortical area I to the doral lateral geniculate nucleus have been preserved.

Acknowledgments

This work was supported by U.S. Public Health Service Grant RO1-EY00607 to R.A.G. and by a Fogarty International Fellowship to A.N.K.

References

Adams, C.W.M. Histochemical mechanisms of the Marchi reaction for degenerating myelin. *Journal of Neurochemistry,* 1958, **2,** 178–186.

Adams, C.W.M. Osmium tetroxide and the Marchi method: Reactions with polar and nonpolar lipids, protein and polysaccharide. *Journal of Histochemistry and Cytochemistry,* 1960, **8,** 262–267.

Albrecht, M.H. Mounting frozen sections with gelatin. *Stain Technology,* 1954, **29,** 89–90.

Albrecht, M.H., & Fernström, R.C. A modified Nauta–Gygax method for human brain. *Stain Technology,* 1959, **34,** 91–94.

Armstrong, J.A. An experimental study of the visual pathways in a reptile (*Lacerta vivipara*). *Journal of Anatomy,* 1950, **84,** 146–167.

Beresford, W.A. An evaluation of neuroanatomical methods and their relation to neurophysiology. *Brain Research,* 1966, **1,** 221–249.

Bielschowsky, M. Die Silberimprägnation der Neurofibrillen. *Journal fuer Psychologie und Neurologie, Leipzig,* 1904, **3,** 169–188.

Bowsher, D., Brodal, A., & Walberg, F. The relative values of the Marchi method and some silver impregnation techniques. *Brain,* 1960, **83,** 150–160.

Carpenter, M.B., & Peter, P. Nigrostriatal and nigrothalamic fibers in the rhesus monkey. *Journal of Comparative Neurology,* 1972, **144,** 93–116.

Chambers, W.W., Liu, C.Y., & Liu, C.N. A modification of the Nauta technique for staining of degenerating axons in the central nervous system. *Anatomical Record,* 1956, **124,** 391–392.

Cottle, M.K.W., & Mitchell, R. Degeneration time for optimal staining by Nauta technique. A study on transected vagal fibers of the cat. *Journal of Comparative Neurology,* 1966, **128,** 209–222.

Cragg, B.G. Centrifugal fibers to the retina and olfactory bulb, and comparison of the supraoptic commissures in the rabbit. *Experimental Neurology,* 1962, **5,** 406–427.

Creel, D., & Giolli, R.A. Retinogeniculate projections in albino and ocularly hypopigmented rats. *Journal of Comparative Neurology,* 1976, **166,** 445–456.

De Olmos, J.S. A cupric-silver method for impregnation of terminal axon degeneration and its further use in staining granular argyrophilic neurons. *Brain, Behavior and Evolution,* 1969, **2,** 213–237.

De Olmos, J.S., & Ingram, W.R. An improved cupric-silver method for impregnation of axonal and terminal degeneration. *Brain Research,* 1971, **33,** 523–529.

Desclin, J.C., & Escubi, J. An additional silver impregnation method for demonstration of degenerating nerve cells and processes in the central nervous system. *Brain Research,* 1975, **93,** 25–39.

De Vito, J.L., & Smith, O.A., Jr. Projections from the mesial frontal cortex (supplementary motor area) to the cerebral hemispheres and brain stem of the *Macaca mulatta*. *Journal of Comparative Neurology,* 1959, **111,** 261–278.

Di Virgilio, A., German, N.I., Horn, R.E., Lewis, R.M., Scharf, R.D., Schwartz, H.J., Warwick, L., & Epstein, R. A modified Nauta–Gygax technique for paraffin sections in the central nervous systems of vertebrates. *Anatomical Record,* 1958, **130,** 403–404.

Eager, R.P. Selective staining of degenerating axons in the central nervous system by a simplified silver method: Spinal cord projections to external cuneate and inferior olivary nuclei in the cat. *Brain Research,* 1970, **22,** 137–141.

Eager, R.P., & Barrnett, R.J. Morphological and chemical studies of Nauta-stained degenerating cerebellar and hypothalamic fibers. *Journal of Comparative Neurology,* 1966, **126,** 487–510.

Ebbesson, S.O.E. The selective silver-impregnation of degenerating axons and their synaptic endings in nonmammalian species. In W.J.H. Nauta & S.O.E. Ebbesson (Eds.), *Contemporary research methods in neuroanatomy.* Berlin & New York: Springer-Verlag, 1970. Pp. 132–161.

Evans, D.H.L., & Hamlyn, L.H. A study of silver degeneration methods in the central nervous system. *Journal of Anatomy,* 1956, **90,** 193–203.

Fink, R.P., & Heimer, L. Two methods for selective silver impregnation of degenerating axons and their synaptic endings in the central nervous system. *Brain Research,* 1967, **4,** 369–374.

Gamble, H.J., Goldby, F., & Smith, G.M.R. Effect of temperature on the degeneration of nerve fibres. *Nature (London),* 1957, **179,** 527.

Gamble, H.J., & Jha, B.D. Some effects of temperature upon the rate and progress of Wallerian degeneration in mammalian nerve fibres. *Journal of Anatomy,* 1958, **92,** 171–177.

Garey, L.J., & Powell, T.P.S. An experimental study of the termination of the lateral geniculo-cortical pathway in the cat and monkey. *Proceedings of the Royal Society of London, Series B,* 1971, **179,** 41–63.

Giolli, R.A. A note on the chemical mechanism of the Nauta–Gygax technique. *Journal of Histochemistry and Cytochemistry,* 1965, **13,** 206–210.

Giolli, R.A., & Creel, D. Inheritance and variability of the organization of the retinogeniculate projections in pigmented and albino rats. *Brain Research,* 1974, **78,** 335–339.

Giolli, R.A., & Guthrie, M.D. Organization of subcortical projections of visual areas I and II in the ation study. *Journal of Comparative Neurology,* 1969, **136,** 99–126.

Giolli, R.A., & Guthrie, M.D. Organization of subcortical projections of visual areas Iand II in the rabbit. An experimental degeneration study. *Journal of Comparative Neurology,* 1971, **142,** 351–376.

Giolli, R.A., & Pope, J.E. The mode of innervation of the dorsal lateral geniculate nucleus and the pulvinar of the rabbit by axons arising from the visual cortex. *Journal of Comparative Neurology,* 1973, **147,** 129–144.

Glees, P. Terminal degeneration within the central nervous system as studied by a new silver method. *Journal of Neuropathology and Experimental Neurology,* 1946, **5,** 54–59.

Glees, P., & Nauta, W.J.H. A critical review of studies on axonal and terminal degeneration. *Monatsschrift fuer Psychiatrie und Neurologie,* 1955, **129,** 74–91.

Grant, G., & Arvidsson, J. Transganglionic degeneration in trigeminal primary sensory neurons. *Brain Research,* 1975, **95,** 265–279.

Grant, G., Ekvall, L., & Westman, J. Transganglionic degeneration in the vertibular nerve. In J. Stahle (Ed.), *Vestibular function on earth and in space.* Oxford: Pergamon Press, 1970. Pp. 301–305.

Gregory, K.M. Central projections of the eighth nerve in frogs. *Brain, Behavior and Evolution,* 1972, **5,** 70–88.

Guillery, R.W. Afferent fibers to the dorso-medial thalamic nucleus in the cat. *Journal of Anatomy,* 1959, **93,** 403–419.

Guillery, R.W. Patterns of fiber degeneration in the dorsal lateral geniculate nucleus of the cat following lesions in the visual cortex. *Journal of Comparative Neurology,* 1967, **130,** 197–222.

Guillery, R.W. Light- and electron-microscopical studies of normal and degenerating axons. In

W.J.H. Nauta & S.O.E. Ebbesson (Eds.), *Contemporary research metbods in neuroanatomy.* Berlin & New York: Springer-Verlag, 1970. Pp. 77-105.

Guillery, R.W., & Ralston, H.J. Nerve fibers and terminals: Electron microscopy after Nauta Staining. *Science,* 1964, **143,** 1331-1332.

Guillery, R.W., Shirra, B., & Webster, K.E. Differential impregnation of degenerating nerve fibers in paraffin-embedded material. *Stain Technology,* 1961, **36,** 9-13.

Heimer, L. Silver impregnation of terminal degeneration in some forebrain fiber systems: A comparative evaluation of current methods. *Brain Research,* 1967, **5,** 86-108.

Heimer, L. Selective siler-impregnation of degenerating axoplasm. In W.J.H. Nauta & S.O.E. Ebbesson (Eds.), *Contemporary research methods in neuroanatomy.* Berlin & New York: Springer-Verlag, 1970. Pp. 106-131.

Heimer, L., Ebner, F.F., & Nauta, W.J.H. A note on the termination of commissural fibers in the neocortex. *Brain Research,* 1967, **5,** 171-177.

Heimer, L., & Peters, A. An electron microscope study of a silver stain for degenerating boutons. *Brain Research,* 1968, **8,** 337-346.

Johnstone, G., & Bowsher, D. A new method for the selective impregnation of degenerating axon terminals. *Brain Research,* 1969, **12,** 47-53.

Joseph, B.S., & Whitlock, D.G. The spatio-temporal course of Wallerian degeneration within the CNS of toads (*Bufo marinus*) as defined by the Nauta silver method. *Brain, Behavior and Evolution,* 1972, **5,** 1-17.

Kalaha-Brunst, C., Giolli, R.A., & Creel, D. An improved silver impregnation method for tracing degenerating nerve fibers and their terminals in frozen sections. *Brain Research,* 1974, **82,** 279-283.

Karnovsky, M.J. A formaldehyde-glutaraldehyde fixative of high osmolarity for use in electron microscopy. *Journal of Cell Biology,* 1965, **27,** 137A.

Land, L.J., Eager, R.P., & Shepherd, G.M. Olfactory nerve projects to the olfactory bulb in the rabbit: Demonstration by means of a simplified ammoniacal silver degeneration method. *Brain Research,* 1970, **23,** 250-254.

Loewy, A.D. Ammoniacal silver staining of degenerating axons. *Acta Neuropathologica,* 1969, **14,** 226-236.

Lund, J.S., & Lund, R.D. The termination of callosal fibers in the paravisual cortex of the rat. *Brain Research,* 1970, **17,** 25-45.

Lund, R.D., & Westrum, L.E. Neurofibrils and the Nauta method. *Science,* 1966, **151,** 1397-1399.

Marchi, V. Sulle degenerazioni consecutive all'estirpazione totale e parziale del cervelletto. *Rivista Sperimentale di Freniatria et Medicina Legale delle Alienazioni Mentali,* 1886, **12,** 50-56.

Marsland, T.A., Glees, P., & Erikson, L.B. Modification of the Glees silver impregnation method for paraffin sections. *Journal of Neuropathology and Experimental Neurology,* 1954, **13,** 587-591.

Mathers, L.H. The synaptic organization of the cortical projection to the pulvinar of the squirrel monkey. *Journal of Comparative Neurology,* 1972, **146,** 43-60.

Merzbacher, L. Untersuchungen an winterschlafenden Fledermäusen. II. Die Nervendegeneration während des Winterschlafes. Die Beziehungen zwischen Temperatur und Winterschlaf. *Archives für des Gesamte Physiologie,* 1903, **100,** 562-585.

Nauta, W.J.H. Über die sogenannte terminale Degeneration im Zentralnerven system und ihre Darstellung durch Silberimprägnation. *Archiv fuer Neurologie und Psychiatrie,* 1950, **66,** 353-376.-

Nauta, W.J.H. Silver impregnation of degenerating axons. In W.F. Windle (Ed.), *New research techniques of neuroanatomy.* Springfield, Illinois: Thomas, 1957, Pp. 17-26.

Nauta, W.J.H., & Gygax, P.A. Silver impregnation of degenerating axon terminals in the Central Nervous System: (2) Chemical notes. *Stain Technology,* 1951, **26,** 5-11.

Nauta, W.J.H., & Gygax, P.A. Silver impregnation of degenerating axons in the central nervous system: A modified technic. *Stain Technology,* 1954, **29,** 91–93.

Nauta, W.J.H., & Ryan, L.F. Selective silver impregnation of degenerating axons in the central nervous system. *Stain Technology,* 1952, **27,** 175–179.

Ohmi, S. Electron microscopic study on Wallerian degeneration of the peripheral nerve. *Zeitschrift fuer Zellforschung und Mikroskopische Anatomie,* 1961, **54,** 39–67.

Powell, E.W., & Brown, G. A critique of silver impregnation methods. *Mikroskopie,* 1975, **31,** 77–84.

Powell, E.W., & Schnurr, R. Silver impregnation of degenerating axons; comparisons of postoperative intervals, fixatives and staining methods. *Stain Technology,* 1972, **47,** 95–100.

Powell, T.P.S., & Cowan, W.M. A note on retrograde fibre degeneration. *Journal of Anatomy,* 1964, **98,** 579–585.

Ramón y Cajal, S. *Degeneration and regeneration of the nervous system.* (Ed. & Trans. by R.M. May) New York: Hafner, 1928.

Schneider, G.E. Retinal projections characterized by differential rate of degeneration revealed by silver impregnation. *Anatomical Record,* 1968, **160,** 423.

Sterling, P., & Kuypers, H.G.J.M. Simultaneous demonstration of normal boutons and degenerating nerve fibres and their terminals in the spinal cord. *Journal of Anatomy,* 1966, **100,** 723–732.

Thompson, J.M., Woolsey, C.N., & Talbot, S.A. Visual areas I and II of cerebral cortex of rabbit. *Journal of Neurophysiology,* 1950, **13,** 277–288.

Tigges, J., & Tigges, M. The accessory optic system and other optic fibers of the squirrel monkey. *Folia Primatologica,* 1969, **10,** 245–262.

Torrey, T.W. Temperature coefficient of nerve degeneration. *Proceedings of the National Academy of Sciences of the U.S.A.,* 1934, **20,** 303–305.

van Crevel, H., & Verhaart, W.J.C. The rate of secondary degeneration in the central nervous system. I. The pyramidal tract of the cat. *Journal of Anatomy,* 1963, **97,** 429–449. (a)

van Crevel, H., & Verhaart, W.J.C. The rate of secondary degeneration in the central nervous system. II. The optic nerve of the cat. *Journal of Anatomy,* 1963, **97,** 451–464. (b)

Velayos, J.L., & Lizarraga, M.F. A simplified silver impregnation method. *Experientia,* 1973, **29,** 135.

Wagor, E., Lin, C.S., & Kaas, J.H. Some cortical projections of the dorsomedial visual area (DM) of association cortex in the owl monkey, *Aotus trivirgatus. Journal of Comparative Neurology,* 1975, **163,** 227–250.

Walberg, F. The early changes in degenerating boutons and the problem of argyrophilia. Light and electron microscopical observations. *Journal of Comparative Neurology,* 1964, **122,** 113–137.

Walberg, F. Does silver impregnate normal and degenerating boutons? A study based on light and electron microscopical observations of the inferior olive. *Brain Research,* 1971, **31,** 47–65.

Walberg, F. Further studies on silver impregnation of normal and degeneration boutons. A light and electron microscopical investigation of a filamentous degenerating system. *Brain Research,* 1972, **36,** 353–369.

Weddell, G., & Glees, P. The early stages in the degeneration of cutaneous nerve fibres. *Journal of Anatomy,* 1941, **76,** 65–93.

White, L.E., Jr. Enhanced reliability in silver impregnation of terminal axonal degeneration-óriginal Nauta method. *Stain Technology,* 1960, **35,** 5–9.

Wiitanen, J.T. Selective silver impregnation of degenerating axons and axon terminals in the central nervous system of the monkey (*Macaca Mulatta*). *Brain Research,* 1969, **14,** 546–548.

Wolfgram, F., & Rose, A.S. Chemical basis of the Marchi method for degenerating myelin. *Neurology,* 1958, **8,** 839–841.

Wong-Riley, M.T.T. Changes in the dorsal lateral geniculate nucleus of the squirrel monkey after unilateral ablation of the visual cortex. *Journal of Comparative Neurology,* 1972, **146,** 519–548.

Chapter 9

The Use of Axonal Transport for Autoradiographic Tracing of Pathways in the Central Nervous System

Anita Hendrickson

Department of Ophthalmology
University of Washington
Seattle, Washington

Stephen B. Edwards

Department of Anatomy
University of Virginia
Charlottesville, Virginia

I.   The Technique and the Problems For Which It Is Appropriate

A.   Development of Autoradiographic Nerve Tracing As a Technique

The rationale for the autoradiographic nerve tracing method (ARNT) can be traced back more than 40 years to observations based on neurons grown in organ or tissue culture (reviewed by Pomerat, Hendleman, Raiborn, & Massey, 1967). Under these specialized conditions, many workers described a considerable traffic of particles or organelles as well as a generalized "bulk flow" of materials away from the cell body along its elongated processes and somewhat less material returning along the same process in the retrograde direction. This movement was called "axonal flow" by Weiss (1961). Technical limitations prevented detailed studies until the introduction in the early 1960's of radioactive isotopes, especially the tritiated (^{3}H) amino acids (Droz, 1965). By 1970 two major new ideas had appeared. The first was that there are at least two rates of movement along the axon, with each rate category containing different materials and each category having somewhat different destinations within the neuron. The second was that radioactively tagged axonal transport might form the basis for a neuroanatomical tracing method.

Information on the mechanisms, substances, rates, and functions of axoplasmic transport form a large body of literature which has been well reviewed (Barondes & Samson, 1969; Grafstein, 1969, 1975; Lasek, 1970; Lubinska, 1975; Ochs, 1972, 1974). For the purposes of this chapter, several pertinent generalizations will be made to establish the theoretical basis of ARNT: the reviews cited above should be consulted for many specific references on which these statements are based.

1. No neuron has yet been found that fails to synthesize proteins for transport, so axoplasmic transport appears to be a universal metabolic function.

2. Amino acid incorporation into transported proteins takes place mainly in the cell body and is undetectable in mammalian axons or synaptic terminals under the conditions of ARNT. This means that within an injection site only the

cell bodies and not the axons passing through, nor the synapses ending within, will synthesize transported protein.

3. Transported protein moves at two general rates: the fast components at 100–450 mm/day and slow components at 1–20 mm/day in mammalian systems. Cold-blooded animals have rates about half those of mammals and birds.

4. The fast-moving material is a complex assortment of glyco- and lipoproteins, nucleic acids, enzymes, membrane components, and transmitters. All of these fast-moving substances reach the synaptic terminal. The axon also becomes labeled in transit, but much more lightly than the terminals. The slow-moving materials are less varied in makeup. They localize mainly in the axon, but some substances such as mitochondria and some enzymes may also reach the synaptic terminal.

The application of this information to create a practical, reliable neuroanatomical tracing technique has been very well accepted (Cowan, Gottlieb, Hendrickson, Price, & Woolsey, 1972; Edwards, 1972; Grafstein, 1967; Graybiel, 1975; Hendrickson, 1975b; Lasek, Joseph, & Whitlock, 1968). The ARNT method answers the question, "Where do the axons from a given collection of neurons terminate?" In practice, obtaining the answer consists of the following steps.

1. Inject a radioactive tracer such as ^{3}H-labeled amino acid into the region of the central or peripheral nervous system which contains the neurons in question.

2. If the distribution of synaptic terminals is desired, a short survival time of 1–2 days is used; if both the axon pathways and synaptic terminals are wanted, a 7–14 day survival is used. The animal is perfused or a small brain is fixed by immersion.

3. The brain is sectioned, and the sections mounted onto glass slides.

4. Under safelight conditions, the slides are dipped into liquid photographic emulsion and dried.

5. The slides are stored in light-tight boxes in the cold for days to months until enough radioactive decays have been recorded in the emulsion—the "exposure" time.

6. The exposed slides are developed and fixed to preserve the decays as silver grains, then washed and stained to create the autoradiograph. The autoradiograph is examined in the microscope, and the pattern of silver grains in the emulsion is determined and then related to the underlying structures in the tissue (Figs. 1 and 2).

B. Where Can Autoradiographic Tracing Be Used?

In what situations can ARNT be used? It is easier to name those situations in which it *cannot* be used, just because it is so universally applicable. First, since

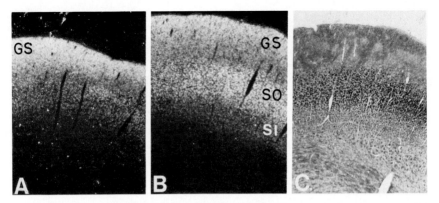

FIG. 1.   Pattern of grain distribution at short (A) and long (B) survival times as seen in rabbit superior colliculus after eye injection of 100 μCi of ^{3}H-proline. (A) At 24 hr after injection the grains are mainly over the stratum griseum superficiale (GS) which contains most of the retinal synapses. These synapses are illustrated at the EM level in Fig. 11B. Darkfield: magnification: × 16. (B) By 21 days after injection the labeling over stratum opticum (SO) is now slightly heavier than over stratum griseum (GS). A diffuse light label has appeared over stratum intermedium (SI). Darkfield; magnification: × 16. (C) Luxol fast blue–cresyl violet-stained section showing the layers in brightfield, magnification: × 16.

the radioactive precursor can be synthesized into the radioactively tagged transported molecule only within the neuron cell body, there must be neurons within the injection site. If only a few neuron cell bodies are present in a nucleus, or they are widely separated, very little labeled material may be synthesized and may be difficult to adequately trace. Second, since axons do not adequately synthesize protein for transport, pathways cannot be labeled by injecting amino acid into axon bundles. This eliminates using peripheral or cranial nerves or end receptors; however, precisely because of this, ARNT does not suffer from a "fiber of passage" problem. If radioactive amino acid is injected into an area, only the nerve cell bodies will incorporate it into transported protein, and so, regardless of how many axons run through the region, the synapses of these axons will not be labeled. Third, since the injection of isotope-tagged precursors is generally quite small, 1 μl or less, and labels 1 mm^2 or less of tissue, the entire nucleus or region will not be labeled. This prevents tracing the total pathway distribution of an area in a single experiment, which is possible with a big lesion. An exception would be intravitreal eye injections which label the entire retinal ganglion cell projection. On the other hand, because of this limitation, ARNT is ideal for topographical organization analysis of systems. Fourth, because axoplasmic transport of proteins is a universal metabolic necessity in neurons, this method works in all systems, unlike silver staining-lesion techniques, which often suffer erratic staining in certain areas such as the hypothalamus. ARNT can

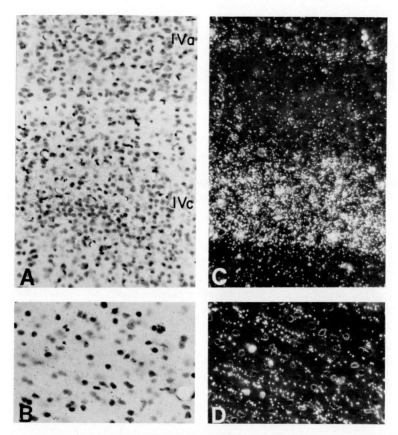

FIG. 2.   Monkey striate cortex photographed in brightfield (A,B) and darkfield (C,D). The monkey dorsal lateral geniculate nucleus was injected 24 hr before sacrifice with ³H-leucine. The grains in (A) and (C) overlie thalamic terminals in layers IVA and IVC while in (B) and (D) the thalamic axons appear lightly labeled in the underlying white matter. A, C, magnification: × 160; B, D, magnification: × 260.

also be used in fetal or immature systems which are difficult to study by lesion techniques owing to unpredictable degeneration rates. Fifth, because the number of silver grains formed is proportional to the exposure time, even a very sparse pathway can be "amplified" by lengthening the exposure.

Before beginning work, the novice in autoradiography should read the classic paper of Kopriwa and Leblond (1962) and browse through Roger's book (1973). Cowan *et al.* (1972), Edwards (1972) and Hendrickson (1975b) delineate the application of autoradiography to nerve tracing. The following discussion will attempt to facilitate each step of ARNT by practical suggestions which have been worked out in the authors' laboratory over the past years. Although they are not the only ways that good results can be obtained, they have worked well for us.

II. Methodology of Autoradiographic Nerve Tracing (ARNT)

A. Step 1: Injecting a Specific Region of Nervous System with Radioactive Amino Acids

1. Choice of Isotope for Autoradiography

Autoradiography is a second-order detecting method in that while the primary event (the radioactive decay) occurs within the tissue, detection of this event is recorded secondarily (as a silver grain) in the overlying emulsion (Fig. 3). The distance between the decay and the resulting silver grain is called *resolution;* it is affected mainly by the energy of the isotope. The closer together the decay and its silver grain occur, the better the autoradiographic resolution, which has led to the use of β emitters almost exclusively because of their low energy range and widespread occurrence. β particles are electrons which are ejected from the nucleus of the atom and have varying amounts of energy so that each β isotope has its own characteristic energy spectrum. Tritium (^{3}H), for instance, has an energy spectrum ranging from 0 to 18.5 keV, but 50% of the electrons have energies below 8 keV. Sulfur-35 (^{35}S) and carbon-14 (^{14}C) are more energetic,

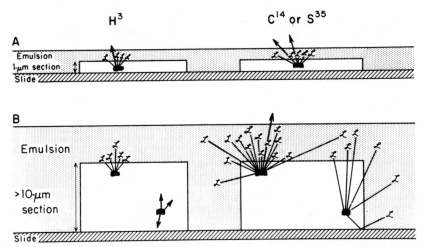

Fig. 3.   Effect of section and emulsion thickness or isotope energy on the pattern and number of silver grains in an autoradiograph. ^{3}H autoradiography is affected very little, but ^{35}S and ^{14}C autoradiography is very sensitive to changes in thickness of either section or emulsion.

having maximum energies of about 167 and 155 keV, respectively, while most of the β particles have energies between 0 and 80 keV (Rogers, 1973). Isotopes with energies greater than this, phosphous-32 at 1600 keV for instance, are not very useful for microscopic autoradiography because most of the β particles travel so far in the tissue and emulsion that the autoradiographic resolution is very poor. ^{35}S, ^{14}C, and ^3H are the isotopes commonly used for pathway labeling, and this discussion will be confined to their advantages and disadvantages for autoradiography. The two largest sources of these isotopes in the United States are Amersham/Searle (2636 So. Clearbrook Drive, Arlington Heights, Illinois 60005) and New England Nuclear (575 Albany Street, Boston, Massachusetts 02118). All individuals involved in ARNT experiments should be thoroughly versed in the handling and disposing of radioactive substances. Local radioactive safety personnel should be consulted before beginning the following techniques.

Tritium is by far the most commonly used isotope for ARNT because of several factors. Because of their low energy, β particles of ^3H travel no more than 2 μm in tissue and 1 μm in emulsion (Rogers, 1973), which means that tritium will yield autoradiographs with a resolution of less than 1 μm under ideal conditions (sections < 1 μm covered by a monolayer of fine grain emulsion, Salpeter, Budd, & Mattimoe, 1974; Fig. 3A). Tritium-labeled substances are available in high specific activities (10–60 Ci/mmole), which means that a large number of labeled molecules are present compared to the unlabeled molecules, and this increases the chance that a labeled molecule will be incorporated into the transported substances. Tritium-labeled substances can be custom-made commercially. Standard ^3H-labeled materials are significantly cheaper than ^{14}C-labeled substances. On the other hand, because of low β particle energy, only the radioactive decays in the 2 μm of the tissue section immediately under the emulsion can reach the emulsion (Fig. 3B). This means that unless the tissue is extremely radioactive, relatively long times will be necessary in order to accumulate enough grains in the emulsion to give a clearly labeled section.

^{14}C and ^{35}S have characteristics that are almost the exact opposite of ^3H. β particles of 20–80 keV penetrate ~ 50 μm in tissue and at least 10 μm in emulsion (Salpeter & Salpeter, 1971), which gives an autoradiographic resolution value ranging from 1–50 μm depending on the thickness of the section and the emulsion coat (Rogers, 1973; Salpeter et al., 1974). The thicker either the section or the emulsion becomes, the poorer the resolution becomes, as can be seen in Figs. 3A and 3B. However, this higher β particle energy also means that a much greater depth of the section will contribute β particles to the emulsion, making exposure times much shorter than for ^3H (Fig. 3B; Bobillier, Séguin, Petitjean, Touret, & Jouvet, 1976; Graybiel, 1975). ^{35}S is available at very high specific activities (> 300 Ci/mmole), but ^{14}C-tagged materials usually have low specific activity. ^{35}S has a half-life of 87 days, which means that tissues must be processed promptly; the long half-lives of ^{3}H (12 years) and ^{14}C (5700 years) allow a more leisurely workup.

2. CHARACTERISTICS OF A GOOD PRECURSOR

Axoplasmic transport involves a diverse collection of materials including proteins, glycoproteins, glycolipids, phospholipids, nucleotides, nucleic acids, and some transmitters (most recently reviewed by Grafstein, 1969, 1975; Lasek, 1970; Lubinska, 1975; Ochs, 1972, 1974). Labeled precursors for each include amino acids for proteins; amino acids, fucose, and glucosamine for glycoproteins; cholesterol, choline, glycerol, or phosphorus-32 for lipids; nucleotides for nucleic acids; and noradrenaline and 5-hydroxytryptamine for transported transmitters. The transmitter precursors are confined to the nerve pathways utilizing these transmitters, which are thoroughly discussed in Hökfelt and Ljungdahl's (1975) recent paper and will not be covered in this article. Lipid precursors are not often used for ARNT because tissue processing for histology causes a severe loss of many labeled lipids, and lipid precursors freely exchange with preformed *in situ* lipids (Droz, 1975). Nucleotides such as ^{3}H-adenosine appear to be transported either as the free nucleotide or within nucleic acids, but there is some disagreement as to whether the transport is in the orthograde, the retrograde, or both directions (Hunt & Kunzle, 1976; Kreutzberg & Schubert, 1975; Wise & Jones, 1976). When this question is resolved, nucleotide precursors could become very useful for ARNT since they might be simultaneous markers of both afferent and efferent pathways.

The most commonly used precursor for ARNT are amino acids labeled with tritium, which are synthesized into radioactive transported proteins. Logically, an amino acid that is commonly found in many proteins should be chosen for initial studies, which led to the choice of leucine and, less often, lysine for early studies. Later it was found that proline has several advantages for ARNT even though it is less commonly found in large amounts in brain proteins. First, for equal amounts injected, more proline is incorporated into the rapid phase of transport than leucine, leading to a much more heavily labeled terminal region at short survival times. Quantitative studies of amino acid effectiveness in ARNT have used the retinal ganglion cell since intravitreal injections are reasonably reproducible and easily done. In the rabbit visual system, proline > methionine > leucine > tryptophan > lysine for labeling of central visual centers 7–8 hr after intravitreal injection (Karlsson & Sjöstrand, 1972), while for fish retinal ganglion cells proline > aspargine (Elam & Agranoff, 1971). Second, proline is more effectively excluded by the blood–brain barrier (Dingman & Sporn, 1959), so that its use for injections into peripheral sites such as the eye results in much lower background of brain tissue (Crossland, Cowan, & Kelly, 1973; Neale, Elam, Neale, & Agranoff, 1974). These results have led to the widespread use of proline as a tracer for many systems, and proline generally has shown greater efficiency per injected dose as compared to leucine. However, there may be a problem in a too rapid generalization of the results in one pathway to all pathways. For instance, in the cat lateral reticular nucleus (Kunzle & Cuénod, 1973) and nucleus gracilis (Berkley, 1975), proline only lightly labels the large cells of the nucleus which are heavily labeled by leucine. In Berkley's paper, the same differential labeling pattern is seen when the terminal areas of these two neuronal populations are compared. A similar result with heavy proline labeling can be seen in the inner plexiform layer of the retina, with only a very light labeling of the protein transported by the photoreceptor to its outer segment (A. Bunt, personal communication). On the other hand, in auditory cortex Sousa-Pinto and Reis (1975) report labeling of cat corticothalamic efferents but *not* cortico-cortico pathways with ^{3}H-leucine. These reports of amino acid discrimination by some feline neurons suggest caution when using only one amino acid to study a pathway. It is much safer to use a mixture of two or three common amino acids like leucine, lysine, proline, and methionine, at least for an initial experimental series.

A second problem with both proline and fucose is the finding that they move transsynaptically (Fig. 4) and can give significant labeling of the axon synapses of a neuron that is postsynaptic to the primarily labeled synapse (Grafstein & Laureno, 1973; Specht & Grafstein, 1973). For instance, after concentrated ^{3}H-labeled proline–fucose mixtures are injected into one eye of mice or monkeys, the thalamic terminals within layer IV of the striate cortex become labeled (Drager, 1974; Kaas, Lin, & Casagrande, 1976; Wiesel, Hubel, & Lam, 1974). There is also significant diffusion of labeled material away from all regions of the primary-labeled neuron (Fig. 1B), which could cause secondary labeling of adjacent but unrelated nuclei. Since this transsynaptic labeling increases the longer the survival time after injection (Grafstein & Laureno, 1973) short survival times reduce but do not altogether eliminate the possibility for detectable transsynaptically labeled pathways.

Intravitreal injections of glucosamine and fucose show that these substances are very effective for labeling the rapid phase of transport and have the additional advantage of showing almost no label in the slow phase (Karlsson & Sjöstrand, 1971; McEwen, Forman, & Grafstein, 1971). This would seem to make them excellent tracers for central nervous system injections, but Graybiel (1975) has briefly reported disturbing results with intracranially injected fucose. Fucose spreads very widely compared to proline, so that 0.2 μl labels half of the brain! As well, when ^{3}H-fucose was injected into the superior cerebellar peduncle (A. M. Graybiel personal communication; also see her Fig. 13, 1975), what appeared to be labeled terminal fields were found in the inferior olive which were the same as those seen after ^{3}H-proline or ^{3}H-fucose injection into the deep cerebellar nuclei. There was no evidence of terminal field labeling when ^{3}H-proline was injected into the peduncle, suggesting that ^{3}H-fucose can be taken up by axons

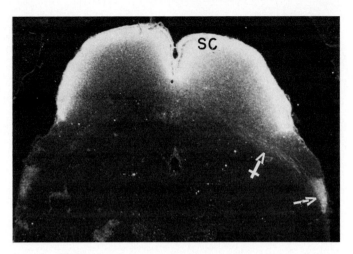

Fig. 4. Superior colliculus (SC) of a squirrel monkey injected in one eye with 1 mCi each of ^{3}H-proline and fucose and sacrificed 14 days later. The section was exposed for 8 weeks. The primary retinal input to the superficial layers of SC is grossly overexposed while the secondary transsynaptic axons (crossed arrow) and terminals from SC to parabigeminal nucleus (arrow) are just apparent. Darkfield; magnification: 22 ×.

and transported to their terminals. If this finding is confirmed, it would mean that unlike amino acids, fucose could show "fiber of passage" problem common to lesion techniques and hence would not be usable for most brain injections.

3. DELIVERY OF ISOTOPES

[3]H-labeled amino acids are usually supplied at a concentration of 1 μCi/μl, although higher concentrations can be obtained by special order. Amino acids are relatively stable molecules, but they are attacked by the free radicals formed by radioactive decay; the higher the specific activity of the amino acid, the greater the autolysis. Most ^{3}H-labeled amino acids are supplied in 0.01 N HCl or 2% ethanol to quench free radical formation. Unless specifically instructed otherwise, they should be stored at 4°–6°C but *not* frozen, since free radical formation is greater in ice than in water. At 4°C in 0.01 N HCl, the rate of decay is 1% per month for a concentration of 1 μCi/μl. If relatively large amounts of material can be injected, e.g., intravitreal injections, sufficient labeling is obtained using the standard 1 μCi/μl concentration so that usable autoradiographs can be produced with 2–3 weeks of exposure. CNS injections must use much smaller volumes ranging from 0.1 to 1.0 μl to avoid tissue damage and to produce discretely labeled sites. This means that in order to achieve sufficient neuronal label (Kunzle, 1975) and to avoid excessively long exposures, ^{3}H-labeled amino acid must be further concentrated to contain 10–100 μCi/μl. If stored at 4°–6°C, the loss in 10× concentrated solutions by autolysis should be about 10% per

month (New England Nuclear, personal communication). Since the alcohol or acid quenching agent is removed during drying, very concentrated (50–100 ×) solutions should be used within a week.

Concentration of the amino acids can be done in a variety of ways. The simplest is by straight heat evaporation in a 55°C oven; time at this temperature should be kept to a minimum, especially in the dry state. Alternatively, the amino acid can be dried at room temperature in a vacuum desiccator, using NaOH pellets as desiccant. This generally is a faster method, but care must be taken that bubbles do not form in the isotope and "boil" the entire amount out of the drying tube. The liquid also can be evaporated under a gentle stream of nitrogen; gentle heating of the isotope during evaporation will speed this process. A rapid but more expensive method is to dry the isotope in a vacuum centrifuge equipped to handle small volumes (SVC 100 Speed Vac, Savant Inst. Inc., Hicksville, N.Y.). When dry, the isotope is redissolved to the desired concentration in sterile water, saline, or buffer.

Concentrated label can be injected into a predetermined region of the brain using a microliter syringe or a glass microelectrode attached to a pressure system. An example of such a microinjector system, which was specifically designed for use in the ARNT and which is now in use in many laboratories, has been described (Edwards & Shalna, 1974). The easiest method is to use a 1-, 5-, or 10-μl syringe (1 μl Hamilton 7000 series, 5 or 10 μl 700 series) with a fixed 26 or 31 gauge needle. The Hamilton microsyringes fill and empty easily and have minimal dead space, which is an important monetary consideration when using very concentrated, expensive isotope solutions. The 26-gauge needle is 2 in. long, making it long and sturdy enough to penetrate to deep regions in cat or monkey brains. Although glass micropipettes probably do less damage at the injection site and seem to produce smaller spots (Schubert & Hollander, 1975), they are much more fragile and liable to plug with tissue at the tip.

For brain injections, the animal should be held securely in a stereotaxic head holder and the syringe or the microelectrode also held securely in a stereotaxic electrode holder. Deep brain regions are usually located by stereotaxic coordinates, but some investigators have also used the injection needle or micropipette as a recording electrode to facilitate finding the correct region (Bunt, Hendrickson, Lund, Lund, & Fuchs, 1975; Edwards, 1977; LeVay & Gilbert, 1976).

Once the needle tip is in the desired region, the isotope is injected as slowly as possible (no faster than 1 μl/5 min) since experience has shown that for equal amounts of isotope, the slower the injection the greater the uptake and the more restricted the injection site. After all of the isotope is deposited, the needle is left in place for at least 15 min to reduce diffusion of labeled precursor along the needle track as it is withdrawn. The isotope should be pulled back slightly from the needle tip both on entry into and on withdrawal from the brain. This air interface helps prevent unwanted isotope deposition along the needle track owing to capillary action leakage at the needle tip. Diffusion along the track may also be prevented by filling the track with oil as the needle is withdrawn (Beitz & King, 1976).

The injection volume and isotope concentration should be varied during the course of an experimental ARNT series. It is advisable to begin with somewhat large volumes of 1–5 μl of 5–10× concentrated ^{3}H- or ^{35}S-labeled amino acid depending on the animal and the region to be injected. This will result in heavy labeling of a large number of fibers and terminals which can easily be detected in autoradiographs. Once all of the labeled pathways have been identified, smaller injections of less than 1 μl can be done. Using too little label can lead to the wrong conclusions. Kunzle (1975) found after injecting 1 μl of concentrated ^{3}H-leucine into area 4 of monkey motor cortex that he was unable to reliably identify the projections to claustrum, caudate, or Darkschewitch nuclei, while the projection to putamen was consistently seen regardless of volume. Injections larger than 3 μl were required to demonstrate the former pathways.

Very small spots or even a single neuron can be labeled by iontophoretic delivery of isotope. The advantage of iontophoresis is that by using current as the delivery vehicle only the charged molecules of tracer are deposited in the tissue. No volume changes occur, minimizing tissue trauma and diffusion of isotope away from the injection site. Several methods of iontophoretic delivery have been published (Graybiel & Devor, 1974; Schubert & Hollander, 1975; Schubert, Kreutzberg, & Lux, 1972), and these papers should be consulted for detailed procedures.

The only consistent difference between these two delivery methods seems to be that iontophoresis gives smaller labeled foci with little or no label along the injection track, but iontophoresis requires much more equipment and experience than a pressure microinjection and is prone to unexplained failures in which no isotope is deposited. The theoretical and technical aspects of these two delivery techniques are further reviewed by Schubert and Hollander (1975).

B. Step 2: Sacrificing the Animal and Fixing Tissue after an Appropriate Survival Time

1. SURVIVAL TIME

Both light and electron microscopy (EM) autoradiography have demonstrated that the rapid phases of transport label mainly synaptic terminals (Figs. 1A, 2, and 11). At longer survival times (Figs. 1B, 4, and 5), axons contain an increasing percentage of the label, although terminals are still radioactive at long survival times, but to a lesser degree (Cuénod, Boesch, Markō, Perisič, Sandri, & Schönbach, 1972; Grafstein, 1967; Hendrickson, 1972; Schönbach, Schönbach, & Cuénod, 1971). Specific survival times depend on the length of the system, the species under study, and whether the terminal fields or the entire pathway is to be

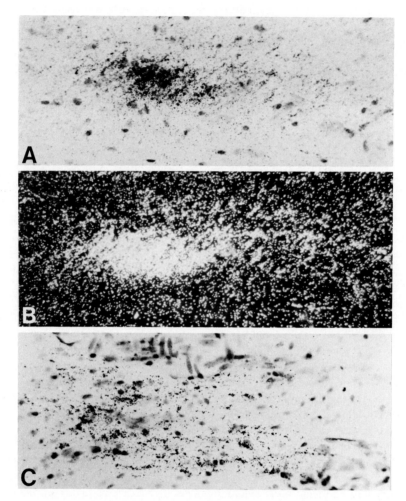

Fig. 5. Focus of terminals in the parageniculate nucleus surrounded by labeled axons ((A), brightfield; (B), darkfield), and the labeled accessory optic axons leaving this region (C). Two-week survival after eye injection in a baboon. Magnification: × 250.

visualized. For instance, the rapid phase of retinal ganglion cell axon transport begins to arrive at the tectal synapse 3 hours after a goldfish eye injection (axon length 6–8 mm) and the slow phase takes 9–11 days (McEwen & Grafstein, 1968), while comparable times for the rabbit retinal ganglion cell (axon length 25–30 mm) are 4–5 hours and 28–30 days, respectively (Hendrickson & Cowan, 1971). In bird and mammalian nervous system, rapid transport travels at least 250 mm/day or approximately 10 mm/hour; the rapid phase moves about one-half to one-third as fast in fish, reptile, and amphibian nervous system (reviewed by Grafstein, 1975). Slow transport average rates usually are 1–5 mm/day in birds and mammals and 0.5–1 mm/day in fish and amphibia, although faster rates of 24 mm/day have been reported for rabbit vagus nerve (McLean, Frizell, & Sjöstrand, 1976). Approximate minimum survival times can be calculated in *mammals* by

$$\frac{\text{length of pathway (in mm)}}{10} = \text{survival time (in hr)}$$

for rapid transport and

$$\frac{\text{length of pathway (in mm)}}{2} = \text{survival time in days}$$

for slow transport.

There are several reasons for thinking that it is better to allow slightly longer survival times for rapid transport than the calculated minimum. There is evidence that a second, slightly slower rapid phase (phase II) contains a considerable amount of protein destined mainly for the terminal (Karlsson & Sjöstrand, 1971; Willard, Cowan, & Vagelos, 1974). If the survival time is long enough for this second phase to reach the terminal, i.e., 4–8 hr after protein first appears, it creates much hotter tissue for autoradiography. Once the rapid and second phases have reached the terminal field, radioactivity remains relatively unchanged in amount for several days (Karlsson & Sjöstrand, 1971; McEwen & Grafstein, 1968; McEwen *et al.*, 1971). Even allowing this extra time, speeds of the rapid and slow components are sufficiently different so that overlap in the region of relatively distant terminal fields is unlikely. As well, a slightly longer survival time will allow slow transport to label the axons near the injection site so that these pathways can be studied.

The type of pathway may also determine survival time. For a relatively concentrated system like many corticothalamic and major sensory pathways, a short survival of 1–2 days, which gives terminal label almost entirely, will produce heavily labeled unequivocal autoradiographs. For sparse or highly diffuse systems like locus coeruleus, raphe nuclei, or reticular formation projections, it may be necessary to use a long survival time so that the axons are labeled in order to be able to definitely identify the labeled pathway (Edwards 1975).

2. Fixation Solutions and Their Use

Proteins containing an isotope are easily and reliably fixed by a variety of aldehyde fixatives, including formalin, paraformaldehyde, and glutaraldehyde. These aldehyde fixatives will retain 92–98% of the labeled protein (Droz, Warshawsky, & Warshawsky, 1963), but it has been shown as well that glutaraldehyde will bind free amino acids in tissue that is fixed minutes after very high doses of radioactive amino acids (Peters & Ashley, 1967). For adequate fixation of many tissues it is necessary to include glutaraldehyde in the fixative. Will this complicate autoradiography? Since the half-life of free amino acid in brain is 1 hr or less (Lajtha, 1975) and survival times for ARNT are usually much longer than 1 hr, the use of glutaraldehyde-containing fixative should present no problems, especially for analysis of terminal localization at a distance from the high amino acid concentration of the injection site. Fixatives containing mercury or lead salts such as Zenker's should be avoided, since they cause chemical reduction of the emulsion (Kopriwa & Leblond, 1962).

If a labeled compound is used that results in something other than protein being the compound under study, careful studies of the effects of histological processing on the tissue will have to be done. The methods for this type of analysis are discussed by Droz (1975).

Once tissue is fixed it can be stored indefinitely in dilute fixative before proceeding to the next steps of autoradiography, although the half-life of the isotope used in the experiment will determine exactly how long. The tissue should be washed thoroughly in water before proceeding to sectioning or embedding.

C. Step 3: Embedding and/or Sectioning Tissue and Mounting Sections on Glass Slides

1. SECTIONING METHODS

Sections for autoradiography can be produced by a variety of methods. Frozen sections have the advantage of speed of preparation and ease of handling for large brains. Frozen sections must be used if the tissue is to be reacted for horseradish peroxidase combined with autoradiography (Colwell, 1975; Graybiel, 1975; Ogren & Hendrickson, 1976). Paraffin has the advantage of holding separate pieces of tissue in place better than the frozen method, and serial sections are easier to prepare. Carbowax preserves much of the lipid and some of the water in the tissue, but can be difficult to section.

For frozen sectioning for ARNT only, tissue is fixed at least overnight and then is held in a solution of 20–30% sucrose in fixative at room temperature until the tissue sinks. This may take 3–4 days for a large whole brain. The sucrose impregnation acts as an antifreeze which prevents ice crystal damage during freezing. The tissue is oriented on the freezing microtome or cryostat and frozen by the method used with the particular machine. Tissue should not be allowed to get too cold or it will shatter during sectioning. Frozen sections are cut serially at 10–50 μm and are kept in dilute (usually 2%) formalin. We use plastic tackle boxes which are divided into 18 sections and have tight-fitting snap lids (Vlchel Plastics, Middlefield, Ohio 44062). The sections can be stored for many months at room temperature, if kept wet with fixative, and still produce satisfactory autoradiographic results. If horseradish peroxidase histochemistry is also to be done, the tissue is handled first for peroxidase (see Chapter 13) and after a sufficient number of frozen sections have been reacted, the rest are put into dilute formalin and then processed for ARNT.

Some brain areas such as cerebellum or cerebral cortex which separate into several pieces when freeze-sectioned are very tedious to mount. This can be eliminated by first encasing the tissue in egg yolk or albumin–gelatin (Ebbesson, 1970) before sectioning. For paraffin or Carbowax, tissue is embedded in these materials using standard methods and is sectioned serially at 5–10 μm.

What effect does section thickness have on autoradiography? With tritium, only the upper 2–3 μm of tissue will influence the emulsion (Boren, Wright, & Harris, 1975; Caviness & Barkley, 1971). Sections thicker than 3 μm will not contribute additional β particles to the autoradiograph and sections thicker than this will have very little increase in grain number (Fig. 1B). ^{35}S or ^{14}C β particles have much higher energy, so the full thickness of a 10–20 μm section will contribute β particles to the emulsion, although not with the same efficiency. Any increase in section thickness at least up to 20 μm will result in an increase in the total number of grains found in the emulsion (compare Figs. 1A and B).

Does the embedding medium have any effect on the number of grains in the final autoradiograph? The denser the medium the more β particles that will be stopped within the section (Boren et al., 1975). To determine this difference, pieces of a uniformly ^{3}H-labeled mouse liver were embedded in Epon or paraffin and frozen-sectioned. Sections were coated with NTB-2, exposed for 3 weeks, and developed together in D19. The 2 μm Epon section had 65 ± 3 grains/900 μm^2, the 10 μm paraffin section 181 ± 6 grains/900 μm^2, and the 25 μm frozen section 229 ± 5 grains/900 μm^2. Even though in theory both paraffin and frozen sections were "infinitely thick" (Rogers, 1973, p. 75) and should have had the same grain counts as the Epon sections, these counts suggest that the density of the Epon embedding medium which was present during autoradiography absorbs a significant number of β particles; however, another explanation can be offered. Tissue shrinks 20–30% in paraffin embedding and during frozen-section dehydration before coating. This shrinkage results in a greater density of radioactive molecules in the upper 2–3 μm of the section compared to the Epon embedded tissue which does not shrink in embedding. Probably both of these factors contribute to the increased number of β particles that reach the emulsion from frozen and paraffin sections over what would be predicted from Epon sections.

Although in theory celloidin sections can be used for autoradiography, the length of processing time, the difficulty of cutting reasonably thin sections, and the problems of obtaining smooth sections tightly affixed to slides for dipping preclude celloidin as a practical embedding medium.

2.   MOUNTING SECTIONS ONTO SLIDES

Paraffin or Carbowax sections are mounted on slides using either gelatin in the water bath or albumin on the slide to adhere the sections to the slides. Paraffin or Carbowax sections are dewaxed, dehydrated, and brought to water, washed, and dried before coating.

Frozen sections can be mounted either from alcoholic gelatin or onto chrome alum–gelatin subbed slides (Hendrickson, Moe, & Nobel, 1972). Frosted-end "write-on" slides are useful in the darkroom because the side of the slide containing the tissue can be identified by touch. Frozen sections which are defatted before coating have been found to give more reliable staining and less background and a more even emulsion coat (Hendrickson *et al.,* 1972). Sections are dried thoroughly after mounting, preferably overnight in a 40°C oven, and then sequentially dehydrated through 70, 95, and 100% ethanol (two 10-min changes each) to xylene where sections are left for 1 hour. They are taken back through descending (100, 95, 70%) concentrations of ethanol, washed for 10 min in running water, and then dried before dipping.

D.   Step 4: Coating Sections with Melted Emulsion in Darkroom

1.   DARKROOM REQUIREMENTS

The darkroom is an essential part of autoradiographic technique. It can be as luxurious as space and money can provide, but if necessary, an adequate darkroom only requires certain minimum facilities. First, if at all possible, the room should be used only for autoradiography and should not be shared for photography. This suggestion is prompted by the need to keep the room as clean and dust-free as possible. Photographic fluids, especially fixer, often kept standing in open pans ready for use, contribute large amounts of particulate matter to the atmosphere of a room, making it a virtual certainty that a clean emulsion in the bottle will be a dirty emulsion when the freshly dipped slide is dried in such a room. Second, the room should have a double door system so that a person can go in and out without exposing the darkroom to light. Once the emulsion is melting in the water bath the room must be under continuous safelight conditions until the coated slides are boxed. Third, the room must *really* be light tight. This should be checked from all angles including the floor by a fully dark-adapted (30 min) observer. At the same time, connections and switches for all electrical equipment should be checked for sparking. Autoradiographic emulsion is *very* sensitive to light flashes. Fourth, temperature-controlled water is desirable but not absolutely essential to regulate the development process. Fifth, the temperature of the room should be relatively stable; better too cold than too warm if a choice

Fig. 6.   The difference in darkroom lighting conditions between (A) the Thomas Duplex-super sodium safelight, and (B) a Kodak 15-W bulb safelight with a Wratten 1A filter.

can be made. Sixth, a temperature-controlled water bath is an essential piece of equipment.

Recommended safelight conditions for Kodak emulsion vary in published reports from "use in total darkness" to a Wratten 1A filter over a 15-W bulb used at 3 ft to the sodium vapor light. Experience shows that working in total darkness is unnecessary and constitutes cruel and unusual punishment. A Wratten 1A filter is better, but not by much. Although it is more expensive than a Kodak safelight, the authors have found the Thomas Duplex-Super sodium safelight (Thomas Inst. Co., 1313 Belleview, Charlottesville, Virginia 22901) to be a valuable asset to both our darkroom and our mental attitude (compare Figs. 6A and B). It is used with the red-tape-bound ortho glass filters in the vanes and the yellow tape bound ortho glass filters inside the lamp. We have our light hanging overhead about 7 ft from the floor. The light should be turned on for 30 min before sensitive materials are exposed to it; if not fully warmed up, the sodium tube can emit wavelengths which will expose the emulsion. The life of the sodium tube will be prolonged if the lamp is switched on and off as little as possible.

2.   Emulsion Handling

In the United States, Kodak NTB-2 and NTB-3 bulk emulsions (Sensitized Products Division, Eastman Kodak, Rochester, N.Y. 14650) are most commonly used for light microscopic autoradiography; in Europe, Ilford K-2, K-5, and G-5 emulsions are frequently used (available in the United States through Polysciences Inc., Warrington, Pennsylvania 18976). This discussion mainly will be concerned with Kodak emulsions, while an excellent discussion of Ilford emulsions and their use can be found in Rogers' textbook (1973).

A photographic emulsion is a suspension of sensitized silver halide crystals in gelatin. All emulsions are unstable, but some, for example NTB-3, are slightly more so than others such as NTB-2 or Ilford L4. Their shelf life is 1–2 months at 4°C. All emulsions are stored at 4°C and should never be frozen. When a new emulsion is received it is tested by melting a small amount, coating a test slide containing a tissue section, drying, and developing immediately. Silver grains which appear in an unexposed emulsion are called "background"; they are due to stress, γ-rays, or spontaneous events. Acceptable background will vary somewhat depending on laboratory criteria, but will probably be between 5–15 grains/oil immersion field. Excessively high background emulsions should be discarded; Rogers (1973) gives some ways they might be restored to a usable level.

For use in ARNT both Kodak and Ilford solutions are diluted so that there is at least a double layer of silver halide crystals over the section. The thickness of the emulsion coat can be determined by coating a test section on a slide, drying it, exposing it to a brief flash of room light, and then immediately developing it (Fig. 7A). For most work, a 1:1 dilution of NTB-2 or -3 is sufficient, but for some tissues and for ^{35}S or ^{14}C label, a thicker 2:1 emulsion/water dilution will be needed. If the emulsion coat is less than a monolayer thick anywhere over the tissue, low and variable grain counts will result in these regions. Rather than being randomly localized, this thin emulsion coat may be related to the underlying tissue organization such as seen in Fig. 7B from retina. Spurious low counts can lead to quite erroneous conclusions about distribution of a labeled molecule.

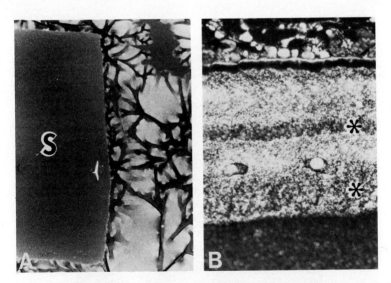

FIG. 7.   (A) Notice how the light-flashed emulsion coat is smooth over a section (S) but badly wrinkled and irregular in thickness over the slide. Emulsion thickness should always be checked using a section. Brightfield: magnification: × 10. (B) A light-flashed emulsion-coated paraffin section of retina. The surface of the section is irregular so that there is a thicker coat of emulsion over the plexiform and ganglion cell layers (*). Brightfield; magnification: × 80.

For ARNT, it is definitely better to lean toward a too thick emulsion rather than a too thin one. This simple light-flash test should always be run when changing mounting or embedding methods or when using different types of tissue.

3. DIPPING METHODS TO COAT THE SECTIONS ON THE SLIDES

Emulsion is melted at 40°–45°C in a temperature-controlled water bath; Kodak emulsion can be melted several times at 45°C without increase in background. If large numbers (300–500) of slides are dipped at the same time, the entire bottle can be melted and the required amount poured out and diluted. The bottle will be used up before background increases excessively. When only small numbers of slides are done at a time, repeated remelting can increase background before the bottle is used up. We conserve emulsion in two ways. First, we have constructed a dipping chamber (Fig. 8) from acrylic plastic (Plexiglas) that is just large enough to hold the glass slide to be used; for instance, for 25×75 mm (1×3 in.) slides, the internal dimensions of the chamber should be $30 \times 5 \times 60$ mm. After this chamber is filled to the dipping line with 10 ml³ of diluted emulsion, 15 slides can be dipped before the chamber needs to be topped off with a fresh emulsion. The dipping chamber stands in a water-filled beaker in the 45°C water bath while slides are being dipped to prevent cooling and resulting thickening of

FIG. 8.   Acrylic plastic dipping chamber for coating slides. The illustrated dimensions are for $25 \times$ 75-mm slides; appropriate adjustments can be made for a larger chamber to dip bigger slides.

the emulsion coat as the level of the emulsion drops down. Second, we melt only the emulsion we will need. With this chamber, we have found that 30 ml of diluted emulsion is needed to coat 100 slides. To produce 30 ml of 1:1 emulsion we put 15 ml of water in a 50-cm^3 graduated glass centrifuge tube or a 50 ml glass pharmaceutical beaker which is sitting in a 45°C water bath, and solid emulsion is added to bring the water level to 30 ml. We have found that if excessive bubbling occurs in the diluted emulsion during coating, it can be suppressed by adding 0.1% sodium lauryl sulfate in distilled water to dilute the emulsion. The emulsion is spooned out of the stock bottle with a disposable plastic spoon which is thrown away after use. This method results in very little wasted emulsion, allows variable dilutions, and keeps the stock bottle at very low background.

When the emulsion is melted, the chamber is filled to the dipping line and the sides dipped one at a time. Slightly less emulsion will be used if two slides are dipped at once, with the empty back sides apposed and the tissue-containing sides facing outward. The slides are separated and dried vertically. A simple device for drying large numbers of slides is illustrated in Figs. 9A, B, and C.

All emulsions contain gelatin, which causes considerable stress within the emulsion as it dries. Slides should be dried slowly and at a relatively high (50–75%) humidity. Very warm and/or very dry darkrooms should be avoided because if the emulsion is dried too rapidly the stress in the gelatin will produce many grains which are often arranged in a cobblestone or straight-line pattern. We cover the bottom of our drying chest (Fig. 8) with wet sponges to create sufficient humidity during this step. Alternatively, a commercial humidifier can be added to the darkroom.

Dry slides are placed in light-tight boxes containing packets of drying medium such as silica gel or Dri-Rite, and the boxes are taped closed with black electrical tape to prevent any light leaks. Black plastic, black cardboard, or wooden slide boxes are satisfactory for exposure containers, but beige plastic slide boxes admit enough light to fog the emulsion. Kodak emulsion must be kept quite dry during exposure to prevent fading of the latent images in the emulsion; long exposures in a humid refrigerator may necessitate changing the drying agent at least once. Ilford emulsions need less drying agent during exposure.

E. Step 5: Exposing Coated Slides for Sufficient Time to Accumulate Detectable Labeling

EXPOSURE CONDITIONS

Autoradiography provides a cumulative record in the emulsion of the radioactive decay within the tissue over the period of time that the section is in contact with the emulsion; this period is called the exposure time. During the exposure time, the β particles given off by the isotopically labeled molecules in the tissue

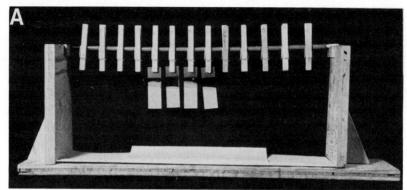

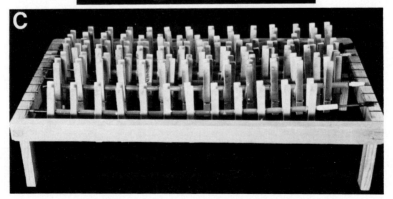

262

interact with silver halide crystals in the emulsion to form an unstable developmental state called "latent images." The "latent images" form silver grains when exposed to developer, thereby creating the final record of the radioactive event. Exposure conditions must be optimal for (a) the creation of latent images caused by radioactive decay in the tissue, (b) the preservation of these latent images so that the silver halide does not revert to the stable state or "fade" before being developed, and (c) the suppression of latent image formation due to factors other than radioactive decay in the tissue, "background." As with most things, what is optimal for one may not be for the other.

Since the work of Kopriwa and Leblond (1962) most exposures have been done at 4°–6°C. Recently, Boren et al. (1975) have reexamined the question of temperature during exposure over the range $-192°$ to 37°C. They find below $-20°C$ that there was a severe loss of emulsion sensitivity and very few grains were produced. Between $-20°$ and 24°C there was an almost linear increase in the number of grains, while above 24°C the number dropped off sharply, presumably due to a combination of high humidity and melting of the emulsion. If Kodak emulsions are exposed under high humidity regardless of temperature, there is severe latent image fade and few grains can be developed; thus the need for fresh desiccant during exposure. Boren et al. agree that 4°C is a good compromise exposure temperature in that the emulsion is reasonably sensitive, and the humidity easily can be kept at low levels by including a desiccant in the slide boxes, eliminating latent image fade. However, if slides are exposed at room temperature in a desiccator or with large amounts of desiccant in the boxes, Boren et al. (1975) claim that exposure at 24°C will produce 50% more grains in the same period of time as compared to exposure at 4°C and with no increase in background. We have repeated this experiment with uniformly ³H-labeled liver-tissue embedded in Epon and found that there was a 19% difference between the

---

FIG. 9. (A) A rack for holding the snap rods which keep the slides in a vertical position while drying. The plywood end posts are 24-cm high and have a notch in the top to hold the rod. Each rod is 52-cm long and is made out of a 7-mm-diameter brass rod. Each rod carries 13 wooden snap clothes pins threaded onto the rod through the central spring mechanism. They are each separated by a 25-mm-long piece of rubber tubing. A piece of rubber tubing is glued to the rod on each end after the last clothes pin to prevent them from slipping off. The slides are dipped into emulsion and then clipped by the nondipped end to the clothes pin, as can be seen in the picture. (B) A drying box for dipped slides constructed from a commercial fiberglass camping ice chest. The slides hang vertically on the snap rods seen in (A), which are resting in grooves in a wooden frame inside the box. There are two layers of slides allowing approximately 200 (25 × 75mm) slides to be dried at the same time. The box has a layer of wet sponges on the bottom to provide humidity during drying. When the top is closed and locked, the box is light tight. This box has the double advantage of protection against accidental light exposure during drying and of providing a compact space to dry a large number of slides in a vertical position without the danger that the emulsion coat will be touched. (C) One level of the double-level wooden frame inside of the drying box. It is made of light wood and has grooves spaced 25 mm apart in the top two sides to hold the rods.

grain counts of tissue exposed at refrigerator (8°C) and at freezer (-15°C) temperatures. This significant difference with no difference in background reinforces the suggestion by Boren et al. (1975) that exposures should be done at the warmest possible temperature.

In addition to heat and humidity, some chemical substances can either produce latent images (positive chemography) or cause fading of formed latent images (negative chemography). When beginning autoradiography or when some major aspect of an established technique is changed, it is wise to include both a light-flashed slide containing a sample section and a coated section of similarly processed but nonradioactive tissue to check for chemography during exposure.

Although it is obvious that the more radioactivity in the tissue the shorter the exposure time, there is no hard and fast rule about the absolute length of exposure when tracing pathways. Generally this is determined by mounting a series of test sections that are taken from areas suspected to contain radioactive terminals or axons, and developing these at regular intervals such as 14, 28, and 56 days after dipping. For instance, our laboratory has found that for an injection of 1 μl containing 10 μCi of tritiated proline into visual cortex, 2 weeks is the shortest exposure time that will adequately reveal sites of terminal labeling in thalamus. The acceptable labeling intensity will also depend on whether the slides are viewed in dark- or brightfield; a labeled terminal field which is easily missed in brightfield will be clearly labeled with the sharper contrast of darkfield (compare A and B in Figs. 2 and 5). It is a good idea to coat several series of slides with exposure times differing by a factor of at least two times, e.g., 2, 4, and 8 weeks, to be sure that sparse as well as dense pathways have been detected.

The most time-consuming step in ARNT is the exposure period, which can be several months long. Although methods have been published which claim to shorten exposure times by incorporating scintillation agents into either the section or the emulsion (the most recent is Durie & Salmon, 1975), these have not proven successful in routine laboratory use for tritium-labeled tissue. Is there any other way that this time can be shortened? The following five modifications to the usual procedures for ARNT as they appear in the literature may be of some help, but no guarantee comes with the suggestions.

1. Inject [14]C- or ^{35}S-labeled amino acid instead of tritium. There will be some loss of resolution compared to tritium, but exposures can be completed in days rather than weeks (Bobillier et al., 1976; Graybiel, 1975; Ogren & Hendrickson, 1976).

2. Use Kodak NTB-3 or Ilford G5 emulsion for tritium. Quantitative studies have shown that NTB-3 gives 15 grains/100 disintegrations compared to 6 grains for NTB-2; G5 gives 12 grains compared to K2 which gives 1 grain/100 disintegrations (Ron & Prescott, 1970). NTB-3 is more highly sensitized and hence may also have a higher background and a shorter shelf life, so it should be tested each time before using.

3. Use an emulsion coat at least three silver grains thick for tritium; a monolayer captures only 60% of the ^{3}H β particles entering it (Rogers, 1973).

4. Develop slightly longer times, for instance, develop 6 min instead of 3 with D19 and 3 min instead of 1.5 with Dektol. This gives both more grains and larger grains (Kopriwa & Leblond, 1962).

5. Expose at the warmest temperature that is practical. This is likely to be in a refrigerator at 4°–8°C, but if possible expose at room temperatures in a desiccator. Exposing at freezer temperatures reduces emulsion sensitivity (Boren *et al.*, 1975).

F. Step 6: Developing, Fixing, and Staining Sections

1. PHOTOGRAPHIC PROCESSING

Slides should be allowed to warm to development temperature in the boxes used for exposure, especially if they have been exposed in a freezer. Slides are developed in fresh developer at 15°–19°C; when developed above 20°C slides have excessive background (Kopriwa and Leblond, 1962). Kodak developers D19 or Dektol (D72) are commonly used for NTB-2 or NTB-3. D19 is used undiluted with a 3-min development. Dektol is diluted 1 part to 2 parts water for a 1.5-min development. Longer development times such as 6 min for D19 and 3 min for Dektol produce significantly larger and slightly more grains with only a very small increase in background. Lengthening development times more than this causes a large increase in background for both developers. Slides should be *gently* agitated once or twice during developing.

Various types of photographic fixer are suitable for autoradiography such as 30% sodium thiosulfate (hypo), undiluted Kodak hypo, and undiluted Kodak Rapid Fix. All of these are used for 2–3 min. A fixer that contains a hardener for the gelatin in the emulsion (Kodak hypo, Rapid Fix with hardener added) will provide greater protection against slippage or scratching of the emulsion but makes no difference in either the size or number of silver grains.

Thorough washing after fixing is absolutely essential. Thiosulfate in fixer becomes adsorbed to the surface of the grains and unless removed will erode them over time and even cause their complete disappearance, especially in acid media (Rogers, 1973, p. 26). Traces of fixer also cause staining irregularities and precipitation.

Since emulsion contains proportionately much more silver than film or paper, solutions of both developer and fixer should always be fresh and should be changed frequently during processing. Distilled water should be used in their preparation. Slides should always be given a thorough rinsing in distilled water after the running water wash. If the tapwater is of poor quality, it may be necessary to use only distilled water for washing; dirty tap water can deposit a great deal of particulate matter on the soft emulsion during washing. The emulsion on the back of the slides can be scraped off more easily if scraping is done after washing but before the slides are dried. Once thoroughly washed, rinsed, and scraped, slides can be dried and stored indefinitely before staining.

2. Staining

A wide variety of stains are available for ARNT and have been discussed in several papers (Hendrickson *et al.*, 1972; Sidman, 1970; Thurston & Joftes, 1963). Staining can be done before coating, after developing, or combined by doing one portion of the stain before coating and another portion after the developing. The most widely used stain for central nervous system (CNS) studies is the basophilic group including cresyl violet, toludine blue, and thionine. These are applied after development and washing.

A second stain which should be more widely used is the myelin stain luxol fast blue. When combined with cresyl violet or thionine, it gives sharp definition of myelinated axon bundles and unmyelinated terminal areas (compare

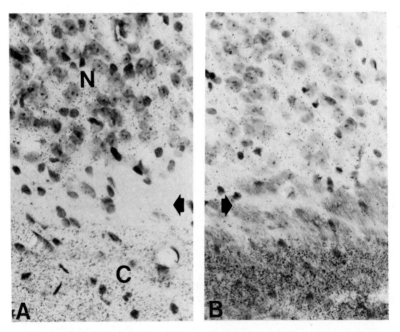

FIG. 10.   A comparison of (A) a cresyl violet and (B) a luxol fast blue–cresyl violet-stained section. The suprachiasmatic nucleus (N) lies just above the optic chiasm (C) and receives a direct retinal projection, as can be seen by the silver grain distribution. In (A) the zone between nucleus and chiasm (arrow) appears empty, while in (B) it is obviously filled with stained but unlabeled axons. Brightfield: magnification: × 225.

Figs. 10A and B). It also has an advantage in that paraffin sections prestained in luxol fast blue before coating consistently show more grains over labeled pathways than adjacent sections coated unstained (Hendrickson *et al.*, 1972). The reason for this is not clear, but it can be useful to reduce exposure time somewhat.

III. Other Uses for Isotopes in Nerve Tracing

A. *Electron Microscope Autoradiography*

One of the advantages of ARNT is that it can be extended to the EM level without significant changes in the techniques (Fig. 11). EM–ARNT has two main advantages over tracing pathways at the EM level using degeneration following a lesion. Since synaptic terminals remain labeled for long periods after an injection

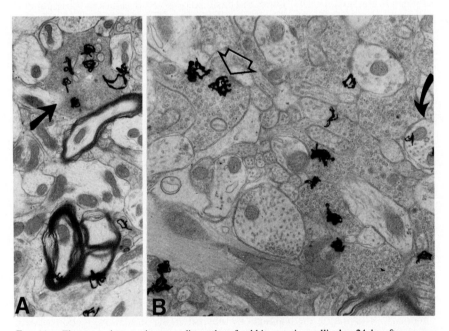

FIG. 11. Electron microscopic autoradiographs of rabbit superior colliculus 24 hr after an eye injection of ³H-lysine. (A) At a level close to stratum opticum the black-twisted silver grains overlie large single synaptic terminals (arrow) and myelinated axons. Magnification: × 10,000. (B) Close to the pial surface grains (open arrow) are almost entirely over clusters of terminals rich in round vesicles and mitochondria or small microtubule-rich pale profiles (arrow) of uncertain origin. Magnification: × 15,000.

(Cuénod *et al.*, 1972; Hendrickson, 1972), survival times have a relatively wide latitude. This is in contrast to degeneration, which may have a very rapid or erratic course, requiring very precise survival times. EM labeling of synapses may also give a more accurate count of the number of terminals originating from a pathway. LeVay and Gilbert (1976) found that 28% of the terminals in layer IV of striate cortex were labeled after ^{3}H-proline injection into dorsal lateral geniculate nucleus, instead of the 5–10% found degenerating after dorsal lateral geniculate lesions. Likewise, Dekker and Kuypers (1976) found that ^{3}H-leucine labeled 14 times the number of terminals originating from mammillary body compared to degeneration after lesions. The second advantage is that normal morphology is retained in EM autoradiography, facilitating identification of a specific type. Unless degenerating terminals are caught at an early stage, they may be very dense or be engulfed by glial cells with loss of their postsynaptic relationships; this makes positive correlations to normal tissue difficult.

The methodology of EM–ARNT has been reviewed several times (Droz, 1975; Fischer & Werner, 1971; Hendrickson, 1975a; Kopriwa, 1973; Rogers, 1971; Salpeter & Bachmann, 1972) with the basic steps given in detail in these papers. Several new procedures have been introduced. Because exposure times for EM are long, the injection amino acid should be as hot as possible, 100–500 μCi/μl, to cut these times down. It is frequently difficult to locate the radioactive pathway in the brain, for instance, a small focus of terminals within the thalamus after a cortical injection. Cutting frozen sections of the tissue has been advocated (Hendrickson, 1975a), but better EM morphology can be obtained by cutting the fixed but unembedded tissue with a Vibratome (T. Pella Co., P.O. Box 510, Tustin, California 92680, USA; also see LeVay & Gilbert, 1976). Alternate 30- and 200-μm sections are cut serially and the 30-μm sections autoradiographed, providing a light microscopic series which identifies the exact focus of activity. The focus in the 200-μm sections is then cut out and processed for EM, followed by EM autoradiography.

Two basic methods are used to coat the sections. The flat substrate method requires putting the thin section on a plastic-coated slide which is then dipped in dilute emulsion. After developing, the emulsion–section–plastic sandwich is floated onto water and picked up onto a grid. This method gives a uniform emulsion coat, and since section thickness can be directly measured in an interferometer, it is the method of choice for rigidly quantitative studies (see Salpeter & Eldefrawi, 1973, for an example of just how much quantitative data can be obtained). The major disdvantage is that often the emulsion–section–plastic layers will not come off of the glass slide, and part or all of an experiment can be lost. Salpeter and Bachmann (1972) mention ways to cope with this problem.

The second EM coating method involves coating the section on the grid. Thin sections are cut as usual and picked up on bare 200 or 300 mesh grids. Two grids are attached at the very edge of the grid to a tiny piece of double-sided Scotch tape which is placed near the end of a glass slide. The grids on the slide are coated with a "soap bubble" of emulsion using a wire loop (Caro & Van Tubergen, 1962; Hendrickson, 1975a). After exposure, the slides are developed and the grids removed for staining. This method has the advantage of a > 95% recovery rate, but the emulsion coat is often not as regular as with the dipping method. This variation introduces some uncertainty about resolution perimeters, which makes the loop method unsatisfactory for rigid quantitative studies, but for pathway identification it is much easier and quite reliable.

Analysis of EM autoradiographs can be straightforward (Hendrickson, 1972) or can be taken to more sophisticated quantitative levels. These methods are beyond the scope of this chapter but are discussed in Salpeter and Bachmann (1972), Salpeter and McHenry (1973), and Parry and Blackett (1976).

B. Combining Autoradiography with the Retrograde Tracer Horseradish Peroxidase or Silver Degeneration Staining

The two tracing methods based on axonal transport can be easily combined by dissolving horseradish peroxidase (HRP) in ^{3}H- or ^{35}S-labeled amino acid, giving twice as much information for each animal. This method of double tracing (Colwell, 1975; Graybiel, 1975; Ogren & Hendrickson, 1976; Trojanowski and Jacobson, 1975) is quite simple in practice with the histochemical processing for HRP done first (see Chapter 13 by LaVail in this volume) followed by autoradiography. Usually a separate series of sections is done for each technique, but if desired, the same section can be processed for both techniques.

Because of the enzymatic nature of the HRP tracer, tissue fixation following the initial perfusion is quite brief and frozen sections must be cut. We have found that if sections are mounted for the autoradiography series immediately after they are cut, we often get poor autoradiographs with high background. This seems to be prevented by letting the frozen sections sit for at least a week in fixative before doing autoradiography. Except for taking this precaution, we use the standard ARNT techniques given in this chapter.

When both techniques are to be done on the same section (Fig. 12), the HRP processing is done first on the loose section and then the section is mounted, defatted, and coated for autoradiography. Diaminobenzidine, used as one capture agent for HRP (Bunt et al., 1975), will chemically reduce the emulsion, causing tremendous background if it is not completely removed from the section. We have not used benzidine. Extensive washing is necessary to remove the unreacted diaminobenzidine; we wash loose 50-μm sections in four changes of distilled water or buffer for 1 hour each, then leave them overnight in fresh distilled water or buffer before mounting. If high background is still a problem, it can be

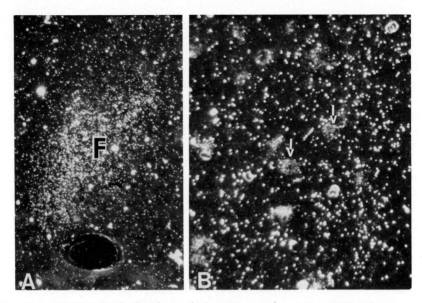

FIG. 12.   (A) A focus (F) of radioactive terminals and horseradish peroxidase (HRP)-labeled neurons in the monkey inferior pulvinar after an injection of both tracers into the striate cortex. The section was first processed for HRP, well washed, and then coated for autoradiography. Darkfield: magnification: × 48. (B) Higher power showing the labeled neurons containing fine HRP granules (arrow) surrounded by silver grains, indicating the pulvinar-striate pathway is topographic and reciprocal. Magnification: × 300. (Modified from Ogren & Hendrickson, 1976, and reproduced by permission of the publisher.)

completely prevented by dipping the slide carrying HRP-reacted, mounted, defatted, dry sections into a 1% parloidin in absolute alcohol–ether (1:1) solution which covers the section with a thin impermeable film. This film will slightly reduce the autoradiographic reaction if ^{3}H-labeled amino acid is used, so longer exposure times should be allowed.

Autoradiography can be combined with the classical Nauta or Fink–Heimer silver staining for degeneration following lesions (Graybiel, 1975). It is most useful when tracing pathways in an experimental situation involving sequential lesions. In the past, when using only silver-stained degeneration, it was necessary to wait for some months between the first and second lesion to be sure that all of the degeneration from the first lesion had disappeared. If the second lesion is replaced by autoradiographic labeling of the pathway under study, this long wait can be eliminated, and no confusion will result. An example of an effective application of this double-label method can be found in Lund and Lund (1976) where it was used to study sprouting of visual pathways during development. The two techniques can be combined in one section with Fink–Heimer done first followed by autoradiography, but results are cleaner when two separate parallel series are done.

C. Other Tracing Methods Involving Isotopes

A radically different tracing method has recently been proposed by Sokoloff and co-workers (reviewed in Kennedym, Des Rosiers, Sakaruda, Shinohara, Reivich, Jehle, & Sokoloff, 1976; Plum, Gjeddes, & Samson, 1976). This involves the autoradiographic visualization of the neurons in the brain that are involved when a pathway or system is functionally active. Labeling is accomplished by the differential uptake of the metabolite ^{14}C-labeled deoxy-D-glucose from the blood into the active neurons. The more metabolically active the neurons, the heavier the uptake. Since deoxyglucose can enter the glucose pathway but cannot be broken down after it is phosphorylated, the resulting ^{14}C-labeled deoxyglucose 6-phosphate accumulates within the cell at a rate proportional to the cell's metabolic activity. In the most recent paper (Kennedy et al., 1976), when one eye of a Macaque monkey is stimulated after the animal has been given a pulse of ^{14}C-labeled 2-deoxy-D-glucose, the striate cortex shows a specific repetitive pattern of label extending from layers I to VI but alternating in heavily and lightly labeled columns about 0.3–0.4 mm wide (Fig. 13). These columns resemble the Macaque eye dominance columns described by Wiesel et al. (1974) using transsynaptic labeling, except that instead of being confined to the thalamic terminals in layer IV, they extend throughout the cortex. This method apparently demonstrates a functional multisynaptic pathway, unlike ARNT, which is confined to the first, and in a few cases the second, synapse formed by the labeled neurons. Unexpected relationships between portions of a system may result from this method; for instance, when one eye is stimulated and the other occluded, the monkey occulomotor nucleus shows reduced activity on the side contralateral to the occluded eye even though both eyes were freely moving. In the olfactory system Sharp, Krauer, and Shepherd (1975) have shown that when a rat smells amyl acetate during ^{14}C-labeled deoxyglucose exposure, two separate bilateral regions in the olfactory bulbs are preferentially labeled. This is in good agreement with previous studies (Pinching & Doving, 1974) using specific odor-induced cell atrophy in the bulb. Even though the exact metabolic rationale for deoxyglucose localization is still not understood, and there are technical problems in soluble substance autoradiography still to be solved, these promising results suggest that the deoxyglucose functional tracing method will be an important addition to both neuroanatomical and neurophysiological methodology.

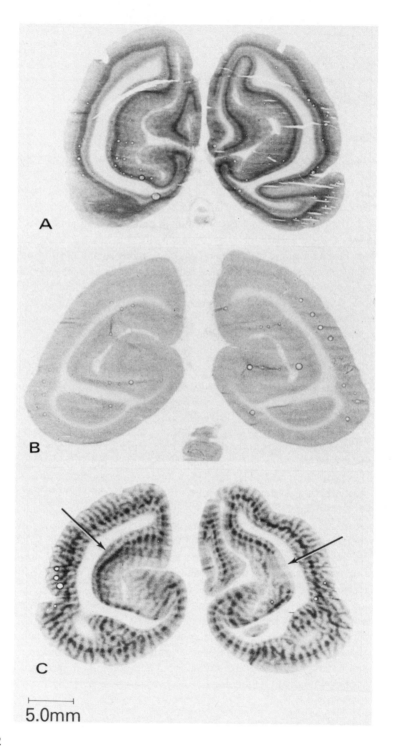

A

B

C

5.0mm

IV. Appearance and Analysis of Data

A. Relating Silver Grains to Their Source

It must again be emphasized that autoradiography is a second-order technique; that is, although the primary radioactive decay takes place in the tissue, it is recorded as a silver grain in the overlying layer of emulsion. A silver grain will look the same whether it is over a cell body, a blood vessel, or an empty slide so that it carries no inherent clues as to where the decay which caused the formation of this grain was localized. Localization is based on two factors—the energy of the isotope and the ability to recognize components in the section.

The process of accurately relating radioactive decay and resulting silver grain involves the limiting factor "resolution." Resolution in autoradiography is the distance between decay and grain and is determined mainly by energy of the β particle (Fig. 3). Low-energy β particles like ^3H do not travel very far from their decay source and so the distance between the two is small, giving "good resolution." Almost all of the grains formed from a ^3H decay will be contained within a circle of 1 μm radius centered over the decay. In the more usual situation resolution is determined for the silver grain, so that the origin will be from the underlying tissue contained in a circle of 1 μm radius centered on the grain. While this is the upper limit of resolution for ^3H regardless of section or emulsion thickness, light microscopic resolution of 0.3 μm can be achieved with very thin sections and a monolayer of emulsion (Salpeter et al., 1974). The good resolution obtained with ^3H makes it possible to reliably relate a silver grain to structures within the neuron or neuropil if those structures can be recognized in the light microscope. Owing to the limitations of light microscopic resolution and to the lack of appropriate staining techniques, under ordinary conditions many components of the CNS cannot be readily identified. While neuronal and glial cell bodies and nuclei are obvious in most methods, neuronal dendrites are more difficult to resolve, and unmyelinated axons, preterminal axons and synaptic endings usually cannot be seen. If appropriate stains are used, myelinated axons are apparent (Fig. 10).

With ^{35}S or ^{14}C the origin of a given grain may be more difficult to determine since in sections 10–50 μm thick, these high-energy isotopes may have a circle of resolution of 10–30 μm radius. This virtually precludes localization within a neuron but allows accurate localization to layers or regions within a structure.

FIG. 13.   Sections of striate cortex of monkeys which were exposed to different types of visual stimulation for 45 min after an intravascular injection of ^{14}C-labeled-2-deoxyglucose. (A) both eyes stimulated; (B) no visual stimulation; and (C) one eye stimulated. Notice in (A) that layer IV is uniformly labeled, while in (C) the "eye dominance columns" of the stimulated eye appear in all layers, but are most prominent in layer IV. Magnification: × 4. (Reproduced by permission of the authors from Kennedy et al., 1976.)

For all isotopes, it is the *pattern* and *number* of silver grains combined with the organization of the tissue within the section that provides the final evidence in ARNT.

B. *Appearance of Labeled Pathways*

A few hours after the radioactive amino acid injection, the entire neuron is labeled by transported and structural nontransported protein, but the relative distribution of these proteins will vary with survival time after injection. Short survivals which are long enough to allow the most rapid phases of transport to reach the synaptic terminals, but which are far too short to allow the slower phases to have moved very far from the neuron cell body will preferentially label the synaptic endings. Since the synaptic terminals have high specific activity compared to their axons, if there is a reasonably dense concentration of synaptic terminals in one area, the high density of terminal label will stand out in sharp contrast to surrounding tissue. The grains will be confined to a specific layer or region of a neuronal structure, or will be found in dense clusters over one area of neuropil (Figs. 1A,2). However, even at these short survival times there is some

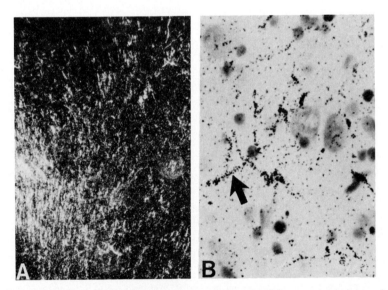

FIG. 14.   (A) Bundles of labeled retina axons coursing through pulvinar-suprageniculate nuclei on their way to pretectum and superior colliculus. The region in the upper right corner is seen in (B). Darkfield; magnification: × 100. (B) A heavily labeled axon (arrow) surrounded by scattered grains, some of which appear in clusters. These could be labeled axons cut in cross section and/or synaptic terminals. Brightfield; magnification: × 455.

light label over axon pathways (Fig. 2), although the ratio of grains is so much in favor of synaptic terminals that there usually is no confusion between the two. A more difficult situation arises when there is only a sparse input to a region so that the number and concentration of synaptic terminals is low. In this case very high levels of injected radioactivity combined with long exposure times may be needed to accumulate enough grains over the terminals to give an unequivocal result. Alternatively, since longer survival times allow more material to accumulate within axons, but do not reduce synaptic labeling very much (Figs. 1B and 5A, B), it may be better to utilize long survival times for sparse projections (Edwards, 1975). Perhaps the most difficult situation for identification of possible synaptic terminals is when a large number of labeled fiber bundles is coursing through a region containing neurons which might also receive synaptic input via collaterals of these bundles (Fig. 14). Resolution of such a case may require myelin stains and judicious juggling of survival time and amino acid, but failing in this, such a problem can be unequivocally resolved by EM autoradiography.

C. Factors in Deciding Whether a Pathway Is Labeled

What constitutes an acceptable level of labeling? Silver grains can be formed from a variety of sources. The most intense source of β particles should be the radioactive molecules within the pathway originating from the cell bodies which were injected; this forms the signal (Fig. 15A). The noise in the system usually comes from two sources. The first source is technical background grains which form in the emulsion due to stress, agitation, aging, or stray radiation (Fig. 15C). This background should be at low levels and should be the same in all areas of the emulsion whether it is on the slide or over nonradioactive tissue. If the number of grains over both slide and tissue is high, there has been some technical failure (light leak, overheating of emulsion, high developer temperature, etc.) which can be identified and eliminated the next time. The second noise source is metabolic background grains over the tissue (Fig. 15B). These are caused by generalized radioactive labeling of neurons throughout the brain by isotope from the injection site conveyed to them in blood or cerebrospinal fluid. This label will be heavier near an injection site and will increase the larger the region, the higher the concentration, and the more energetic the isotope injected. It may also vary depending on whether neuron cell bodies, fiber bundles, and neuropile are examined.

There are two methods of deciding whether a region is or is not labeled. The first is qualitative in that the observer takes into account the total background of the section (technical plus metabolic) and then decides whether there are enough additional grains in a specific region over this background to justify calling that region "labeled." The factors that enter into this judgment are grain pattern, repetition, and reasonableness. A labeled terminal field will usually be marked

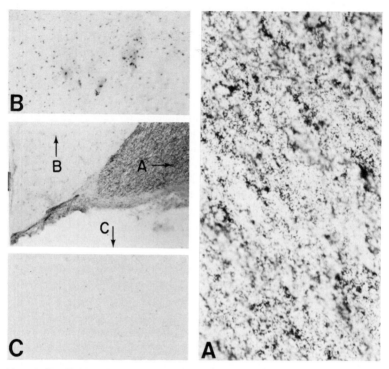

FIG. 15.   (A) Heavily labeled optic tract seen after eye injection; this is the "signal." (B) Grains over neighboring, unrelated tissue; these make up "metabolic background." (C) Grains over the slide outside of the section; this is "technical background." Brightfield; magnification: × 520.

by concentrated radioactivity which is related to some underlying structure such as a specific layer of the cortex or a definite region within a nucleus. In most cases terminal fields are not diffusely and evenly spread throughout an entire region; almost always there will be some type of focal concentration of grains. If labeled axons are seen running through a region, with a few scattered grains surrounding them, although the grains *may* be over a few scattered terminals, it is more likely these are axons *en passage* (Fig. 14). The same result should be found in every experiment and usually in several slides for each experiment. A single slide with a "labeled focus" could be the result of some very localized contamination of the emulsion or some peculiar local stress factor. The final judgment factor is reasonableness, that is, is this result likely, or is it just too wild to be believed? As more and more heretofore unsuspected connections are documented, this factor will always have to be flexible, but a totally unexpected "labeled" region should always be given a skeptical reception until it has been verified by several experiments.

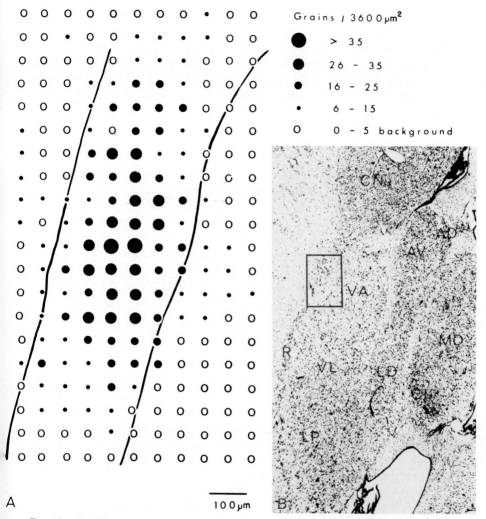

FIG. 16. Graphic representation of the grain counts found within the outlined square over the reticular nucleus of the thalamus. (Reproduced by permission from Jones, 1975, Fig. 11.)

The second method is a quantitative one using actual grain counts. This involves counting at 1000× magnification the number of grains in a unit area throughout the suspected labeled region and then counting the grains in the same unit area over tissue which is outside the suspected labeled region. This background count is then subtracted from the count over the labeled region. Generally a level of at least three times background is needed to demonstrate labeling (for examples, see Jones, 1975, or Price & Wann, 1975). Grain counting inyolves considerable time and is tedious work. It is debatalable whether it provides any more convincing evidence for labeling than the qualitative "considered judgement" method to an individual observer, but for purposes of graphic presentation, grain count distributions present an excellent overview of the extent and intensity of label (Fig. 16). In cases of diffuse light label this may be the only way to convincingly demonstrate input. Wider availability of automatic grain counting devices (Goldstein & Williams, 1974; Price & Wann, 1975; Rogers, 1972; Wann, Price, Cowan, & Agulnek, 1974) should make the quantitative method more feasible as a routine procedure.

V. Limitations and Artifacts of ARNT

A. Technical Artifacts of Autoradiography

The aggravating artifacts listed below have already been mentioned in Sections II, D, E, and IV, A, but are discussed together here, in detail, because they all will occur at some time or the other when using ARNT and must be recognized for what they are and how they can be prevented.

1. HIGH GENERAL BACKGROUND

Many silver grains evenly distributed over both section and slide is probably the most common artifact and is the most easily recognized. The causes are diverse: old emulsion, a light leak in the darkroom safelight, overheating of emulsion during melting, drying the dipped slide at too low humidity or too high temperature, or sheer perversity of the Fates. Ordinarily the series is redone after the fault is corrected; if, however, there are no more sections left—and this capricious artifact is the reason that you never coat every single section at one time—Rogers (1973, p. 29) gives a possible method of removing the bad emulsion so that the slides can be recoated.

2. SPECIFIC LOCALIZED GRAINS UNRELATED TO THE LABELED PATHWAY

These more subtle artifacts fall into two categories. The first consists of a very localized pattern of silver grains distributed along the edges of structures within the section such as the lining of blood vessels, the pial surface, or along prominent fiber tracts (Fig. 17A). These edge artifacts are common in thick frozen sections but are not seen in Epon sections. They seem to be caused by stress within the emulsion during drying as it passes over elevated abrupt tissue edges. It must be emphasized that paraffin and frozen sections are not uniformly flat, but in side view resemble alternating valleys and mountain ranges, especially at the interface of large cells to neuropil, and blood vessels to lumen. Plastic sections from which the plastic is not removed are smooth surfaced.

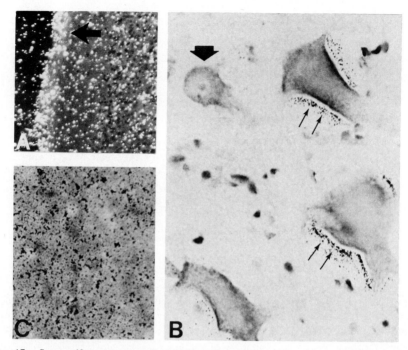

FIG. 17.   Some artifacts encountered in ARNT. (A) A row of artifactual silver grains along the pial edge of a section. Darkfield; magnification: × 360. (B) Artifactual grains of uncertain origin (small arrows) which lie along the edges of the large but not the small (large arrow) neurons in red nucleus. Magnification: × 780. (C) Stain deposit on a cresyl violet-stained section which simulates silver grains in density and size. Brightfield; magnification: × 900.

The second category is localized deposits of silver grains related to a specific region or type of cell. This appears to be caused by chemical reduction of an emulsion by a substance within the section (positive chemography). This is more commonly seen in the unfixed frozen sections which are used for histochemistry, but has been described in ARNT material (Daniels & Gilmore, 1975; Hackman & Vapaatalo, 1972; Hendrickson *et al.*, 1972). Lines of silver grains found along the periphery of the large but not the small neurons of the red nucleus (Fig. 17B) could be a form of both artifact categories. The specificity of such local grain deposits can be tested by exposing an identically processed but nonradioactive section.

3. LATENT IMAGE FADE

This artifact results in the disappearance of the formed latent image during exposure so that few or no silver grains appear over radioactive regions (negative chemography). It can be caused by a chemical in the section suppressing latent image formation or by high humidity during exposure. This artifact can be checked by exposing a coated radioactive section that has been light-flashed so that the maximum number of latent images are present from the beginning. If there is a general or localized reduction of grains over such slides, negative chemography is taking place. If ARNT experiments routinely produce lightly labeled slides even though reasonable amounts of radioactivity are being injected, this artifact could be the answer.

4. UNEVEN EMULSION COAT

If the Kodak emulsions are diluted 1:1, certain regions may be covered by less than a monolayer. Paraffin-embedded 1:1 coated eyes can have a very thin emulsion coat over the cell body layers of retina and a much thicker coat over the synaptic layers (Fig. 7B). Grain counts over such regions give quite erroneous results when compared to counts from a 2:1 coated section.

5. STAIN ARTIFACTS

The first type involves some interaction of the stain with the emulsion to either cause or suppress grain formation. It is very important to use stains in their proper sequence for autoradiography (Hendrickson et al., 1972; Thurston & Joftes, 1963). Since most all of the basophilic stains cause chemical reduction of emulsion, it is important to use clean alcohols when defatting sections before coating so that sections are not contaminated by stains in these solutions. The second artifact type is precipitation of the stain itself, causing deposits which resemble silver grains (Fig. 7C). This seems to be especially bad with cresyl violet which can form on an emulsion a red-brown particle which is close to a silver grain in size and density. Thorough washing to remove all traces of fixer is *absolutely* essential. A preliminary rinse before staining in buffer of the same pH as the stain and use of freshly filtered staining solutions will also help prevent stain artifacts.

B. Limitations to Obtaining Definitive Results Using ARNT

This section will list all of the possible complications to clear interpretation of ARNT results, and some suggestions for their resolution.

1. THE SELECTIVE LABELING OR LACK OF LABELING BY A SPECIFIC CELL TYPE FOR A GIVEN AMINO ACID

Amino acid discrimination does occur in some CNS neurons and has already been discussed in Section II,A,2. Mixtures of several common amino acids should be used for at least some of the injections to eliminate this effect. If quantitative comparisons are being made about numbers of synaptic terminals in different synaptic sites, the possibility of unsymmetrical labeling from a mixed neuronal population should always be considered (Berkley, 1975).

2. THE UPTAKE AND TRANSPORT OF RADIOACTIVE TRACERS BY AXONS WITHIN OR NEAR THE INJECTION SITE

This problem appears to have been ruled out for the amino acids proline, leucine, and methionine and for both myelinated and unmyelinated axons (Cowan *et al.*, 1972; Fink, Kennedy, Hendrickson, & Middaugh, 1972; Graybiel, 1975), but other amino acids have not been well studied. Fucose probably is transported by axons (Graybiel, 1975) while nucleotides (Kreutzberg & Schubert, 1975) have not been tested.

3. RETROGRADE TRANSPORT OF RADIOACTIVE TRACERS

Amino acids are not transported from synapse to cell body, at least in levels detectable using ARNT (Cowan *et al.*, 1972). Nucleotides do move retrogradely (Hunt & Kunzle, 1976; Wise & Jones, 1976) while other tracers have not been well studied.

4. TRANSSYNAPTIC LABELING

Movement of material from the labeled synapse to the postsynaptic neuron with subsequent labeling of its synapse (Fig. 4) has been well studied in the visual system (Grafstein & Laureno, 1973; Specht & Grafstein, 1973) and deliberately used as a valuable tracing method (Drager, 1974; Kaas *et al.*, 1976, Wiesel *et al.*, 1974). Transsynaptic or second-order synaptic labeling appears to be maximized by (a) injections of very large amounts of label, (b) the use of proline and fucose in contrast to leucine, (c) long survival times, (d) a short axon on the postsynaptic neuron, and (e) long exposure. Generally, short survivals of 12–24 hours appear to be less likely to show this problem, but in both mouse olfactory system (Barber & Field, 1975; Barber & Raisman, 1974) and goldfish tectum (Neale *et al.*, 1974) sacrificed 6–12 hours after injections, movement of label into deeper layers was seen that suggested transsynaptic label. This deeper labeling increased with longer survivals. In the case of the mouse olfactory bulb, Parry and Blackett (1976) have shown using EM autoradiography that these deeper grains are mainly over the postsynaptic mitral cell cytoplasm and are not over synapses, so the appearance of grains over a nearby layer or region does not necessarily mean that the grains are over labeled synapses. In all of the identified cases of transsynaptic transport studied thus far, there is a tremendous difference in labeling intensity between the first- and second-order synapses; for instance, in Fig. 4 the first-order retinal synapse in superior colliculus was clearly seen after 1 week of exposure, while the second-order transsynaptic parabigeminal nucleus label required 6–8 weeks of exposure to appear. The combination of short survival time and short exposure time may be the best way to minimize unwanted transsynaptic labeling.

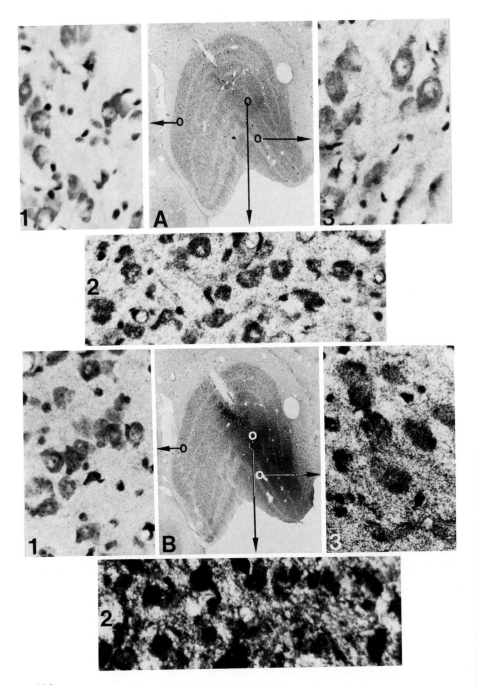

282

5. EXACT LOCALIZATION OF THE INJECTION SITE

This is one of the most difficult problems of ARNT. The apparent size of the injection site depends on the concentration of the isotope, the volume injected, the rate of injection, the survival time after injection, and the length of exposure. The smallest sites in total extent are produced by small volumes of concentrated isotope injected very slowly (Schubert & Hollander, 1975). The apparent size of any injection volume is reduced when the survival time is prolonged, or when the sections are exposed for very short times (Fig. 18). The problem in knowing exactly what neurons have transported protein and those that have not derives from not knowing exactly how radioactive a cell must be in order for it to have radioactive terminals. Obviously, since axons do not transport amino acids, surrounding fiber pathways can be ignored, but what is the likelihood that neurons in adjacent layers or nuclei contribute to the labeled pathway?

Each injection site contains two areas. The central region has neuron cell bodies that are completely covered with silver grains even at short exposure times [Fig. 18 (A2)]. It seems reasonable that these neurons received the most amino acid and therefore would be most likely to have labeled projection pathways. Around these hot neurons in sections with short exposure times is a zone of more lightly labeled neurons at or just over the background levels of the surrounding neuropil [Fig. 18 (A3)]. If the exposure time is lengthened from 7 days to 28 days, cells in this zone can change from lightly labeled (A3) to heavily labeled (B3), so probably their own axon pathways would also be seen with a similar 4 × increase in exposure. When neurons on the edge of a focus are still at or barely above surrounding neuropile labeling after a long exposure [Fig. 18 (A1), (B1)], it is unlikely that these cells will have labeled pathways. One exception might be if a particular type of neuron avidly utilizes the injected amino acid for transport so that most of the labeled protein has already left the cell body by the time of sacrifice. Then even a relatively lightly labeled cell in or near the focus might have quite heavily labeled pathways, but as yet there is no evidence for such neurons.

The size and shape of the injection site is also determined by the region injected. In addition to not transporting amino acid, fiber pathways tend to

---

FIG. 18.   The effect of exposure time on the apparent size of the injection focus. (A) ³H-proline injection (0.8 μl containing 8 μCi) in dorsal lateral geniculate nucleus of monkey sacrificed 24 hr after injection; the section was exposed for 7 days. Magnification: × 20. (1) Unlabeled region of nucleus, (2) the heavily labeled neurons in the focus, and (3) lightly labeled neurons some distance from the center of the focus. Magnification: × 320. (B) An adjacent section from the same monkey, but exposed for 28 days. Magnification: × 20. (1) The neurons are now very lightly labeled in this previously uninvolved region, (2) the center of the focus with intensely labeled neurons, and (3) the previously lightly labeled neurons in A3 are now heavily labeled and would probably be considered a part of the focus if this section were the only one examined. Magnification: × 320.

restrict amino acid diffusion. For instance, injections into gray matter of cortex tend to run parallel to the white matter rather than cross it. If a nucleus is limited by white matter, a larger injection volume can be used with less chance that adjacent, unrelated nuclei will be involved.

6. AN ENTIRE PATHWAY CANNOT BE LABELED IN A SINGLE
    EXPERIMENT

With the exception of the retinal ganglion cell, only a portion of the total projection pathway can usually be labeled by one or more injections of ^{3}H-labeled amino acid. On the other hand, this localized characteristic of injection sites does facilitate accurate topographic mapping of connections.

7. SILVER GRAINS CONTAIN NO INHERENT INFORMATION
    ABOUT THEIR ORIGIN

ARNT is a second-order technique with the radioactive event occurring in the tissue while the record is preserved in the emulsion as the silver grain. Silver grains all look alike, so it is necessary to maximize recognition of structures within the section to facilitate their probable origin (see discussion in Section IV,A regarding resolution). In addition to basophilic stains, myelin stains (Fig. 10) help detect labeled bundles of axons and are simple to do routinely (Hendrickson *et al.*, 1972). Myelin staining may help resolve questions arising from grain patterns in labeled regions such as seen in Figs. 5 and 14. As is often the case in degeneration techniques as well, it may be necessary to go to EM for a definitive answer as to exactly what the silver grains overlie.

8. THE TECHNIQUE TAKES A LONG TIME TO OBTAIN
    RESULTS

In a typical experiment using rapid transport localization, the animal is injected, sacrificed 24 hr later, the brain is fixed overnight, sunk in sucrose for 24 hr and frozen sectioned, giving a processing time of 3–4 days. The sections are mounted and dried overnight, then defatted and coated which takes 2–3 days longer, so the total minimum processing time from injection to slide coating is 5–7 days. If slow transport localization is needed, this may add 7–21 days to the survival time, and if the brain is embedded in paraffin it may take an additional 7–10 days. This means that even using the longest procedures that the slides will be coated 5 weeks after injection. These times are not too different from classic silver degeneration staining methods.

ARNT takes an additional increment of time for exposure of the coated slides. This can be relatively short if ^{35}S-methionine is used (Graybiel, 1975; Ogren & Hendrickson, 1976), but even with concentrated isotopes it takes at least 2 and often 4 weeks to adequately demonstrate labeled projections. In Section II,E, some technical changes are suggested that might help reduce exposure times somewhat.

Even with this additional time exposure, total processing for ARNT from start to finish can be done in 4–6 weeks. It is always a good idea to run a test series through the suspected injection site and expose it for a week, by which time the center of the focus will be apparent. If the injection was in the wrong place, only 2 weeks will be lost rather than the 4–6 weeks needed to demonstrate pathways.

Acknowledgments

This work was supported by National Institutes of Health Research Grant Nos. EY01208 and NS11254, Research Career Development Award No. EY39039, and by a grant from the National Society for the Prevention of Blindness, Inc. Some facilities were supported in part by National Institutes of Health Grant Nos. RR00166 and HD02274. Discussions with Ann M. Graybiel, Diane Majors, and W. M. Cowan and the skilled technical assistance of Ms. Ludelle Moe and Bente Noble contributed immeasurably to the successful development of these techniques.

References

Barber, P.C., & Field, P.M. Autoradiographic demonstration of afferent connections of the accessory olfactory bulb in the mouse. *Brain Research,* 1975, **85,** 201–203.

Barber, P. C., & Raisman, G. An autoradiographic investigation of the projection of the vomeronasal organ to the accessory olfactory bulb in the mouse. *Brain Research,* 1974, **81,** 21–30.

Barondes, S. H., & Samson, F. E., Jr. Axoplasmic transport. *Neurosciences Research Programs Bulletin,* 1969, **3,** 191–300.

Beitz, A.J., & King, G.W., An improved technique for the microinjection of horseradish peroxidase. *Brain Research,* 1976, **108,** 175–179.

Berkley, K. J. Different targets of different neurons in nucleus gracilis of cat. *Journal of Comparative Neurology,* 1975, **163,** 285–304.

Bobillier, P., Séguin, S., Petitjean, F., Touret, M., & Jouvet, M. The raphe nuclei of the cat brain stem: A topographical atlas of their efferent projections as revealed by autoradiography. *Brain Research,* 1976, **113,** 449–486.

Boren, H. G., Wright, E. C., & Harris, C. Quantitative light microscopic autoradiography emulsion sensitivity and latent image fading. *Journal of Histochemistry and Cytochemistry,* 1975, **23,** 901–909.

Bunt, A., Hendrickson, A., Lund, J. S., Lund, R. D., & Fuchs, A. Monkey retinal ganglion cells: Morphometric analysis and tracing of axonal projections, with a consideration of the peroxidase technique. *Journal of Comparative Neurology,* 1975, **164,** 265–286.

Caro, L. G., & Van Tubergen, R. P. High resolution autoradiography. I. Methods. *Journal of Cell Biology,* 1962, **15,** 173–188.

Caviness, V. S., Jr., & Barkley, D. S. Section thickness and grain count variations in tritium autoradiography. *Stain Technology,* 1971, **46,** 131–135.

Colwell, S. A. Thalamocortical-corticothalamic reciprocity: A combined anterograde-retrograde tracer technique. *Brain Research,* 1975, **92,** 443–449.

Cowan, W. M., Gottlieb, D. I., Hendrickson, A., Price, J. L., & Woolsey, T. A. The autoradiographic demonstration of axonal connections in the central nervous system. *Brain Research,* 1972, **37,** 21–51.

Crossland, W. J., Cowan, W. M., & Kelly, J. P. Observations on the transport of radioactively labeled proteins in the visual system of the chick. *Brain Research,* 1973, **56,** 77–105.

Cuenod, M., Boesch, J., Marko, P., Perisič, M., Sandri, C., & Schönbach, J. Contributions of axoplasmic transport to synaptic structure and functions. *International Journal of Neuroscience,* 1972, **4,** 77–87.

Daniels, J. S., & Gilmore, S. A. Chemography associated with specific anatomic areas in autoradiographs of brain stems from adult rats. *Brain Research,* 1975, **98,** 343–347.

Dekker, J. J., & Kuypers, H. G. J. M. Quantitative E. M. Study of projection terminals in the rat's AV thalamic nucleus. Autoradiographic and degeneration techniques compared. *Brain Research,* 1976, **117,** 399–422.

Dingman, W., & Sporn, M. B. The penetration of proline and proline derivatives into the brain. *Journal of Neurochemistry,* 1959, **4,** 148–153.

Drager, U. C. Autoradiography of tritiated proline and fucose transported transneuronally from the eye to the visual cortex in pigmented and albino mice. *Brain Research,* 1974, **82,** 284–292.

Droz, B. Fate of newly synthesized proteins in neurons. In C. P. Leblond & K. Warren (Eds.), *The use of radioautography in investigating protein synthesis.* New York: Academic Press, 1965. Pp. 159–175.

Droz, B. Autoradiography as a tool for visualizing neurons and neuronal processes. In W. M. Cowan & M. Cuénod (Eds.), *The use of axonal transport for studies of neuronal connectivity.* Amsterdam: Elsevier, 1975. Pp. 127–154.

Droz, B., Warshawsky, B., & Warshawsky, H. Reliability of the radioautographic technique for the detection of newly synthesized protein. *Journal of Histochemistry and Cytochemistry,* 1963, **11,** 426–435.

Durie, B. G., & Salmon, S. E. High speed scintillation autoradiography. *Science,* 1975, **190,** 1093–1095.

Ebbeson, S. O. E. The selective silver-impregnation of degenerating axons and their synaptic endings in non-mammalian species. In W. J. H. Nauta & S. O. E. Ebbeson (Eds.), *Contemporary research methods in neuroanatomy.* Berlin & New York: Springer-Verlag, 1970. Pp. 132–161.

Edwards, S. B. The ascending and descending projections of the red nucleus in the cat: An experimental study using an autoradiographic tracing method. *Brain Research,* 1972, **48,** 45–63.

Edwards, S. B. Autoradiographic studies of the projections of the midbrain reticular formation: Descending projections of nucleus cunniformis. *Journal of Comparative Neurology,* 1975, **161,** 341–358.

Edwards, S. B. The commissural projection of the superior colliculus in the cat. *Journal of Comparative Neurology,* 1977, **173,** 23–40.

Edwards, S. B., & Shalna, E. J. Microinjector for use in the autoradiographic neuroanatomical tracing method. *Pharmacology, Biochemistry, and Behavior,* 1974, **2,** 111–113.

Elam, J. S., & Agranoff, B. W. Rapid transport of protein in the optic system of the goldfish. *Journal of Neurochemistry,* 1971, **18,** 375–387.

Fink, R. B., Kennedy, R. D., Hendrickson, A., & Middaugh, M. E. Lidocaine inhibition of rapid axonal transport. *Anaesthesia,* 1972, **36,** 422–432.

Fischer, H. A., & Werner, G. A contribution to electron microscopic autoradiographic technique. *Histochemie,* 1971, **29,** 44–53.

Goldstein, D. J., & Williams, M. S. Quantitative assessment by photometric reflectance microscopy. An improved method using polarized light. *Histochemical Journal,* 1974, **6,** 223–230.

Grafstein, B. Transport of protein by goldfish optic nerve fibers. *Science,* 1967, **157,** 196–198.

Grafstein, B. Axonal transport: Communication between soma and synapse. *Advances in Biochemical Psychopharmacology,* 1969, **1**, 11–25.

Grafstein, B. Principles of anterograde axonal transport in relation to studies of neuronal connectivity. In W. M. Cowan & M. Cuénod (Eds.), *The use of axonal transport for studies in neuronal connectivity.* Amsterdam: Elsevier, 1975. Pp. 47–68.

Grafstein, B., & Laureno, R. Transport of radioactivity from eye to visual cortex in the mouse. *Experimental Neurology,* 1973, **39**, 44–57.

Graybiel, A. M. Wallerian degeneration and anterograde tracer methods. In W. M. Cowan & M. Cuénod (Eds.), *The use of axonal transport for studies of neuronal connectivity.* Amsterdam: Elsevier, 1975. Pp. 173–216.

Graybiel, A. M., & Devor, M. A microelectrophoretic delivery technique for use with horseradish peroxidase. *Brain Research,* 1974, **68**, 167–173.

Hackman, C. R., & Vapaatalo, H. Reproducible chemography in autoradiographs of the rat brain. *Experientia,* 1972, **28**, 492–493.

Hendrickson, A. Electron microscopic distribution of axoplasmic transport. *Journal of Comparative Neurology,* 1972, **144**, 381–397.

Hendrickson, A. Technical modifications to facilitate tracing synapses by electron microscopic autoradiography. *Brain Research,* 1975, **85**, 241–247. (a)

Hendrickson, A. Tracing neuronal connections with radioisotopes applied extracellularly. *Federation Proceedings, Federation of American Societies for Experimental Biology,* 1975, **34**, 1612–1615. (b)

Hendrickson, A., & Cowan, W. M. Changes in the rate of axonal transport during post-natal development of the rabbit's optic nerve and tract. *Experimental Neurology,* 1971, **30**, 403–422.

Hendrickson, A., Moe, L., & Nobel, B. Staining for autoradiography of the central nervous system. *Stain Technology,* 1972, **47**, 283–290.

Hökfelt, T., & Ljungdahl, A. Uptake mechanisms as a basis for the histochemical identification and tracing of transmitter specific neuron populations. In W. M. Cowan & M. Cuénod (Eds.), *The use of axonal transport for studies of neuronal connectivity.* Amsterdam: Elsevier, 1975. Pp. 249–306.

Hunt, S. P., & Kunzle, H. Bidirectional movement of label and transneuronal transport phenomena after injection of [H^3] adenosine into the central nervous system. *Brain Research,* 1976, **112**, 127–132.

Jones, E.G. Some aspects of the organization of the thalamic reticular complex. *Journal of Comparative Neurology,* 1975, **162**, 285–308.

Kaas, J. H., Lin, C. S., & Casagrande, V. A. The relay of ipsilateral and contralateral retinal input from the lateral geniculate nucleus to striate cortex in the owl monkey: A transneuronal transport study. *Brain Research,* 1976, **105**, 371–378.

Karlsson, J. O., & Sjöstrand, J. Synthesis, migration and turnover of protein in retinal ganglion cells. *Journal of Neurochemistry.* 1971, **18**, 749–767.

Karlsson, J. O., & Sjöstrand, J. Axonal transport of proteins in retinal ganglion cells. Amino acid incorporation into rapidly transported proteins and distribution of radioactivity to the lateral geniculate body and the superior colliculus. *Brain Research,* 1972, **37**, 279–285.

Kennedy, C., Des Rosiers, M. H., Sakaruda, O., Shinohara, M., Reivich, M., Jehle, J. W., & Sokoloff, L. Metabolic mapping of the primary visual system of the monkey by means of the autoradiographic C^{14} deoxyglucose technique. *Proceedings of the National Academy of Sciences of the U.S.A.,* 1976, **73**, 4230–4234.

Kopriwa, B. M. A reliable standardized method for ultrastructural electron microscopic radioautography. *Histochemie,* 1973, **37**, 1–18.

Kopriwa, B. M., & Leblond, C. P. Improvements in the coating technique of radioautography. *Journal of Histochemistry and Cytochemistry,* 1962, **10**, 269–284.

Kreutzberg, G. W., & Schubert, P. The cellular dynamics of intraneuronal transport. In W. M. Cowan & M. Cuénod (Eds.), *The use of axonal transport for studies of neuronal connectivity.* Amsterdam: Elsevier, 1975. Pp. 83–112.

Kunzle, H. Notes on the application of radioactive amino acids for the tracing of neuronal connections. *Brain Research,* 1975, **85,** 267–271.

Kunzle, H., & Cuénod, M. Differential uptake of [H³] proline and [H³] leucine by neurons: Its importance for the autoradiographic tracing of pathways. *Brain Research,* 1973, **62,** 213–217.

Lajtha, A. Transport and incorporation of amino acids in relation to measurement of axonal flow. In W. M. Cowan & M. Cuénod (Eds.), *The use of axonal transport for studies of neuronal connectivity.* Amsterdam: Elsevier, 1975. Pp. 25–46.

Lasek, R. J. Protein transport in neurons. *International Review of Neurobiology,* 1970, **13,** 289–324.

Lasek, R. J., Joseph, B. S., & Whitlock, D. G. Evaluation of a radioautographic neuroanatomical tracing method. *Brain Research,* 1968, **8,** 319–336.

LeVay, S., & Gilbert, C. D. Laminar patterns of geniculocortical projections in the cat. *Brain Research,* 1976, **113,** 1–19.

Lubinska, L. On axoplasmic flow. *International Review of Neurobiology,* 1975, **17,** 241–296.

Lund, R. D., & Lund, J. S. Plasticity in the developing visual system: The effects of retinal lesions made in young rats. *Journal of Comparative Neurology,* 1976, **169,** 133–154.

McEwen, B., & Grafstein, B. Fast and slow components in axonal transport of protein. *Journal of Cell Biology,* 1968, **38,** 494–508.

McEwen, B. S., Forman, D. S., & Grafstein, B. Components of fast and slow axonal transport in the goldfish optic nerve. *Journal of Neurobiology,* 1971, **2,** 361–377.

McLean, N. S., Frizell, M., & Sjöstrand, J. Slow axonal transport of labeled proteins in sensory fibers of rabbit vagus nerve. *Journal of Neurochemistry,* 1976, **26,** 1213–1216.

Neale, J. H., Elam, J. S., Neale, E. A., & Agranoff, B. W. Axonal transport and turnover of proline- and leucine-labeled protein in the goldfish visual system. *Journal of Neurochemistry,* 1974, **23,** 1045–1055.

Ochs, S. Fast transport of materials in mammalian nerve fibers. *Science,* 1972, **176,** 252–260.

Ochs, S. Fast axoplasmic transport of materials in mammalian nerve and its integrative role. *Annals of the New York Academy of Sciences,* 1974, **193,** 43–58.

Ogren, M., & Hendrickson, A. Pathways between striate cortex and subcortical regions in *Macaca mulatta* and *Saimiri sciureus:* Evidence for a reciprocal pulvinar connection. *Experimental Neurology,* 1976, **53,** 780–800.

Parry, D. M., & Blackett, N. M. Analysis of electron microscope autoradiographs using the hypothetical grain analysis method. *Journal of Microscopy (Oxford),* 1976, **106,** 117–124.

Peters, T., Jr., & Ashley, C. A. An artifact in radioautography due to binding free amino acids to tissues by fixatives. *Journal of Cell Biology,* 1967, **33,** 53–60.

Pinching, A. J., & Doving, K. B. Selective degeneration in the rat olfactory bulb following exposure to different odours. *Brain Research,* 1974, **82,** 195–204.

Plum, F., Gjeddes, A., & Samson, F. E., Jr. Neuroanatomical functional mapping by the radioactive 2-deoxy-D-glucose method. *Neurosciences Research Program Bulletin,* 1976, **14,** 457–518.

Pomerat, C. M., Hendleman, W. J., Raiborn, C. W., Jr., & Massey, J. F. Dynamic activities of nervous tissue *in vitro.* In H. Hydén (Ed.), *The neuron.* Amsterdam: Elsevier, 1967. Pp. 119–178.

Price, J. L., & Wann, D. F. The use of quantitative autoradiography for axonal tracing experiments and an automated system for grain counting. In W. M. Cowan & M. Cuénod (Eds.), *The use of axonal transport for the studies of neuronal connectivity.* Amsterdam: Elsevier, 1975. Pp. 155–172.

Rogers, A. W. Recent developments in the use of autoradiographic techniques with electron microscopy. *Philosophical Transactions of the Royal Society of London, Series B.*, 1971, **261**, 159–171.

Rogers, A. W. Photometric measurements of grain density in autoradiographs. *Journal of Microscopy (Oxford)*, 1972, **96**, 141–153.

Rogers, A. W. *Techniques of autoradiography* (2nd ed.) Amsterdam: Elsevier, 1973.

Ron, A., & Prescott, D. M. *In* D. M. Prescott (Ed.), The efficiency of tritium counting with seven radiographic emulsions. *Methods in Cell Physiology*, Vol. VI, 1970, **6**, 231–240.

Salpeter, M. M., & Bachmann, L. Electron microscope autoradiography. *In* M. Hayat (Ed.), *Principles and Techniques of Electron Microscopy*, Biological Applications. New York: Van Nostrand-Reinhold, 1972, **2**, 221–278.

Salpeter, M. M., & Eldefrawi, M. E. Sizes of end plate compartments, densities of acetylcholine receptors and other quantitative aspects of neuromuscular transmission. *Journal of Histochemistry and Cytochemistry*, 1973, **21**, 769–778.

Salpeter, M. M., & McHenry, F. A. Electron microscope autoradiography, analysis of autoradiograms. In J. K. Koehler (Ed.), *Advanced techniques of biological electron microscopy*. Berlin & New York: Springer-Verlag, 1973. Pp. 114–151.

Salpeter, M. M., & Salpeter, E. E. Resolution in electron microscope radioautography, II. Carbon[14]. *Journal of Cell Biology*, 1971, **50**, 324–332.

Salpeter, M. M., Budd, G. C., & Mattimoe, S. Resolution in autoradiography using semithin sections. *Journal of Histochemistry and Cytochemistry*, 1974, **22**, 217–222.

Schönbach, J., Schönbach, C., & Cuénod, M. Rapid phase of axoplasmic flow and synaptic proteins: An electron microscopical autoradiographic study. *Journal of Comparative Neurology*, 1971, **141**, 485–498.

Schubert, P., & Hollander, H. Methods for the delivery of tracers to the central nervous system. In W. M. Cowan & M. Cuénod (Eds.), *The use of axonal transport for studies of neuronal connectivity*. Amsterdam: Elsevier, 1975. Pp. 113–126.

Schubert, P., Kreutzberg, G. W., & Lux, H. D. Use of microelectrophoresis in the autoradiographic demonstration of fiber projections. *Brain Research*, 1972, **39**, 274–277.

Sharp, F. R., Krauer, J. S., & Shepherd, G. M. Local sites of activity-related glucose metabolism in rat olfactory bulb during olfactory stimulation. *Brain Research*, 1975, **98**, 596–600.

Sidman, R. L. Autoradiographic methods and principles for study of the nervous system with thymidine-H[3]. In W. J. H. Nauta & S. O. E. Ebbeson (Eds.), *Contemporary research methods in neuroanatomy*. Berlin & New York: Springer-Verlag, 1970. Pp. 252–274.

Sousa-Pinto, A., & Reis, F. F. Selective uptake of [H[3]] leucine by projection neurons of the cat auditory cortex. *Brain Research*, 1975, **85**, 331–336.

Specht, S., & Grafstein, B. Accumulation of radioactive protein in mouse cerebral cortex after injection of H[3]-fucose into the eye. *Experimental Neurology*, 1973, **41**, 705–722.

Thurston, J. M., & Joftes, D. L. Stains compatible with dipping autoradiography. *Stain Technology*, 1963, **38**, 231–235.

Trojanowski, J. Q., & Jacobson, S. A combined horseradish peroxidase-autoradiographic investigation of reciprocal connections between superior temporal gyrus and pulvinar in squirrel monkey. *Brain Research*, 1975, **85**, 347–353.

Wann, D. F., Price, J. L., Cowan, W. M., & Agulnek, M. A. An automated system for counting silver grains in autoradiographs. *Brain Research*, 1974, **81**, 31–58.

Weiss, P. The concept of perpetual neuronal growth and proximo-distal substance convection. In S. S. Kety & J. Elkes (Eds.), *Regional neurochemistry*. Oxford: Pergamon Press, 1961. Pp. 220–242.

Wiesel, T. N., Hubel, D. H., & Lam, D. M. K. Autoradiographic demonstration of ocular dominance columns in the monkey striate cortex by means of transneuronal transport. *Brain Research*, 1974, **79**, 273–279.

Willard, M., Cowan, W. M., & Vagelos, P. R. The polypeptide composition of intra-axonally transported proteins: Evidence for four transport velocities. *Proceedings of the National Academy of Sciences of the U.S.A.*, 1974, **71**, 2183–2187.

Wise, S. P., & Jones, E. G. Transneuronal or retrograde transport of [H^3]adenosine in rat somatic sensory system. *Brain Research*, 1976, **107**, 127–131.

Chapter 10

The Use of Somatofugal Transport of Horseradish Peroxidase for Tract Tracing and Cell Labeling

Hugh J. Spencer
Gary Lynch

Department of Psychobiology
University of California
Irvine, California

R. Kevin Jones

Department of Biology
University of Wollongong
Wollongong. N.S.W. Australia

I.   Introduction

The introduction of new tracing techniques in neuroanatomy has caused an almost explosive increase in knowledge about the organization of the vertebrate brain; a prominent example of this has been the development of the horseradish peroxidase (HRP) histochemical procedure (Kristensson & Olsen, 1971; LaVail & LaVail, 1972). This method has virtually replaced earlier retrograde degenerative procedures for tracing afferents of brain regions and has been used to demonstrate circuits, the existence of which was hardly suspected 5 years ago. Shortly after the appearance of reports illustrating the use of HRP as a retrograde tracing method (LaVail, Winston, & Tish, 1973) the authors found that the enzyme also traveled in an anterograde direction and suggested that it might be useful in following efferent as well as afferent connections (Lynce, Gall, Mensah, & Cotman, 1974b; Lynch, Smith, Mensah, & Cotman, 1973). Initially, there was some controversy regarding the orthograde movement of the enzyme but many laboratories have now provided ample evidence of its existence (e.g., Colman, Scalia, & Cabrales, 1976; Sotelo & Riche, 1974; Winfield, Gatter, & Powell, 1975).

This observation that HRP will move in a somatofugal as well as somatopetal direction also suggested that it might be used for purposes other than the tracing of connections. Specifically, it occurred to the authors that if injected into the region of neuronal cell bodies, the enzyme might be incorporated into the somata and transported through dendritic and axonal arborizations, thereby providing a means of staining entire neurons in a manner similar to the Golgi method. This possibility was realized when HRP histochemistry was coupled with iontophoresis from recording microelectrodes (Lynch, Gall, & Deadwyler, 1974a; Lynch, Smith, Browning, & Deadwyler, 1975).

This chapter will review the methods used to demonstrate sites of HRP activity with particular emphasis on recent improvements in the technique. The advantages, disadvantages, and pitfalls in using HRP to trace efferent projections will then be discussed. Finally, efforts to exploit somatofugal transport of HRP for use in the labeling of central neurons will be reviewed briefly.

II. Histochemical Procedures for Demonstrating Horseradish Peroxidase: The "Blue," "Brown," and "Green" Methods

A. Comparison

HRP pathway-mapping experiments involve the injection of solutions of the enzyme at selected areas in brain and, after various survival times, (typically 1 to 2 days) sacrifice of the animal, fixation, sectioning, and finally incubation of the sections in reagent solutions to produce reaction products at sites of enzyme activity. This incubation medium consists of a chromagen (typically a benzidine derivative) and a substrate, hydrogen peroxide. In the presence of enzymatically active HRP, the chromagen is oxidized and becomes visible. Three procedures, differing primarily in the nature of the chromagen used, have been employed for demonstrating enzyme activity. Benzidine dihydrochloride gives an intense "blue" reaction product, which is both electron dense and highly refractile (thus highly visible under dark-field illumination) (Lynch et al., 1973, 1974a). The reaction product produced by diaminobenzidine (DAB) (3,3'-diaminobenzidine tetrachloride, LaVail et al., 1973) is "brown," electron dense, but considerably less visible than that produced by benzidine dihydrochloride (Mesulam, 1976). The "green" reaction product of o-dianisidine hydrochloride (another benzidine derivative) is as intense as that produced by the benzidine dihydrochloride technique, but is water soluble (de Olmos, 1977). The main benefit claimed for this "green" technique is sensitivity and the use of a relatively safe chromogen, o-dianisidine; the method itself, however, is complex. The reported carcinogenic tendencies of benzidine dihydrochloride led to the general adoption of the diaminobenzidine procedure derived by LaVail and LaVail (1972) from Graham and Karnovsky's (1966) electron microscope procedure, which yields the brown reaction product. The use of this procedure has become standard practice in many anatomical laboratories. In the author's studies the benzidine "blue" method developed by Straus (1964), which used benzidine dihydrochloride was modified. Mesulam (1976) has recently performed a significant experiment in which he compared the sensitivities of the two procedures and found that the "blue" method provided much better visualization of HRP than did the DAB procedure, particularly in areas in which enzyme concentrations were low. This probably accounts for the very limited orthograde movement of the enzyme often reported by various workers using the diaminobenzidine procedure. Mesulam (1976) was also able to significantly increase the sensitivity of the "blue" method over the levels that the authors had achieved.

In discussing the various stages involved in HRP histochemistry as applied to neuroanatomical work it must be appreciated that only the incubation step has been investigated in anything approaching a methodical fashion and while Straus (1964) has performed a series of studies on HRP methodology using nonneural tissue, the applicability of his findings to brain is open to questions. Consequently any protocol for the method must necessarily be somewhat tentative.

B. Injection

Typically, for pathway tracing, the HRP is administered by pressure through a microsyringe or broken glass micropipette, usually over a period of 10–15 min at a rate of 0.1 to 2 μl/min. The enzyme concentration and the media in which it is dissolved varies from laboratory to laboratory. Many workers prefer using HRP solutions with concentrations of 15–30%, but we have often used concentration of around 8% and obtained good results. The less concentrated version has the advantage of causing less cellular destruction and provides a more discrete injection. Distilled water serves as an adequate solvent for the enzyme but the pH of the solution might be profitably manipulated, particularly if pinocytosis of HRP is pH dependent. Type VI HRP is widely used but it is very possible that some purification steps will increase the sensitivity of the procedure. HRP consists of several isozymes, and Bunt (personal communication) has suggested that only certain of these are actively transported by neurons. This, in turn, could lead to a significant increase in the intensity of labeling.

C. Survival Time

HRP appears to be transported (somatofugally), at about 100 μm/min. Most workers have allowed survival period of 1 to 2 days postinjection. When injecting into brain tissue slices *in vitro*, postinjection times of 10–30 min appear to be quite adequate for labeling cell processes including axons.

D. Fixation

Most workers use a 0.01 M phosphate buffered (pH 7.4) solution of paraformaldehyde or paraformaldehyde–glutaraldehyde, at typical concentrations of 0.5–4.0% paraformaldehyde, 0.5–5.0% glutaraldehyde. Straus (1964) recommends not using buffers during fixation and the authors' results have agreed, in general, with this. The optimal duration of postperfusion fixation has not been studied. Forty-eight hours gives acceptable tissue preservation and enzymatic activity, but *in vitro* slices can be sectioned within 6 hr. It is advisable to keep the material at 4°C during this time as formalin slowly deactivates the enzyme.

E. Sectioning

The brain is sliced on a freezing microtome or in a cryostat. Section thickness is a variable that conceivably could influence the results of the incubation step: adequate results are achieved with 30–50 μm sections, but thinner sections might be considered.

F. Processing

Sections can be processed either unmounted (using small-cavity ice cube trays) or mounted on the slide with albumin. The albumin used must be diluted (1 mg dried egg albumin/10 ml water) and should be applied to carefully cleaned slides by dipping or brushing, followed by drying for about 10 min at 50°–60°C, which ensures that the albumin film is dry but not denatured. A thick albumin coating will result in poor penetration of the reagents into that part of the section adjacent to the glass and thus cause partial development. If no albumin is used, the sections will strip off the slides during the color development procedure owing to gas evolution by the peroxidase. Sections are floated onto these coated slides and allowed to air dry before processing. While sections can be stained individually before mounting, they tend to get damaged during handling, and we have found that the method described above gives good results and is considerably less tedious.

G. Color Development

The following solutions are used.

Solution A: 0.2% solution of benzidine dihydrochloride in 50% ethanol (stock)
Solution B: 0.06% H_2O_2 in water (fresh)
Solution C: 25% Ethanol (rinse) (stock)
Solution D: Sodium nitroferricyanide (= nitroprusside) 6% solution in 20% ethanol (stock)

Color development (as presently employed in this laboratory) must be carried out at 1°–5°C either in cooled Coplin jars of reagent, or with the slides lying flat on ice in a tray and the reagents applied with a dropping pipette. Benzidine solution (equal parts of solution A and B) is applied for 1–3 min. Color development should be rapid; if it is prolonged with these concentrations, crystalline artifacts appear (Mesulam, 1976; Straus, 1964). The sections are then washed with 25% ethanol and slides covered for 1 hour or so with sodium nitroprusside at 4°C, which stabilizes the "blue" reaction product (Mesulam, 1976; Straus, 1964).

1. INCUBATION

Incubation under acidic conditions is necessary for success with "blue" method; beyond this, however, it is not clear that any particular pH is optimal. Straus and Mesulam use acetate buffers to keep the pH at 5.0 and if for no other reason than standardization of the method from one experiment to the next, this is probably advisable. However, in our laboratory we often leave out the buffers and incubate at the pH of the incubating media (about 4.5).

Mesulam (1976) has reported that the two modifications described below greatly improve the sensitivity of the method.

2. PRESOAKING

In this step the sections are soaked in a bath of benzidine, sodium nitroferricyanide, ethanol, acetate buffer, and distilled water. This serves to allow the benzidine to penetrate into the sites of HRP activity. After presoaking, the chemical reaction leading to oxidation of the benzidine is then initiated by adding H_2O_2 to the presoaking solution.

3. PROLONGATION OF THE INCUBATION PERIOD

In our protocol we have used relatively high concentrations of benzidine and H_2O_2 with a very brief incubation step (less than 1 min); Mesulam (1976) reported obtaining much better results by reducing the concentrations of the incubating media and extending the reaction time to 25 min.

H. Results

After the reaction product has been stabilized by nitroprusside the sections should be washed in 50% alcohol, counterstained in eosine or saffranin, and mounted in a neutral mounting medium. HRP labeled axon terminals can be

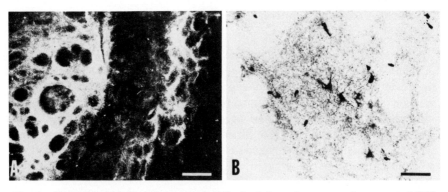

FIG. 1.   Examples of labeling of axon terminals in the thalamus by anterograde transport of HRP after injection into cortex. A. Low-power dark-field photomicrograph showing intense labeling in the thalamic reticular nucleus (to the right) and in the ventral basal complex. B. Medium-power bright-field photomicrograph of the ventral basal complex showing a few heavily labeled neurons, surrounded by a field of labeled axon terminals. (From Robertson, 1977; reproduced here by permission of Elsevier Press.)

identified as fields of dusty appearing reaction product, whose density varies with the particular neural system involved, the amount of HRP injected, survival time, incubation parameters, and probably other factors. It is usually easy to distinguish anterograde-transported from retrograde-transported HRP, as shown in Fig. 1. The reaction product may fade over a period of days to months, and photography and camera lucida studies should be carried out with reasonable haste. (See also p. 311)

III. Advantages, Disadvantages, and Pitfalls in Using HRP Histochemistry to Trace Efferent Projections

HRP histochemistry possesses a number of features that make it an attractive method for the tracing of efferent projections in the nervous system; chief among these are the simplicity of the method and its potential for combined light/ electron microscopy experiments.

The method is uncomplicated, in fact it is not much more difficult than the standard Nissl stains and with a little practice produces reliable results. Furthermore, the entire procedure, from injection of the enzymes to analysis of the results, can be carried out in only a few days as compared to the weeks required for autoradiography. These characteristics make it reasonable to conduct pilot studies of the method's utility in almost any tracing experiment.

Probably the greatest potential advantage of HRP histochemistry is the possibility of using it in both light and electron microscope experiments. The "brown" reaction product is electron dense, and it appears that the same is true of the "blue" material (H. Karten, personal communication); this opens the way for experiments in which a terminal population can be localized to a given brain region and then the ultrastructural details of the endings analyzed with the electron microscope. In this way unambiguous descriptions of the characteristics of terminals from known sources could be accomplished. Previous efforts at this type of work have involved autoradiography, which is laborious and not entirely satisfactory, or the use of degeneration studies, which result in the distortion of the postsynaptic terminal. Some authors have already reported on the use of anterograde HRP to identify specific terminal populations (e.g., Winfield et al., 1975), but clearly much more work will be required before optimal EM/light microscopic methods are developed.

The orthograde HRP procedure has an additional advantage in that it can be used with very localized injections, and in many cases labeled fibers can be readily followed from injection site to their terminal fields. This can be of real advantage in analyzing intrinsic connections of only a few millimeters length. Autoradiography often gives too high a background exposure (because of diffusion from the injection site) to be used for these types of experiments.

There are several problems which must be taken into consideration when deciding whether to use HRP to trace efferent projections and when interpreting experimental results. First and foremost is the problem of the false negative, that is, failure to observe transport from an injection site. In fact, this cannot be taken as conclusive evidence that the brain regions involved are not connected with each other. Some efferent and afferent projections do not appear to be detected by the HRP procedure; there is no obvious reason for this, but the fact remains. A second problem in using HRP is that the extent to which damaged axons become filled with enzyme and transport it bidirectionally has not been satisfactorily resolved. The pipettes and microsyringes used to inject the HRP create considerable local damage including, presumably, the severing of axons of passage. There is a possibility that these broken fibers might take up the HRP and transport it to their terminals and cells of origin. In our studies of hippocampus it appeared that uptake by broken fibers did not result in transport for more than a few millimeters in either direction, but the issue has not been adequately resolved.

Potential problems in distinguishing between retrograde and orthograde transport also exist (see Fig. 1B). It is conceivable that the enzyme might be transported in a retrograde direction and then enter a collateral; this could be interpreted wrongly as an efferent projection originating in the injection site.

In a similar fashion, confusion may occur in distinguishing between HRP that has been transported in retrograde fashion to neurons and thence into their dendrites, and misinterpreted as HRP in axon terminals. Effects of these types have not yet been reported (we are hopeful that they are nonexistent or rare), but experimenters should at least be aware of the possibilities. Orthograde transport also represents a possible source of confusion in interpreting the results of retrograde experiments involving transport since axon terminals on the cell bodies of its target neurons could appear under light microscopy to be retrogradely transported enzyme within the cell. It is essential to establish that the reaction product is located *within* the cell body in retrograde experiments.

The availability of benzidine dihydrochloride has been restricted in the United States on the basis of its reported carcinogenic tendencies. Thus, the substance ought to be handled with a great degree of care. Unfortunately, benzidine derivatives other than the benzidine base do not appear to be as satisfactory for producing the reaction product.

Two recent developments in HRP technology may help alleviate problems with histochemical procedures. The use of radioactively labeled HRP combined with standard autoradiographic procedures provides a tool for tracing both retrograde and anterograde connections, and may be more sensitive than those using histochemical procedures (Geisert, 1976). The development of an immunological approach is also potentially helpful. In this, the HRP is inactivated as an enzyme, injected, and then detected using labeled antibodies. This has a number of potential advantages. The molecule can be manipulated in any number of ways, and as long as the antigenic sites are preserved, it can still be detected. Thus, it may be possible to develop a form of the enzyme which is readily taken up, causes minimal cellular disruption, and is transported. The authors have done some preliminary testing of an immunohistochemical procedure of this type and while it is clear that chemically manipulated enzymatically inactive versions of HRP are transported in an orthograde direction, we have no data on the relative sensitivities of the procedures so far developed.

IV.   Labeling of Individual Neurons by Using HRP Histochemistry

A.   Overview

The use of HRP as a marker in neurophysiological experiments originated from work in this laboratory, in which it was found that HRP could be ejected from extracellular recording electrodes by iontophoresis. The ejected HRP in some cases produced exquisite labeling of one or more cells located near the electrode tip (Lynch et al., 1974a). This suggested that it might be possible to use the method to identify cells that had been recorded from by intracellular and extracellular electrodes. The power of such a technique, in which the anatomy of a cell can be directly correlated with its physiological function as recorded by a microelectrode, can be readily appreciated.

Iontophoretically applied HRP is taken up by nerve cells and their dendrites, and to a lesser degree by axons, then is rapidly transported bidirectionally throughout the cell by fast axoplasmic and dendritic flow (Lynch et al., 1974 , 1975). When successful, extracellular and intracellular (Kitai, Kocsis, Preston, & Sugimori, 1976; Snow, Rose, & Brown, 1976) HRP cell marking produces a labeled neuron with cell body, dendrites, and axons clearly delineated. More frequently, however, extracellular marking attempts with HRP have resulted in unsatisfactory results; often we found a diffuse stain not associated with particular cells or a large clump of stained cell bodies with no, or very little, dendritic impregnation. In other cases, too large a population is labeled to allow any attempt at correlation of physiological results with particular cell morphologies.

In Section IVB–H we will review our efforts to achieve a more reproducible and tractable extracellular marking technique and discuss its merits and shortcomings. This is presented with the hope that the material will be a useful guide to those interested in using and modifying the technique for both extra- and intracellular marking. These results have been obtained using in vitro slices of hippocampus (Lynch et al., 1975; Spencer, Gribkoff, Cotman, & Lynch, 1976), and although we have no reason to suspect that different conditions will prevail under in vivo conditions, the possibility should not be ignored.

B. *Varieties of HRP*

For the results reported here we used Sigma type II HRP or a fractionated varient of it produced by ultrafiltration (discussed later). During our initial experiments there appeared to be no appreciable difference between the Sigma type VI and the less expensive type II peroxidase. For filling the electrodes, the HRP was made up as a 0.5% solution (1 mg/200 μl giving a straw-colored solution); the exact concentration does not appear to be particularly critical. Initially, the HRP was dissolved in saline (0.1–0.2 M NaC1); higher molarities appear to cause precipitation of the peroxidase in the electrode. Phosphate buffer, pH 6.2 (KH_2PO_4/K_2HPO_4) + 0.1 M NaCl, was also used as a solvent. Both solutions had a sufficiently low resistivity to ensure that the microelectrodes (2–3 μm tip diameter) had an adequately low (2–10 Mohm) series resistance for extracellular recording and current passage. The phosphate buffered HRP appears to be taken up more readily by cells than is that dissolved in the saline alone.

We routinely purify commercial type II HRP using ultrafiltration (Amicon diaflo filters). The fraction lying between 10,000 and 300,000 M.W. when lyophilized, is almost white in color, and accounts for the bulk of the material. The activity of the 10,000–300,000 M.W. fraction was comparable or greater than the original type II HRP, and also appeared to be released more readily from micropipettes.

Horseradish peroxidase is an unusually stable enzyme, and solutions of it can be made up days or weeks before use. Provided they are stored below 5°C they do not appear to deteriorate appreciably. In fact, we suspect that aging the peroxidase solutions for 4 or more days improves their performance. Filtration of peroxidase solutions is not necessary if reasonable care is exercised to avoid contamination of the solution and electrodes.

C. *Electrodes*

Electrodes have proved to be the major source of variability. Initially we used, with some success, a quantity of 0.6 mm o.d. borosilicate capillaries (Owens-Illinois) that are employed routinely in the department for other procedures. Further testing, however, showed this type of glass failed to give consistent positive results. We then began using glass tubing containing an integral glass fiber (American Glass Co., Omega-Dot tubing), which performed equally erratically at first, there being variability in release of HRP even between electrodes pulled from the same length of glass.

This problem may stem from distributed fixed charges on the glass resulting from the pulling operation. There is a charge separation during the pulling process, and these charges become fixed when the molten glass solidifies. The presence of such charges becomes evident if one leaves a freshly pulled electrode in a dusty environment for a day or so as they are revealed by the adhesion of dust particles on the electrode tip.

Various attempts to counter or neutralize this surface charge by filling the electrode with various dilute detergent solutions (sodium lauryl sulfate; Igepon TC., hexadecyldimethyl amidosulfonate) after pulling, and then emptying and drying the electrodes under vacuum, did not materially improve the performance. To date our best success has been with old (1 year) 1 mm o.d. capillary-filled tubing (American Glass Co., Omega-Dot), which has been rinsed with acetone and dried by suction prior to pulling and filling with HRP. We emphasize that it is highly probable that a variety of electrode glass will have to be evaluated by the prospective HRP user. We have also on some occasions observed precipitation of the HRP solution on contact with glass, causing a plug of brown precipitate to be carried forward by the advancing meniscus into the tip and blocking it, despite the use of filtered HRP and clean glass tubing.

The electrode tip should not be broken back to the desired diameter (1–3 μm) until required for use. We perform this breaking-back operation by running the electrode into a block of metal or glass positioned under a microscope ($\times$ 50), using a longitudinally grooved 1 $\times$ 3 in. plexiglass slide mounted on the mechanical stage as the electrode carrier. Following this operation the electrode is pressure filled with HRP by using a 32-gauge filling needle.

D. HRP Ejection Assay

The ejection assay represents a real advance in the technique, as it permits immediate assessment of the HRP-releasing characteristics of the electrode. We have modified the o-dianisidine–HCl assay described in the Worthington Enzyme Manual (Worthington Biochemical Corporation, 1972) for use in microliter volumes, and can readily detect released HRP in amounts of approximately 1 ng. The test is a simple spot test, but with the use of suitable microcuvette (50 μl capacity) it can be adapted for use in a recording spectrophotometer if an accurate assay of the HRP release is desired.

The necessary assay reagents are

1. Stock dye solution, o-dianisidine in absolute methyl alcohol (75 mg/10 ml). Keep in stoppered bottle

2. Potassium-phosphate buffer, 0.01 M, pH 6

For the substrate (test) solution, add 50 μl of 3% H_2O_2 and 50 μl of o-dianisidine solution to 6.0 ml of phosphate buffer. This substrate solution lasts for weeks if kept cold. For the spot test the discoloration that appears over time does not affect the sensitivity of the test.

The electrode is filled, mounted on a suitable electrode carrier, and dipped into a drop ($\sim$10 μl) of saline ($\sim$1%). Silver wires (30 gauge) are used to make contact with the solution in the electrode and with the saline drop (Fig. 2B). The

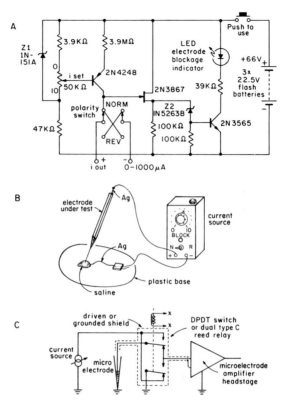

FIG. 2. A. Constant current source with electrode blockage/open circuit indicator. The current set potentiometer should be calibrated using a current meter having a 0–1 μA range, or the voltage developed across a 1-MΩ resistor (0–1.0V) should be monitored with a voltmeter having a high (100 MΩ) input resistance. All resistors are 5% tolerance. Substitute semiconductors should be chosen carefully: 2N4248, PNP high voltage, low leakage, high gain; 2N3867, low leakage, 200 μA IDSS, J-Fet; 2N3565, high gain; NPN; 1N751A, 5.8 V, 100 mW; 1N5263B, 50 V, 100 mW, zener diodes. Other decade resistance values can be substituted for the 3.9 MΩ resistor to give higher current ranges (390 kΩ = 1–10 μA, etc.) for making lesions. B. Suggested electrode test setup. The lower silver wire should not lie flat on the plastic base so as to preclude the possibility of capillary flow of the saline along its length. C. Suggested arrangement for the "breakaway" switch or relay, which must be installed at the input of the electrometer probe. So long as low-leakage components are used, the exact components used are not critical. The second contact pair is used to ground the amplifier input during current injection and to ground the current source line to prevent AC hum from entering the amplifier during recording. The relay coil connections marked x-x can be switched by an additional pair of contacts on the constant current-box switch (which must also be a low-leakage type). The box containing the breakaway switch must be either grounded or connected to the amplifier's guard terminal (in which case it *must not* be grounded).

electrode should not be exposed to the air for more than a few minutes after filling since the HRP may dry and coagulate, blocking the tip. A positive current of 500 nA is passed for 1 min, the electrode removed, and the drop transferred to a white plastic or porcelain spotting plate. When an equal size drop of the test solution is applied to the drop on the place, a color should develop within 60 sec if the electrode has ejected a sufficient quantity of HRP. While the test is being completed, the electrode should be moved into another saline drop to prevent coagulation at its tip. Color development may be slower if only a small amount of peroxidase is ejected but should be complete within 3 min. The intensity and rate of the initial color development will indicate the relative amount of HRP ejected. Good HRP release gives an initial purple color that changes to a clear straw-brown; weaker (marginal) releases produce only a slowly developing straw-brown color. A 10 ng/ul standard solution (1 mg HRP/100 ml water) should be used to check the reagent activity and rate of colour development as discussed above. This solution must be kept refrigerated.

Electrodes that show better than marginal color development should be used immediately if possible. We have found that electrodes can often be used for 20 or more injections, but the longer the period between insertion into the tissue and injection, the lower the success ratio. It has also proven wise to repeat the test procedure between injections. This may be most easily accomplished by a miniaturized modification of the above-described test system positioned above the preparation.

E. Ejection Systems and Parameters

We have concentrated on using constant-current injection systems, since they are reliable and easy to use.

Currents of $+50$ to $+500$ nA appear to effectively eject peroxidase from the electrode. Some method of monitoring the current passage should be provided since electrode blocking during current passage is not uncommon. The circuit in Fig. 1A provides a linear current calibration, a high compliance voltage (60 V), and a means of detecting a blocked electrode or open circuit.

Some form of low leakage breakaway switch or reed-relay (Type C, Fig. 2C) must be used at the amplifier input to prevent the applied current-driving voltage from destroying the electrometer amplifier stage. We have found that for ejection, current of $+50$ to $+150$ nA are sufficient when applied for 1–4 min. This seems to avoid the problems of damaging the cell by high local current densitites, although cells generally stop firing following injection, apparently regardless of whether the injection is fully successful. Electrodes, especially those with fine (less than 1 μm) tips, tend to block during positive current ejection, and periodically reversing the polarity sometimes unblocks the electrode. After injection it is wise to allow at least 10 min after the last HRP ejection in the preparation for axon and dendrite transport of the HRP to be completed before fixation.

Since HRP appears to be taken up by the cell by pinocytosis, we began to look for agents which would increase directly the uptake of HRP by this process. It has been reported that adenosine increases pinocytosis in macrophage type cells (C. Gall, personal communication), and recently, we have tried adding this nucleoside (~0.05% w/v) to the HRP solutions. The addition of adenosine, in trials to date, resulted in a dramatic increase in the probability of getting satisfactory release and uptake of HRP. Further parametric studies on the effect of adenosine are clearly required, but these initial findings are most encouraging.

F. Processing

Following buffered formaldehyde–glutaraldehyde fixation (10–20 hr in the cold), sections are cut on a conventional freezing microtome, and serial sections are floatation mounted directly onto albumin-coated slides, then drained and allowed to air dry. Serial sections are absolutely necessary to allow reconstruction of dendritic and axonic fields. The sections are incubated in benzidine dihydrochloride as described above and then counterstained with saffranin.

G. Marking Specificity

As with extracellular recording in general, the actual relationship between the electrode and the cell whose activity is being recorded can only be approximated, since nearby cells can be "silent." Electrotonic paths can be unpredictable in a mixed population of varied cell types, and the amplitude of the extracellular spike observed may be both a function of the cell type and of the position of the electrode tip with respect to that cell.

The same considerations appear to apply to the probability of cells being marked by ejected HRP. However, using the 2 μm electrode tip size, we are able to approach very closely the cell soma of pyramidal and granular cells in tissue slices of the rat hippocampus. The extracellular spike amplitude of the cells in the pyramidal cell body zone (presumably pyramidal cells) were of the order of 50 μV to 4 mV prior to injection. In many cases it appeared that the cell being recorded was, in fact, that being marked. In other instances however, HRP injection has resulted in the labeling of two or more cells, often very clearly, which were not necessarily adjacent. It is likely that retrograde axonal or dendritic transport is involved in this effect.

It is possible, since most of the recordings from subsequently labeled cells were made close to the soma, that some degree of current disruption of the soma had occurred since the cells were generally silent following ejection. Presently we have no way of verifying this hypothesis.

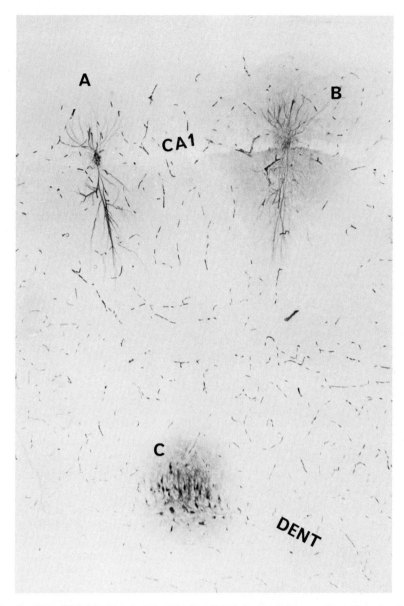

FIG. 3. Three HRP injections carried out in the cell body layers of the same tissue slice, showing the three characteristic degrees of uptake. A is presumably a pyramidal neuron in the CA1 cell layer that has been selectively labeled. B, also in CA1, is a group of cells which have all become labeled together with some extraneous label surrounding the cells. C is a nonspecific label following injection in the dentate granule cell layer (DENT). Injection parameters for all injections were 150 nA for 2 min; the HRP was dissolved in pH 6.2 phosphate buffer with adenosine. The black dotted lines are blood vessels that stain because of the endogenous peroxidase activity of erythrocytes.

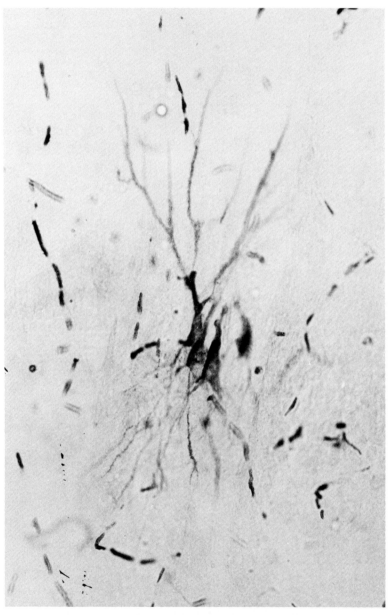

FIG. 4. Four CA4 polymorph neurons labeled as a result of a single injection. In this example the cell bodies have taken up the peroxidase while there has been little axon or dendritic labeling. 250 nA, 1 min, pH 6.2, no adenosine.

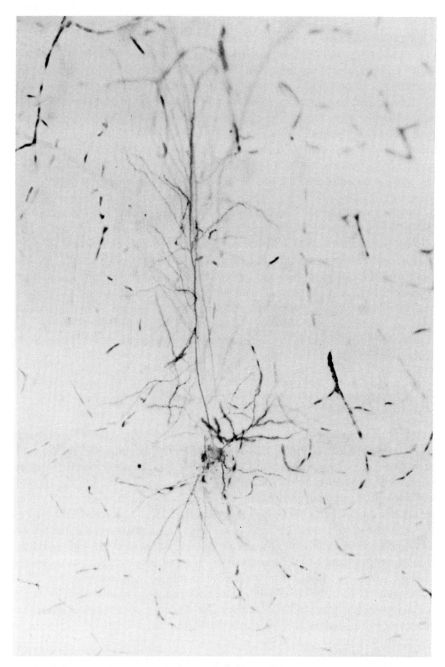

FIG. 5. Labeling of two cells in the CA1 pyramidal cell layer. The upper cell (in focus) is morphologically distinct from the pyramidal cells (one of which is below the focal plane) in that it has a single apical dendrite. This neuron was also unusual in that it alone of all the cells labeled in this study produced a ''brown'' reaction product with benzidine.

Even with the use of adenosine, about 20% of the marks were of the type illustrated in Fig. 3C—diffuse labeling of a large number of cells. The remainder were single labeled cells (Fig. 3A) or single or multiple labeled cells labeled together with a quantity of extraneous "nonspecific" marking (Fig. 3B).

The cells that take up and transport a sufficient amount of the peroxidase label demonstrate a satisfying amount of detail, which in many aspects is comparable to that of a Golgi preparation (Fig. 4,5, and 6). Dendritic spinelike processes are clearly evident and can be readily distinguished from the needlelike artifacts produced by overdevelopment in benzidine (Straus, 1966). Axons can also be traced for a considerable distance in the more heavily labeled cells (Figs. 6 and 7).

The observed degree of uncertainty of marking and the uncertain specificity of the cell marked can be partly overcome at least by increasing the sample size. Continual association of a particular labeled cell type with a particular response pattern could in some situations be regarded as adequate evidence that the cell labeled is the cell producing the observed physiological response. Consistent quasi-intracellular recording, at least just before injection, would be an even better method of assuring that the recorded cell will be labeled. Intracellular HRP labeling, however, must still be regarded as the ultimate marking technique and is at present still subject to the same vagaries of ejection discussed above (S. T. Kitai, personal communication).

H. Morphological Characteristics of HRP-Labeled Neurons

Although there appears to be considerable similarity between the appearance of cells labeled with HRP and those impregnated by the Golgi technique, there are significant differences that must be considered when examining HRP-labeled neuronal material.

Golgi silver impregnation appears to be an impregnation of the extracellular membrane (and possibly the intracellular material as well), whereas HRP is taken up and transported by the intracellular contents alone. Thus, on the assumption that neither technique results in the thickening or increase in volume of the impregnated or labeled process, the HRP material displays the "core" of the neuron, while the Golgi displays the membrane-bound entirety. For gross anatomical reconstructions, the differences between the two techniques probably are not particularly important. However, for detecting such morphological features as dendritic spines, this intracellular labeling characteristic of HRP must be taken into account.

Well-labeled neurons, such as that illustrated in Fig. 5, provide an excellent example for this. The observer's first reaction on seeing the decorated dendrites is to assume that the protuberances on the shafts are spines, which, in fact, they

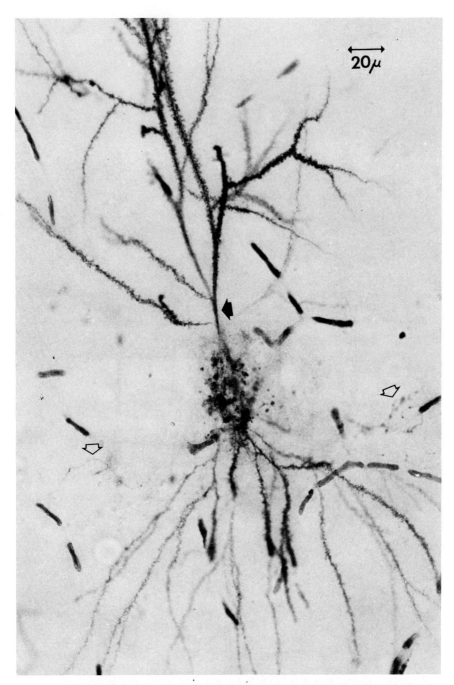

FIG. 6.   This is an enlargement of the neuron illustrated in Fig. 3A showing the labeled dendrites in detail. The spinelike processes are clearly evident in both the basal and apical dendrites, while the proximal apical dendrites (solid arrow) are free of processes. Two axon processes (open arrows) can be seen on either side of the cell body. The soma of this neuron is only partially filled with HRP.

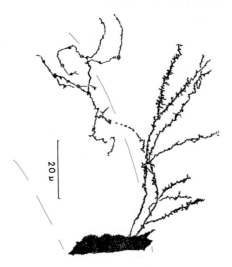

FIG. 7.   *Camera lucida* drawing of the axon process which appears on the righthand side of the cell in Fig. 6. Part of the cell body is shown together with some of the apical dendrites. Evident in this drawing are the triangular axonal swelling which may constitute *en passant* terminals. The dashed lines represent the extent of the cell body layer.

may be. But, examination under high power reveals that the HRP labeling consists of discrete fine globules like ferritin particles, and in the more weakly marked processes these particles appear to be arranged around a "core" to give a lampbrush chromosomelike appearance. Whether these irregular aggregations of particles, in fact, do represent the intracellular aspect of the dendritic spine lumens remains to be demonstrated. In the cell illustrated, the proximal segments of the apical dendrites (solid arrow, Fig. 6) appear to be free of such brushlike labeling.

Axons of pyramidal cell-layer neurons, when adequately labeled, demonstrate what could be considered synaptic swellings (*en passant* terminals) as the *camera lucida* sketch of the presumed axons (in Fig. 7) illustrates. The axon in this picture, which seems to originate from this cell, appears to make a number of *en passant* terminations with the proximal basal dendrites of an adjoining (unlabeled) neuron.

Thus, until the anatomy of a large sample of HRP-labeled cells is compared with high-quality Golgi impregnations (and possibly with EM studies as well) the user of this technique should be cautioned against drawing hasty morphological and functional conclusions on the basis of HRP-labeled material alone.

V. Addendum

Since the initial preparation of this chapter, a number of procedural modifications have been published or transmitted to the authors, which we have felt justified inclusion in this chapter.

A. Modified Benzidine Incubation Procedure (Blue Reaction Product)

Mesulam (1976) has published a modification of Straus's (1964) technique in which the majority of incubation steps are carried out at room temperature. Only the sodium nitroferricyanide stabilizing bath is processed at $0°$ to $4°C$. In our laboratory this procedure has been further modified to allow all steps to be carried out at room temperature (procedure courtesy of C. Gall). The following solutions are required.

1. Benzidine stock solution. This solution should contain these final concentrations of ingredients in water
     0.1% Benzidine w/v
     5% Sodium acetate to give the solution oa pH of 5
     15% Ethanol
2. H_2O_2 solution. Must be made up fresh
     0.8 ml of 3% H_2O_2 in 90 ml of 30% ethanol
     0.6 ml of solution 3
3. Stabilizing bath
     9% Sodium nitroferricyanide in 50% ethanol

Tissue is perfused with 10% unbuffered formalin followed by a wash and subsequent storage in 10% sucrose at $4°C$. The length of time that the tissue is exposed to formalin should be as short as possible since formalin does denature HRP, and weakly labeled cells or processes could become unobservable.

Preincubate sections in a 1:1 mixture of solution 1 and 30% ethanol for 20 min. Transfer sections to a 1:1 mixture of solutions 1 and 2 for 20 min. Stabilize the reaction product in solution 3 for 20 min followed by rinsing in several changes of water. Mount sections on slide (may also be done prior to histochemical processing as described earlier). Allow to air dry. Counterstain with safranin or neutral red.

Sections prepared in this manner appear to be free of the crystalline artifacts that readily appear in sections prepared by the earlier method. They also appear to demonstrate a higher sensitivity, in that anterograde transport is more frequently observed (G. Rose, personal communication).

In a recent publication, Adams (1977) has described a number of technical refinements in the use of HRP as a neuronal tracer. He draws attention to the importance of removing the fixative as rapidly as possible following adequate tissue fixation to prevent denaturation of the HRP label. A postfixation perfusion with a sucrose–dimethyl sulfoxide solution is recommended. The result is a severalfold increase in sensitivity of the procedure. Adams also recommends the addition of hypochlorite bleach (e.g., Chlorox) to waste solutions containing benzidine to oxidize it to a nontoxic state.

B. Electron Microscopy

Drs. H. Karten and Christine Laverack (Departments of Psychiatry and Behavioral Science, SUNY, Stony Brook, N.Y.) have kindly supplied the following protocol for EM examination of HRP-containing neural tissue.

1. Following the desired survival time after HRP injection, the animal is killed and perfused with Reese–Karnovsky 1% PFA + 1.5% glutaraldehyde; the brain is stereotaxically blocked into 2- to 10-mm-thick slabs, *in situ*.

2. The tissue is mounted on a Vibratome stage with Eastman 910 or similar cyanoacrylate adhesive. Do not freeze the tissue for sectioning.

3. The Vibratome trough is filled with 0.1 M phosphate-buffered sucrose and the solution is maintained at 4°C with a circulating cooler.

4. Cut serial sections at 40 to 100 μm and collect in phosphate-buffered sucrose.

5. Individual sections are incubated for either brown or blue reaction product. A modified version of Mesulam's blue protocol (1976) is used, having a much reduced alcohol concentration, since the high alcohol concentration in the original produces poor EM results.

6. Individual sections of the HRP-containing cells are treated with osmium, dehydrated, and embedded in Epon 812 in the usual manner, in either a flat mold or in the flat end of a BEEM capsule.

7. The osmic section is light brown, the labeled cells brown to black. Individual cells can be localized within the untrimmed block by using a dissecting microscope having substage illumination. Alternatively, the use of a conventional microscope with a "high-dry" objective for use without coverslips allows detailed visualization of dendrites.

8. Cut 1-μm sections and stain with Richardson's or a similar stain. Examine with a bright-field/dark-field light microscope to further identify and localize the labeled cell. Polarization and Nomarski optics are of even greater value.

9. Trim the selected area to form a mesa. Cut thick (1-μm) sections from the mesa to further confirm the cell and then cut thin (1000-Å) sections.

10. Sections of 1000 Å on a grid can be examined directly with Nomarski optics to confirm the location of HRP. The same grid is then examined under an electron microscope. Alternatively, the section on the grid can be examined

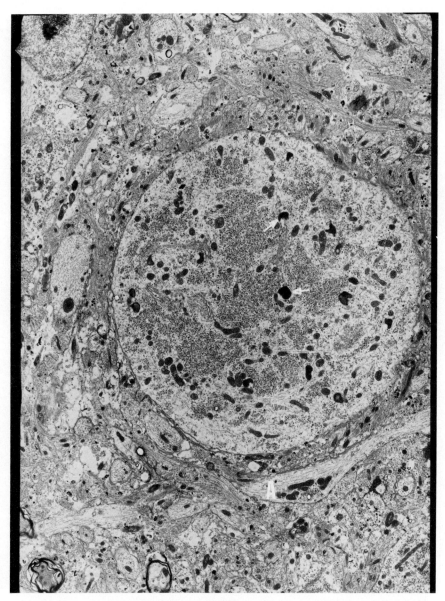

FIG. 8. Soma of a labeled retrogradely with HRP from the vagal dorsal motor nucleus of the pigeon. This cell shows numerous densely staining lysosomes containing HRP, distributed throughout the cytoplasm (arrows). Note the HRP labeled terminal (A). Magnification: × 10,000. (Photograph courtesy of Drs. H. Karten and C. Laverack.)

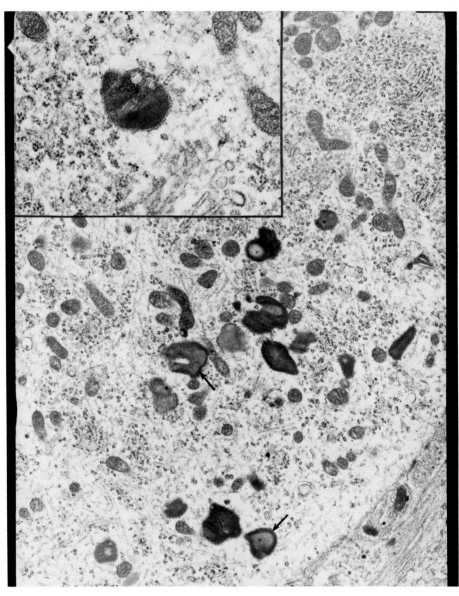

FIG. 9. A view at higher magnification (× 20,000) of another cell showing the HRP-containing lysosomes in more detail. A laminar structure is evident in several of the lysosomes (arrows). Inset: HRP-containing lysosomes showing the confining membrane (× 50,000). (Photograph courtesy of Dr. H. Karten and C. Laverack.)

under a light microscope and then under a scanning-transmission electron microscope.

There are several important points that must be observed if one hopes to obtain optimal results:

1. Do not use frozen sections. They convert the brain to Swiss cheese.

2. The blue reaction is not as consistently electron dense as the brown reaction product. This may, however, be due to the presence of ethanol; work is in progress to attempt to rectify this.

3. In general Karten and Laverack urge caution in the interpretation of an electron-dense profile. A lysosome always looks like a lysosome, no matter what it contains. Therefore, the correlation with the light microscope picture is absolutely mandatory. This too may be ambiguous, but at least it reduces potential errors.

References

Adams, J. C. Technical considerations on the use of horseradish peroxidase as neuronal marker. *Neuroscience*, 1977, **2**, 141–145.

Colman, D. R., Scalia, F., & Cabrales, E. Light and electron microscopic observations on the anterograde transport of horseradish peroxidase in the optic pathway in the mouse and rat. *Brain Research*, 1976, **102**, 156–163.

de Olmos, J. An improved HRP tract tracing technique. *Exp. Brain Res.*, 1978, in press.

Geisert, E. E. The use of horseradish peroxidase for defining neuronal pathways: A new application. *Brain Research*, 1976, **117**, 130–135.

Graham, R. C., & Karnovksy, M. J. The early stages of absorption of injected horseradish peroxidase in the proximal tubules of the mouse kidney: Ultra-structural cytochemistry by a new technique. *Journal of Histochemistry and Cytochemistry*, 1966, **14**, 291–302.

Kitai, S. T., Kocsis, J. D., Preston, R. J., & Sugimori, M. Monosynaptic inputs to caudate neurone identified by intracellular injection of horseradish peroxidase. *Brain Research*, 1976, **109**, 601–606.

Kristensson, K., & Olssen, Y. Retrograde axonal transport of protein. *Brain Research*, 1971, **29**, 363–365.

LaVail, J. H., & LaVail, M. M. Retrograde axonal transport in the central nervous system. *Science*, 1972, **176**, 1416–1417.

LaVail, M. M., & LaVail, J. H. Retrograde intraaxonal transport of horseradish peroxidase in retinal protein for identification of cell bodies of origin of axons terminating within the CNS. *Brain Research*, 1973, **58**, 470–477.

Lynch, G., Gall, C., & Deadwyler, S. Labelling of central nervous system neurones with extracellular recording microelectrodes. *Brain Research*, 1974, **66**, 337–341. (a)

Lynch, G., Gall, C., Mensah, P., & Cotman, C. Horseradish peroxidase histochemistry: A new method for tracing efferent projections in the central nervous system. *Brain Research*, 1974, **65**, 373–380. (b)

Lynch, G., Smith, R. L., Browning, M. D., & Deadwyler, S. Evidence for the bidirectional dendritic transport of horseradish peroxidase. *Advances on Neurology*. 1975, **12**, 297–314.

Lynch, G., Smith, R. L., Mensah, P., & Cotman, C. Tracing the dentate gyrus mossy fiber system with horseradish peroxidase histochemistry. *Experimental Neurology*, 1973, **40**, 516–524.

Mesulam, M. M. The blue reaction product in horseradish peroxidase neurohistochemistry: Incubation parameters and visibility. *Journal of Histochemistry and Cytochemistry*, 1976, **24**, 1273–1280.

Robertson, R. T. Bidirectional movement of horseradish peroxidase and the demonstration of reciprocal thalamocortical connections. *Brain Research*, 1977, **129**, 538–544.

Snow, P. J., Rose, P. K., & Brown, A. G. Tracing axons and axon collaterals of spinal neurons using intracellular injections of horseradish peroxidase. *Science*, 1976, **191**, 310–311.

Sotelo, C., & Riche, D. The smooth endoplasmic reticulum and the retrograde and fast orthograde transport of horseradish peroxidase in nigro striatonigral loop. *Anatomy and Embryology*, 1974, **146**, 209–218.

Spencer, H. J., Gribkoff, V. K., Cotman, C., & Lynch, G. S. GDEE antagonism of iontophoretic amino acid excitations in the intact hippocampus and in the hippocampal slice preparation. *Brain Research*, 1976, **105**, 471–481.

Straus, W. Factors affecting the cytochemical reaction of peroxidase with benzidine and the stability of the blue reaction product. *Journal of Histochemistry and Cytochemistry*, 1964, **12**, 462–469.

Winfield, D. A., Gatter, K. C., & Powell, T. P. S. An electron microscopic study of retrograde and orthograde transport of horseradish peroxidase to the lateral geniculate of the monkey. *Brain Research*, 1975, **92**, 462–467.

Worthington Biochemical Corporation. *Worthington enzyme manual*. Worthington Biochemical Corporation, Freehold, New Jersey, 1972.

Chapter 11

Neurochemical Effects of Lesions

Ronald M. Kobayashi

Neurology Service
Veterans Administration Hospital
San Diego, California
  and
Department of Neurosciences
University of California
La Jolla, California

I. Introduction

The study of the effects of localized destructive lesions is a broadly applied technique for the investigation of both functional and anatomical organization of

317

the central nervous system (CNS). Destruction of the cell body produces loss of all axonal projections arising from the cell. Destruction of the axon produces anterograde degeneration of the distal axon and degeneration of terminals at a distance from the cell body. If the axon is severed proximal to sustaining collateral branches, the cell body will also undergo retrograde degeneration. Thus, neuronal degeneration will result from either destruction of cell bodies or transection of axons. This process may be systematically examined to elucidate neurochemical properties, as well as neural structure.

For normal neurotransmitter function to occur, it is necessary that neuronal integrity be preserved. Lesions may affect certain neurotransmitter characteristics or "markers," which identify the presence and integrity of neural pathways associated with specific neurotransmitter agents. Biochemical markers of special interest include neurotransmitter concentration, specific enzymes associated with particular neurotransmitters, and the presence of a high affinity uptake system for the neurotransmitter or related compounds.

To be most useful, any observed neurochemical effects should be interpreted in conjunction with neuroanatomical studies indicating a specific relationship of cell bodies to axonal projections and terminals. This adheres to the concept of localization of function, which holds that specific components of the nervous system control specific functions (Moore, 1970). Present anatomic maps of CNS catecholamine pathways are largely the result of formaldehyde-induced histofluorescent studies after lesions or drug treatment (Dahlstrom & Fuxe, 1964; Ungerstedt, 1971b). More recently, studies based on glyoxylic acid fluorescence have permitted improved visualization of catecholaminergic terminals and cell bodies (Lindvall & Bjorklund, 1974; see Moore & Loy, this volume).

The purpose of this chapter is to illustrate the neurochemical effects of lesions as a tool in tracing axonal projections and function, as well as in elucidating neurotransmitter properties. Major emphasis will be placed upon the catecholamines, norepinephrine (NE) and dopamine (DA). Only representative studies will be presented, since it is not within the scope of this review to comprehensively include all related work.

II. Lesions and Anatomic Methods

A. Stereotaxic Technique

A variety of methods are available for producing localized lesions of the nervous system, and these will be briefly reviewed. For a comprehensive and timely review, the reader is referred to Moore (1978). Production of reproducible lesions which are accurately localized to a particular site is based on the stereotaxic technique. Simply stated, three important elements in this technique are the stereotaxic instrument, anatomic references, and method of creating the lesion. The stereotaxic instrument is designed to rigidly fix the head of the animal at three or more points. Rigid fixation of the skull to the rigid stereotaxic apparatus then results in a constant relationship of the brain to the instrument. A series of calibrated adjustments on the instrument permits precise localization in three different dimensions. Anatomic references in three dimension have been developed for many animal species, and these stereotaxic atlases serve as aids to accurately direct the lesion-producing device to the desired location. The more frequently utilized methods for lesion production are electrolysis, heat produced by radio-frequency current, aspiration, surgical knife, and injection of neurotoxic amine derivatives. Only the latter method is selective; the former methods are nonselective in that they destroy, more or less, all neural elements within the lesion site. Specific details of these methods have been recently reviewed (Breese, 1975; Moore, 1978).

1. SPECIFIC LESION TECHNIQUES

Before one can fully evaluate the neurochemical effects of lesions, some features of the lesion techniques need to be considered. Electrolytic lesions utilize fine needle electrodes capable of more discrete lesions than those produced by the larger radio-frequency electrode. However, the lesions produced by radio frequency tend to be more reproducible and more spherical. Aspiration is particularly well suited to destruction of cortical rather than of subcortical structures. Surgical knife methods are principally useful for transection of fiber bundles, such as the medial forebrain bundle or dorsal noradrenergic bundle. A variation of this technique, developed by Halasz and Pupp (1965), permits a sweeping cut around the base of the brain which totally or partially isolates the hypothalamus from afferent (and efferent) impulses. This lesion is particularly suited for the analysis of neuroendocrine function involving the hypothalamus.

Two chemical analogs of catecholamines, 6-hydroxydopa (6-OHDOPA) and 6-hydroxydopamine (6-OHDA), selectively destroy catecholamine elements (Breese, 1975). This method has the special feature of producing widespread effects on brain catecholamines after parenteral, intraventricular, or intracisternal administration. Direct intracerebral injections of 6-OHDA have been utilized to study effects of more restricted damage. Since 6-OHDOPA passes the blood–brain barrier to a greater extent than 6-OHDA, parenteral administration of 6-OHDOPA may be utilized to study effects on both peripheral and central catecholamine neuron systems. Preferential depletion of brain NE may be achieved by successive injections of small doses of intraventricular 6-OHDA; DA is preferentially depleted if the 6-OHDA is preceded by pretreatment with desmethylimipramine, which prevents uptake of the neurotoxin into NE terminals (Breese & Traylor, 1971). Besides lesions produced by these techniques, spontaneously occurring lesions in humans may be analyzed for neurochemical effects. This approach is exemplified by the neurochemical studies of the nigro-striatal system in patients with Parkinson's disease. The finding of reduced striatal DA served as the rationale for therapy with L-dopa.

2. ASSESSMENT OF LESIONS

Once the lesion is produced, the localization and extent of the damage must be appropriately studied before neurochemical effects are evaluated. Different methods may be required to assess lesions produced by different techniques. For most lesion-producing techniques, conventional histological methods adequately demonstrate the tissue damage. However, when the neurotoxins are utilized, histological evidence of degeneration in catecholamine pathways or terminal areas may be absent, despite significant reductions in NE levels (Hedreen & Chalmers, 1972).

Regardless of the lesion technique employed, a zone of mechanical damage is produced by passage of the electrode, needle, or knife. If the instrument tract interrupts important structures other than the target site, a significant source of error may arise. To control for this potential problem, one can approach the target site from different angles, and thus avoid consistently damaging other areas. For example, Moore and Heller (1967) transected the medial forebrain bundle by a conventional direct approach and compared these results with a contralateral approach directed along an oblique path. No difference in neurochemical effects were observed, suggesting that the electrode tract was not responsible for the observed reduction in amines.

Interpretation of lesion effects should also take into consideration the post-lesion survival period. In the peripheral nervous system, sympathetic denervation results in an abrupt decrease in tissue catecholamines within several days (Kirkepar, Cervoni, & Furchgott, 1962). However, in the CNS, the decrease is delayed and gradual. When the medial forebrain bundle was unilaterally lesioned, maximal decrease in brain NE and serotonin required 12 days and was not further reduced when examined at 35 days after the lesion (Moore & Heller, 1967). More rapid terminal degeneration has been observed in the cerebral cortex, where degeneration of terminals was complete within 9 days after transection of axons (Colonnier, 1964).

The analysis of the effects of lesions in the acute period, that is less than 2 weeks after the lesion, must consider effects of the anesthetic used, nonspecific trauma of the procedure such as intracerebral hemorrhage and altered cerebral spinal fluid dynamics (Moore, 1976), altered food and water consumption, or involvement of other organ systems (Amaral & Foss, 1975). After the first 2 weeks, one must consider tissue reparative processes, possible retrograde or transneuronal degeneration, and regenerative sprouting in the denervated area (Moore, 1978).

B. Anatomic Methods

1. REGIONAL DISSECTION

After verification of the lesion, the next step is the removal of brain parts for the desired neurochemical studies. In simplest outline, the entire brain may be taken for study after a bilateral lesion of a paired structure, or one-half of the brain may be compared to the other side after a unilateral lesion. Regional dissection methods separating the brain into parts such as hypothalamus or cortex offer the advantage of assessing effects in smaller, more homogeneous brain parts (Glowinski & Iversen, 1966). Enhanced reproducibility is also possible with a "brain slicer" composed of fine wires which transect the fresh brain at fixed intervals of thickness (Segal & Kuczinski, 1974) and with use of a McIlwain tissue chopper (R. E. Zigmond and Ben-Ari, 1976).

2. MICRODISSECTION

Recently, a microscopic method to remove discrete nuclei from frozen microtome-sectioned brains has been described (Palkovits, 1973; Kobayashi, Palkovits, Jacobowitz, & Kopin, 1975; Kobayashi, Palkovits, Kopin, & Jacobowitz, 1974). Hollow stainless-steel needles are used with a dissecting microscope to remove pellets of tissue from tissue slices, and these may be examined for neurotransmitters, enzymes, uptake, and turnover properties (see below). Anatomic landmarks have been described for removal of nuclei from the hypothalamus (Palkovits, Brownstein, Saavedra, & Axelrod, 1974b), brainstem (Palkovits, Brownstein, & Saavedra, 1974a), and limbic system (Palkovits, Saavedra, Kobayashi, & Brownstein, 1974c). Neurochemical effects of lesions thus may be examined in specific nuclei and selective effects may be evident in one nucleus, but not in adjacent ones (Kobayashi et al., 1975; Kobayashi, Palkovits, Kopin, & Jacobowitz, 1974; Kobayashi, Lu, Moore, & Yen, 1976a; Kobayashi, Palkovits, Kizer, Jacobowitz, & Kopin, 1976b). At these more precise levels of dissection, it has become increasingly possible to attempt neurochemical studies in an amount of brain tissue which is anatomically less heterogeneous than was permitted by other dissection methods.

III. Effects of Lesions on Specific Neurotransmitter Properties

A. Neurotransmitter Concentrations

1. MEASUREMENT OF CATECHOLAMINES

Probably the most widely used neurochemical technique for mapping neuronal projections is the measurement of amine reduction after lesions. Three different

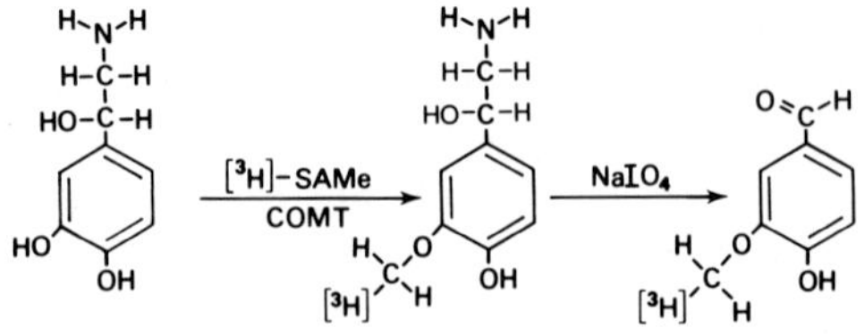

Reaction for the assay of Norepinephrine by enzymatic O-methylation

[³H]-SAMe = S-Adenosyl methionine-methyl-[³H]

COMT = Catechol-O-methyl transferase

FIG. 1. Reaction for the assay of norepinephrine by enzymatic O-methylation. NE is O-methylated in the presence of [³H]-S-adenosylmethionine ([³H]-SAMe) to normetanephrine which is oxidized to vanillin. This product is separated by organic extraction and radioactivity is measured by scintillation spectrometry. COMT, Catechol-O-methyltranferase.

biochemical methods have been employed to quantify the effects of lesions on amines: fluorometry, enzymatic–radioisotopic determination, and mass fragmentography. Sufficient sensitivity to permit measurement of amine change in small brain parts is achieved particularly by the last two methods. The enzymatic–radioisotopic method is based on the conversion of NE and DA by the enzyme catechol-O-methyltransferase (COMT) or the enzyme phenylethanolamine-N-methyltransferase (PNMT) to radioisotopically labeled methylated derivatives, which are then separated by organic solvent extraction (Coyle & Henry, 1973; Henry, Starman, Johnson, & Williams, 1975) (see Fig. 1). This is a sensitive, specific, and relatively rapid method. It is sensitive to levels of approximately 100 femtomoles (fmoles) for NE and 330 fmoles for DA. The assay for DA is contaminated by NE (if present in large amounts approaching the level of DA), and epinephrine and NE are not distinguishable by the COMT method (Coyle & Henry, 1973), but are separable by the PNMT method (Henry et $al.$, 1975). Mass spectroscopy combined with gas chromatography is the most sensitive and specific method, detecting levels of amines and metabolites to the femtomole level in tissue containing as little as $10 \mu g$ protein (Abramson, McCaman, & McCaman, 1974; Koslow, Cattabeni, & Costa, 1972). The major limitations of the technique are inability to assay large numbers of samples and expense of the instrumentation. As these limitations become reduced, mass fragmentography is likely to become a more widely applied method, especially where maximal sensitivity and specificity are desired.

2. MODEL FOR BIOCHEMICAL MAPPING

One of the most widely studied catecholamine systems arises from the locus coeruleus. This is a paired nucleus located in the roof of the pons (Fig. 2) and is

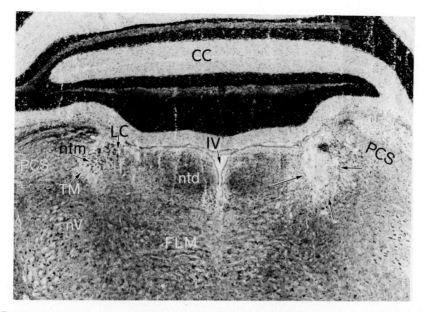

FIG. 2. Frontal section through locus coeruleus. Right half received a lesion (arrows) and normal structures are indicated on left half. Luxol fast blue–cresyl violet stain. Magnification: × 21. CC, cerebellar cortex; FLM, fasciculus longitudinalis medialis; LC, locus coeruleus; ntd, nucleus tegmenti dorsalis; ntm, nucleus tractus mesencephalicus nervi trigemini; nV, nucleus motorius nervi trigemini; PCS, pedunculus cerebellaris superior; TM, tractus mesencephalicus nervi trigemini; IV, fourth ventricle.

the principal source of noradrenergic neurons innervating the cerebral and cerebellar cortices and hippocampus, as well as diencephalic nuclei. A biochemical model to study axonal projections has been described based on neurotransmitter concentrations after a unilateral lesion of the locus coeruleus (Kobayashi *et al.*, 1974, 1975). This model is illustrated in Fig. 3. If innervation to a particular region is totally from the ipsilateral locus coeruleus, the regions on the side of the lesion should be depleted of NE and the corresponding contralateral region should contain the normal amount of the transmitter. If innervation is totally crossed, the NE contralateral to the lesion should be depleted while NE ipsilateral to the lesion remains intact. If innervation is equally distributed bilaterally to a particular region, NE on both sides should be reduced to one-half of normal. Under any of these conditions, complete destruction of one of the paired nuclei should reduce by 50% the sum of the NE content of regions receiving their total innervation from the locus coeruleus. Variables in this model include the completeness of the lesion, innervation by more than one nucleus or by mixed unilateral and bilateral projections, and time course of effects. Appropriate con-

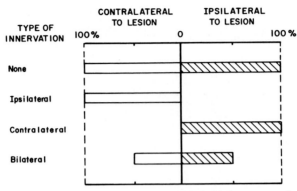

FIG. 3.   Model to indicate innervation pattern based on norepinephrine concentration after unilateral lesion of the putative locus of origin. Corresponding regions contralateral and ipsilateral to the lesion are represented as percentage of normal (control) values for the specific region. (Reprinted from *Neurology*, 1975, **25**; 231 with permission from The New York Times Media Co., Inc.).

trols for a unilateral locus coeruleus lesion include the contralateral nonlesioned side, sham-operated animals, and nonoperated animals. The last two types of controls are particularly important to reveal contralateral effects, which might otherwise be inapparent. Statistically, the effects of lesions on corresponding regions on operated and nonoperated sides of the same animal may be analyzed by a t test for paired comparison or by analysis of variance (Sokal & Rohlf, 1969).

3.   Locus Coeruleus Lesions and Catecholamines

Application of the model for biochemical mapping is illustrated by Fig. 4. After unilateral electrolytic lesions of the locus coeruleus, NE was measured radioisotopically in nuclei removed by microdissection (Kobayashi *et al.*, 1974, 1975). NE levels declined significantly in the ipsilateral hypothalamic paraventricular and periventricular nuclei, thalamic ventral and anterior ventral nuclei, the habenular nuclei, cerebellum, hippocampus, and cerebral cortex with no change in regions contralateral to the lesion, suggesting unilateral innervation to these nuclei and regions. Bilateral decrease in NE levels consistent with partial bilateral innervation was observed for the medial geniculate body, the inferior colliculus, and the posterior half of the cerebellar cortex. Regions in which NE levels did not change, suggesting no innervation from the ipsilateral locus coeruleus, included the medial preoptic nucleus, nucleus interstitialis stria terminalis (ventralis), hypothalamic dorsomedial nucleus, medial forebrain bundle, and the inferior olive.

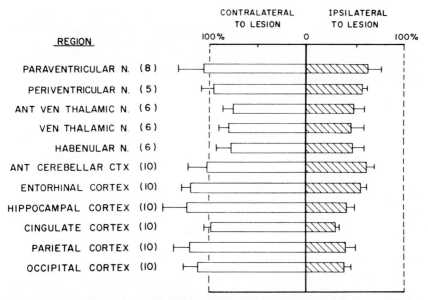

FIG. 4.   Brain norepinephrine concentration in 11 regions after unilateral locus coeruleus lesion. Brain regions contralateral and ipsilateral to the lesion are represented as percentage of normal values (nonoperated controls) for the corresponding region. Each bar represents the mean ± SEM for the numbers of pairs of samples indicated in parentheses. (Reprinted from *Neurology*, 1975, **25**, 229 with permission from The New York Times Media Co., Inc.).

4. AXONAL LESIONS AND AMINES

Besides placing lesions in the region of cell bodies, axonal tracts may be selectively destroyed and neurotransmitter levels examined. The effects of unilateral transection of the medial forebrain bundle in the rat lateral hypothalamus on both serotonin and NE levels in the brain half ipsilateral to the lesion were investigated by Moore and Heller (1967). Lesions were induced electrolytically; brains were dissected into cortex, septum, striatum, amygdala, hippocampus, and brainstem; amines were measured fluorometrically. The monoamine decrease first appeared on the postoperative days 2 and 3, progressed thereafter, and was complete by day 12.

Unilateral injection of 6-OHDA into the dorsomedial reticular formation in the caudal mesencephalon (dorsal noradrenergic bundle) interrupted the axonal projection from the locus coeruleus (Lidbrink & Jonsson, 1974). At postoperative day 2, the NE in the ipsilateral cortex increased about 75%, fell at day 5, and at day 15 was only about 15% of control. Hypothalamic NE did not change over the

2-week period. The initial increase was suggested as being due to increased synthesis, since the rise was prevented by α-methyl-p-tyrosine, a potent inhibitor of tyrosine hydroxylase activity.

Unilateral dorsal bundle lesions have been stereotaxically produced with a knife (Roizen, Kobayashi, Muth, & Jacobowitz, 1976). At day 38 postlesion, regions on each side of the brain were removed by microdissection, and NE levels were measured isotopically. In agreement with a previous study in which the locus coeruleus was lesioned (Kobayashi et al., 1975), ipsilateral NE reduction was demonstrated in the hypothalamic periventricular nucleus, cingulate cortex, and hippocampal cortex. Failure to demonstrate changes in nuclei previously shown to be innervated by the locus coeruleus suggested that alternate routes for axons or recurrent collaterals or sprouting may have occurred in the 38 days since the lesion was produced. This study indicates that the site of lesion placement along the neural pathway and the time effects may be important in interpreting data about the biochemical effects of lesion.

5. Intraventricular 6-OHDA and Hypothalamic Catecholamines

Injection of the catecholamine toxin 6-OHDA into the lateral ventricles produces less prominent degeneration of the hypothalamus as compared to other regions. The most widely accepted interpretation is inaccessibility of 6-OHDA to this area. More direct accessibility of the 6-OHDA to the hypothalamus is accomplished by injection into the third ventricle. When measured 15 days after third ventricular injection, hypothalamic NE was reduced to 17% of control value while DA was normal (Cuello, Shoemaker, & Ganong, 1974). In contrast to these results, a marked depletion to the range of 14–20% of control values was observed for both DA and NE in the arcuate nucleus and median eminence 10 days after third ventricular injection of 6-OHDA (Kizer, Arimura, Schally, & Brownstein, 1975).

B. Enzyme Activity

1. Locus Coeruleus Lesions and Enzymes

Brain lesions may alter enzymatic activity and serve as a biochemical marker of axonal projections. The noradrenergic enzyme dopamine β-hydroxylase (DNH) catalyzes the conversion of DA to NE and is a specific marker of noradrenergic neurons. The enzymatic assay of DBH activity is based on the sequential conversion of either phenylethylamine or tyramine by DBH to a radioactive N-methyl derivative by reaction with the enzyme phenylethanolamine-N-methyltransferase in the presence of [^{14}C]-S-adenosylmethionine (Molinoff, Weinshilboum, & Axelrod, 1971). This reaction sequence is summarized in Fig.

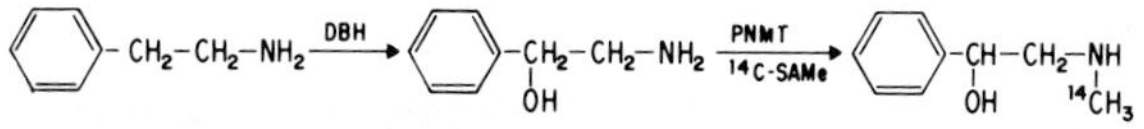

FIG. 5. Dopamine β-hydroxylase (DBH) assay reaction sequence. PNMT, Phenylethanolamine-N-methyltransferase; [^{14}C]-SAMe; [^{14}C]-S-adenosylmethionine.

5. The N-methyl derivatives are separated by solvent extraction and their radioactivity is determined. DBH activity was examined following unilateral electrolytic lesions of the rat locus coeruleus (Ross & Reis, 1974). At day 21 postlesion, DBH activity in the ipsilateral olfactory bulb, frontal cortex, and hippocampus was less than 20% of nonoperated control animals. Reduction in DBH activity was observed in the ipsilateral thoracic spinal cord, and in hypothalamus and cerebellum, portions of the mesencephalon, thalamus, and telencephalon. Failure to observe complete loss of DBH activity was ascribed to incomplete destruction of the locus coeruleus, or in the case of the cerebellum, innervation by neurons within the superior cerebellar peduncle which escaped destruction. DBH activity in the lower brainstem was unchanged. When the time course of effects was examined from 3 to 40 days after the operation, maximal depletion of enzyme activity had occurred by 12 days with no further reduction at 21 days and 40 days for most regions.

Reversible increase in tyrosine hydroxylase (TH) activity in the opposite locus coeruleus was observed after a unilateral electrolytic lesion of the locus coeruleus (Buda, Roussel, Renaud, & Pujol, 1975). Nuclei were removed by microdissection. This increase appeared on day 2 postlesion, was maximal (77% rise) at day 4, and returned to normal by day 13. Altered enzyme kinetics suggested that this was related to an increase in the number of active enzyme sites. This effect was selective for the locus coeruleus and was not present in dopaminergic cell bodies or the serotonergic cell bodies of the dorsal raphe nucleus on either side of the lesion.

2. LATERAL HYPOTHALAMIC LESIONS AND ENZYMES

Posterolateral hypothalamic lesions destroying axons ascending from the locus coeruleus in the medial forebrain bundle were utilized to study the dynamics of anterograde and retrograde changes in DBH activity (Reis & Ross, 1973). In the ipsilateral cerebral cortex, DBH activity decreased to 30% of control by day 14 and remained low for up to 45 days. In the ipsilateral cerebellum and brainstem, DBH activity initially rose during the first 2 days, then fell below control from 7–14 days, and returned to normal in 21 days. In a block of pontine tissue containing the locus coeruleus, DBH activity fell to 52% of control by day 12 postlesion, but was normal at day 28. More anteriorly placed lesions in the lateral hypothalamus did not alter DBH activity in the brainstem, indicating that the retrograde changes are a function of the distance of cell bodies from the site of axonal transection. Cortical DBH activity fell, indicating that ascending noradrenergic axons were interrupted. In these studies, the changes caudal to the lesion site were of particular interest, since they indicated a reversible change in DBH activity as part of the retrograde reaction.

Further neurochemical studies (Ross, Joh, & Reis, 1975) indicated that activity of tyrosine hydroxylase (TH) was reduced along with DBH, and these reductions were attributable to reduced accumulation of specific enzyme protein. This reduction was demonstrated by the technique of immunotitration, in which increasing amounts of tissue enzyme are added to fixed amounts of antibody. As more tissue enzyme is added, it binds the available antibody and increasing amounts of enzyme remain in the supernatant. The specific DBH protein from the locus coeruleus of lesioned animals at day 14 was decreased to 56% of controls, which was identical to the percentage fall in DBH activity. Similarly, the specific TH protein in the locus coeruleus in lesioned animals was reduced to 67% of controls, in close correspondence to the 64% fall in enzyme activity. No changes in the activities of the nonspecific enzymes dopa decarboxylase and monoamine oxidase in the ipsilateral locus coeruleus of lesioned animals was demonstrated at either day 14 or 21. This supports the use of TH and DBH as specific aminergic markers. There was no reduction in the number of locus coeruleus neurons, indicating that the reduced activity of DBH and TH is due to reduction of enzyme protein. The possibility was raised that reaction to axonal injury in some CNS neurons produces reversible neurochemical changes in the absence of classical chromatolysis. These results suggest that during the retrograde reaction selective priority is given for protein metabolism to resynthesize structural components at the expense of enzyme proteins involved in neurotransmission.

3.   VENTRAL NORADRENERGIC BUNDLE LESIONS AND ENZYMES

It is generally held that DA in the median eminence is related to the tuberoinfundibular DA system while NE arises from brainstem neurons via the ventral noradrenergic pathway. The contribution by locus coeruleus neurons to the median eminence and to the hypothalamus is considered to be minor. Hypothalamic catecholamine innervation was examined by measurement of DBH and TH activity in hypothalamic nuclei after bilateral knife lesions of the ventral noradrenergic bundle (Kizer, Muth, & Jacobowitz, 1976). Two weeks later, DBH activity was markedly depleted to 14% of control in the ventromedial nucleus, to 8% in median eminence, and was undetectable in the arcuate nucleus. Significant reduction of DBH activity to 40–60% of control occurred in the medial preoptic, periventricular, paraventricular, supraoptic, and dorsomedial nuclei and posterior medial forebrain bundle. TH activity in the mediobasal hypothalamic nuclei was unchanged. These results provide supporting evidence that noradrenergic innervation to the mediobasal hypothalamus originates from lateral medullary neurons. The incomplete reduction in DBH activity in the other hypothalamic nuclei is consistent with mixed innervation from both the locus coeruleus and from other brainstem neurons.

4. INTRAVENTRICULAR 6-OHDA AND ENZYMES

In addition to reducing catecholamine levels, intraventricular 6-OHDA significantly reduced the activity of the enzymes TH and DBH in terminal and cell body regions of both dopaminergic and noradrenergic neuronal systems (Sorimachi, 1975).

Selectivity of this neurotoxin toward catecholaminergic neurons was utilized to examine whether the enzymes dopa decarboxylase and 5-hydroxytryptamine decarboxylase could be differentiated. It is generally held that decarboxylation in both the catecholaminergic and serotonergic pathways is catalyzed by a nonspecific enzyme, but the extent to which these enzymes are similar in unclear. After intracisternal 6-OHDA in rats pretreated with pargyline, dopa decarboxylase activity was significantly reduced to 30–60% of control values in all ten brain regions examined at day 15 postinjection (Sims & Bloom, 1973). By contrast, 5-hydroxytryptamine decarboxylase activity did not decrease significantly in these same ten regions. These results suggest that the two enzymes are distinctive.

Histochemical data corroborate the biochemical evidence that 6-OHDA causes degeneration of noradrenergic cell bodies in the locus coeruleus. By fluorescence microscopy, disappearance of the locus coeruleus was observed after intraventricular 6-OHDA even when examined 3 months later (Descarries & Saucier, 1972). This change was attributed to retrograde degeneration of cell bodies after axonal injury.

5. INTRAPARENCHYMAL 6-OHDA AND ENZYMES

Although susceptible to intraventricular 6-OHDA, the NE cells in the locus coeruleus are resistant to direct intraparenchymal injections, while DA cell bodies in the substantia nigra are very sensitive (Ungerstedt, 1971a). Noradrenalin fluorescence in the locus coeruleus persisted after direct injection of 6-OHDA, even with large amounts (8 μg) of the catecholeamine injected into the locus coeruleus (Ungerstedt, 1971a). This suggested greater resistence to 6-OHDA of cell bodies in the locus coeruleus as compared to cell bodies located elsewhere and as compared to terminals. Since the destructive effects of 6-OHDA appeared dependent on uptake of the neurotoxin, the much greater uptake activity in terminal sites may explain the greater destruction in these regions as compared to the cell body regions.

6. CORRELATION OF NEUROCHEMICAL EFFECT WITH EXTENT
   OF LESION

Correlation of the neurochemical effect with placement of the lesion is
exemplified by the work of Moore, Bhatnagar, and Heller (1971). In studying the
cat nigrostriatal pathway, electrolytic lesions of the medial internal capsule were
observed to reduce caudate DA, TH, and dopa decarboxylase by 42–49%. When
the lesion was larger and invaded both the medial internal capsule and the lateral
hypothalamus, the reductions in caudate DA, TH, and dopa carboxylase were
much more prominent, ranging from 72 to 89%. These chemical data indicate
that the neurochemical effects of lesions of the nigro-neostriatal projection are
proportional to the amount of the projection destroyed.

7. TIME COURSE OF EFFECTS AFTER HEMITRANSECTION

Changes with time in the activities of several different enzyme systems of the
rat substantia nigra and striatum after brain hemitransection at the mid-
hypothalamic level have been studied (McGeer, Fibiger, McGeer, & Brooke,
1973). TH activity declined rapidly in both the striatum and the substantia nigra
after hemitransection. However, compared to the nigra, reduction in the striatum
was more rapid (to 10% of control value vs 60% by day 2 postoperation) and
more complete (virtually absent compared to 30% at day 30 postoperation).
These results suggest that some nigral TH activity may be associated with
neurons whose axons were not transected, and that these axons project caudally
since rostrally directed axons would be interrupted by the extensive hemitransection.

C. Uptake

1. UPTAKE AS A BIOCHEMICAL MARKER

Neurons associated with specific transmitters are characterized by selective
uptake of that agent. This uptake system is specific for the neurotransmitter
agent, requires intact innervation, is related to the tissue content of the agent,
exhibits high affinity, and is saturated at increasing concentrations adhering to
Michaelis–Menten enzyme kinetics, and is temperature and ion dependent (Iversen, 1975; Kuhar, 1973). High affinity neuronal uptake occurs predominantly in
axonal terminals. Thus, uptake of the neurotransmitter serves as a useful marker
for the study of the biochemical organization of neuronal projections. Uptake
may be measured using synaptosomes (which are pinched off nerve endings
composed of presynaptic terminals with synaptic vesicles and a portion of the
postsynaptic membrane) collected by selective centrifugation (Kuhar, 1973).
These synaptosomal preparations are homogenized and incubated in a buffered
medium containing radioactively labeled neurotransmitter. After termination of

the reaction by immersion in an ice bath, the mixture is centrifuged, the pellet is repeatedly washed, and the supernatant is discarded. The pellet is then solubilized and the accumulated radioactivity, which reflects the amount of uptake, is measured by scintillation spectrometry.

2. UPTAKE OF NE AFTER LATERAL HYPOTHALAMIC LESIONS

The *in vitro* uptake of ^{3}H–NE by synaptosomes prepared from rat telencephalon was reduced by 40–50% 1 month after electrolytic lesions of the lateral hypothalamus (M. J. Zigmond, Chalmers, Simpson, & Wurtman, 1971). Normal uptake was preserved in homogenates from the contralateral side and from the brains of sham-operated animals. ^{3}H–NE uptake was reduced as early as 24 hr after the operation and was 32% reduced at day 3. No significant further decline was seen after 3 days, and the effect was persistent at days 70 and 140 postoperation. Reductions in NE concentrations paralleled the reduction in uptake, but the amount of decline, 60%, exceeded the reduction of uptake. This pattern of disparity has been commented on by Kuhar (1973) and was ascribed to nonspecific low-affinity uptake. The reduced uptake of ^{3}H–NE is consistent with loss of uptake sites as a consequence of anterograde degeneration of terminals after transection of noradrenergic axons in the lateral hypothalamus.

3. UPTAKE OF NE AFTER 6-OHDA

The uptake of ^{3}H–NE was reduced after 6-OHDA induced lesions of the dorsal NE bundle at the caudal mesencephalic level and by hemitransection (Lidbrink & Jonsson, 1974). After bilateral lesions, uptake in cerebral cortex slices was compared to untreated controls. After a unilateral lesion, uptake ipsilateral and contralateral to the lesion were compared. Comparison only to the contralateral side raises the possibility that bilateral effects, such as resulting from crossed innervation, might not be appreciated (Kobayashi *et al.*, 1975; Reis & Ross, 1973). At 5 days after a unilateral lesion, uptake in the ipsilateral cortex was reduced to 70% of the contralateral side. At 15 days, the uptake was 30–40% of the contralateral side and remained at this level for the next 3 months. This time course is in close correspondence to the reduction of endogenous NE and of DBH after destruction of ascending NE projections (Moore & Heller, 1967; Reis & Ross, 1973). The maximum reduction of ^{3}H–NE uptake of about 65% achieved by unilateral 6-OHDA lesions of the dorsal bundle was not further increased by bilateral lesions. This suggests that crossed innervation is not present. However, this indicates uptake in structures other than NE terminals. Possible alternative uptake sites include DA terminals and extraneuronal elements such as glia, pericytes, and endothelial cells (Lidbrink & Jonsson, 1974). Possible uptake of ^{3}H–NE into DA terminals is consistent with the observation that NE uptake was reduced only 30–50% in the entorhinal cortex and cingulate gyrus, limbic regions which have been shown recently to receive both DA and NE innervation

(Lindvall, Bjorklund, Moore, & Stenevi, 1974; Thierry, Blanc, Sobel, Stinus, & Glowinski, 1973).

4. UPTAKE OF DA IN MICROPELLETS

Recently ^{3}H–DA uptake has been measured in pellets of striatum removed by the microdissection method of Palkovits (1973) (Tassin, Cheramy, Blanc, Thierry, & Glowinski, 1976). Frozen brains are sectioned at $-7°C$ and homogenized in 0.25 M sucrose to produce crude synaptosomal preparations. Several details of these experimental techniques should be mentioned. ^{3}H–DA uptake was detectable only when the micropellets were homogenized; uptake was not detectable in nonhomogenized micropellets. It is generally held that only unfrozen fresh tissue may be used for uptake studies (Iversen, 1975; Kuhar, 1973). In this study, ^{3}H–DA uptake was accomplished on frozen tissue, but temperature is critical since cryostat microtome temperatures below $-7°C$ abolished uptake. Uptake into the frozen striatal preparations resembled the process in fresh tissues since it was unaffected by desmethylimipramine, inhibited by benztropine, and abolished by lesioning of the substantia nigra. Unilateral lesions of the substantia nigra produced by 6-OHDA abolished ^{3}H–DA uptake in the striatum 1 month later.

D. *Metabolism*

1. LOCUS COERULEUS LESIONS AND MHPG

The major metabolite of NE in the CNS is 3-methoxy-4-hydroxyphenylglycol (MHPG). Measurement of MHPG serves as an index of turnover and is considered to be a more accurate reflection of noradrenergic activity than is NE concentration. MHPG may be measured by gas chromatography (Walter & Eccleston, 1973) or fluorometrically after adsorption chromatographic separation (Meek & Neff, 1972). MHPG was measured in neocortex 3 weeks after electrolytic lesions of the locus coeruleus (Arbuthnott, Christie, Crow, Eccleston, & Walter, 1972). Unilateral lesions resulted in a 61% reduction of MHPG in the ipsilateral cortex and no change in the contralateral side, compared to nonoperated controls. Bilateral lesions reduced MHPG by 54–61%. Parallel reductions of NE and MHPG were observed 10 days after a unilateral electrolytic lesion of the locus coeruleus (Korf, Aghajanian, & Roth, 1973b). In the cortex and hippocampus ipsilateral to the lesion, NE was reduced by 78% and MHPG by 69%. No change in the contralateral side occurred, indicating no significant crossed innervation.

2. LOCUS COERULEUS LESIONS AND STRESS-INDUCED RISE IN MHPG

The role of the locus coeruleus in mediating the increased turnover of NE in the rat cerebral cortex after stress has been examined (Korf, Aghajanian, & Roth,

1973a). Application of electric foot shock increased the conversion of NE to MHPG. Acute lesions of the locus coeruleus (25 min before the shock) prevented the rise of MHPG on the side ipsilateral to the lesion, but not on the contralateral side. These results are consistent with activation of locus coeruleus neurons by the stress of electric foot shock.

IV. Conclusions

This chapter has provided a brief overview of techniques presently available for the study of neurochemical effects of lesions. Appropriate assessment of the neurochemical effects of lesions provide important information regarding both neurotransmitter properties and biochemical organization of the nervous system, particularly in relation to its basic structure. Correlation of these data with results derived from physiological and behavioral studies are likely to contribute to enhanced understanding of neural function.

Acknowledgments

This work was supported by a Veterans Administration Clinical Investigatorship and MH 26072.

References

Abramson, F. P., McCaman, M. W., & McCaman, R. E. Femtomole level of analysis of biogenic amines and amino acids using functional group mass spectrometry. *Analytical Biochemistry*, 1974, **57**, 482–499.

Amaral, D. B., & Foss, J. A. Locus coeruleus lesions and learning. *Science*, 1975, **188**, 377–378.

Arbuthnott, G. W., Christie, J. E., Crow, T. J., Eccleston, D., & Walter, D. S. The effect of unilateral and bilateral lesions in the locus coeruleus on the levels of 3-methoxy-4-hydroxyphenylglycol (MHPG) in neocortex. *Experientia*, 1972, **29**, 52–53.

Breese, G. R. Chemical and immunochemical lesions by specific neurotoxic substances and antisera. In L. L. Iversen, S. D. Iversen, & S. H. Snyder (Eds.), *Handbook of psychopharmacology*. Vol. 1. New York: Plenum, 1975. Pp. 137–189.

Breese, G. R., & Traylor, T. D. Depletion of brain noradrenaline and dopamine by 6-hydroxydopamine. *British Journal of Pharmacology*, 1971, **42**, 88–99.

Buda, M., Roussel, B., Renaud, B., & Pujol, J.-F. Increase in tyrosine hydroxylase activity in the locus coeruleus of the rat brain after contralateral lesioning. *Brain Research*, 1975, **93**, 564–569.

Colonnier, M. Experimental degeneration in the cerebral cortex. *Journal of Anatomy*, 1964, **98**, 47–53.

Coyle, J. T., & Henry, D. Catecholamines in fetal and newborn rat brain. *Journal of Neurochemistry*, 1973, **21**, 61–67.

Cuello, A. C., Shoemaker, W. J., & Ganong, W. F. Effect of 6-hydroxydopamine on norepinephrine and dopamine content, ultrastructure of the median eminence, and plasma corticosterone. *Brain Research*, 1974, **78**, 57–69.

Dahlstrom, A., & Fuxe, K. Evidence for the existence of monoamine-containing neurons in the central nervous system. *Acta Physiologica Scandinavica*, 1964, **62**, Supplementum 232, 1–55.

Descarries, L., & Saucier, G. Disappearance of the locus coeruleus in the rat after intraventricular 6-hydroxydopamine. *Brain Research*, 1972, **37**, 310–316.

Glowinski, J., & Iversen, L. L. Regional studies of catecholamines in the rat brain. I. The disposition of [³H] norepinephrine, [³H] dopamine and [³H] DOPA in various regions of the brain. *Journal of Neurochemistry*, 1966, **13**, 655–669.

Halasz, B., & Pupp, L. Hormone secretion of the anterior pituitary gland after physical interruption of all nervous pathways to the hypophysiotropic area. *Endocrinology*, 1965, **77**, 553–562.

Hedreen, J. C., & Chalmers, J. P. Neuronal degeneration in rat brain induced by 6-hydroxydopamine: A histological and biochemical study. *Brain Research*, 1972, **47**, 1–36.

Henry, D. P., Starman, B. J., Johnson, D. G., & Williams, R. H. A sensitive radioenzymatic assay for norepinephrine in tissues and plasma. *Life Sciences*, 1975, **16**, 375–384.

Iversen, L. L. Uptake processes for biogenic amines. In L. L. Iversen, S. D. Iversen, & S. H. Snyder (Eds.), *Handbook of psychopharmacology*. Vol. 3. New York: Plenum, 1975. Pp. 381–442.

Kirkepar, S. M., Cervoni, P., & Furchgott, R. F. Catecholamine content of the cat nictitating membrane following procedures sensitizing it to norepinephrine. *Journal of Pharmacology and Experimental Therapeutics*, 1962, **135**, 180–190.

Kizer, J. S., Arimura, A., Schally, A. V., & Brownstein, M. J. Absence of luteinizing hormone-releasing hormone (LH-RH) from catecholaminergic neurons. *Endocrinology*, 1975, **96**, 523–525.

Kizer, J. S., Muth, E., & Jacobowitz, D. M. The effect of bilateral lesions of the ventral noradrenergic bundle on endocrine-induced changes of tyrosine hydroxylase in the rat median eminence. *Endocrinology*, 1976, **98**, 886–893.

Kobayashi, R. M., Lu, K. H., Moore, R. Y., & Yen, S. S. C. Effects of ovariectomy and estrogen on regional concentrations of hypothalamic LRF and pituitary LH secretion in the rat. Abstract of the Annual meeting of the Society for Neuroscience, 1976. (a)

Kobayashi, R. M., Palkovits, M., Kizer, J. S., Jacobowitz, D. M., & Kopin, I. J. Selective alterations of catecholamines and tyrosine hydroxylase activity in the hypothalamus following acute and chronic stress. In E. Usdin, R. Kvetnansky, & I. J. Kopin (Eds.), *Catecholamines and stress*. Oxford: Pergamon, 1976. Pp. 29–38. (b)

Kobayashi, R. M., Palkovits, M., Jacobowitz, D. M., & Kopin, I. J. Biochemical mapping of the noradrenergic projection from the locus coeruleus: A model for studies of brain neuronal pathways. *Neurology*, 1975, **25**, 223–233.

Kobayashi, R. M., Palkovits, M., Kopin, I. J., & Jacobowitz, D. M. Biochemical mapping of noradrenergic nerves arising from the rat locus coeruleus. *Brain Research*, 1974, **77**, 269–279.

Korf, J., Aghajanian, G. K., & Roth, R. H. Increased turnover of norepinephrine in the rat cerebral cortex during stress: Role of the locus coeruleus. *Neuropharmacology*, 1973, **12**, 933–938. (a)

Korf, J., Aghajanian, G. K., & Roth, R. H. Stimulation and destruction of the locus coeruleus: Opposite effects on 3-methoxy-4-hydroxyphenylglycol sulfate levels in the rat cerebral cortex. *European Journal of Pharmacology*, 1973, **21**, 305–310. (b)

Koslow, S. H., Cattabeni, F., & Costa, E. Norepinephrine and dopamine: Assay by mass fragmentography in the picomole range. *Science*, 1972, **176**, 177–180.

Kuhar, M. J. Neurotransmitter uptake: A tool in identifying neurotransmitter-specific pathways. *Life Sciences*, 1973, **13**, 1623–1634.

Lidbrink, P., & Jonsson, G. Noradrenaline nerve terminals in the cerebral cortex: Effects on noradrenaline uptake and storage following axonal lesion with 6-hydroxydopamine. *Journal of Neurochemistry*, 1974, **22**, 617–626.

Lindvall, O., & Bjorklund, A. The organization of the ascending catecholamine neuron systems in the rat brain as revealed by the glyoxylic acid fluorescence method. *Acta Physiologica Scandinavica*, 1974 Supplementum, **412**, 1–48.

Lindvall, O., Bjorklund, A., Moore, R. Y., & Stenevi, U. Mesencephalic dopamine neurons projecting to neocortex. *Brain Research*, 1974, **81**, 325–331.

McGeer, E. G., Fibiger, H. C., McGeer, P. L., & Brooke, S. Temporal changes in amine synthesizing enzymes of rat extrapyramidal structures after hemitransections or 6-hydroxydopamine administration. *Brain Research*, 1973, **52**, 289–300.

Meek, J. L., & Neff, N. H. Fluorometric estimation of 4-hydroxy-3 methoxyphenylethyleneglycol sulphate in brain. *British Journal of Pharmacology*, 1972, **45**, 435–441.

Molinoff, P. B., Weinshilboum, R., & Axelrod, J. A sensitive enzymatic assay for dopamine-β-hydroxylase. *Journal of Pharmacology and Experimental Therapeutics*, 1971, **178**, 425–531.

Moore, R. Y. Brain lesions and amine metabolism. *International Review of Neurobiology*, 1970, **13**, 67–91.

Moore, R. Y. Surgical and chemical lesion techniques. In L. L. Iversen, S. D. Iversen, & S. H. Snyder (Eds.), *Handbook of pyschopharmacology*. New York: Plenum, 1978, in press.

Moore, R. Y., Bhatnagar, R. K., & Heller, A. Anatomical and chemical studies of a nigro-neostriatal projection in the cat. *Brain Research*, 1971, **30**, 119–135.

Moore, R. Y., & Heller, A. Monoamine levels and neuronal degeneration in rat brain following lateral hypothalamic lesions. *Journal of Pharmacology and Experimental Therapeutics*, 1967, **156**, 12–22.

Palkovits, M. Isolated removal of hypothalamic or other brain nuclei of the rat. *Brain Reserach*, 1973, **59**, 449–450.

Palkovits, M., Brownstein, M., & Saavedra, J. M. Serotonin content of the brain stem nuclei in the rat. *Brain Research*, 1974, **80**, 237–249. (a)

Palkovits, M., Brownstein, M., Saavedra, J. M., & Axelrod, J. Norepinephrine and dopamine content of hypothalamic nuclei of the rat. *Brain Research*, 1974, **77**, 137–149. (b)

Palkovits, M., Saavedra, J. M., Kobayashi, R. M., & Brownstein, M. Choline acetyltransferase content of limbic nuclei of the rat. *Brain Research*, 1974, **79**, 443–450. (c)

Reis, D. J., & Ross, R. A. Dynamic changes in brain dopamine-hydroxylase activity during antero-grade and retrograde reactions to injury of central noradrenergic axons. *Brain Research*, 1973, **57**, 307–326.

Roizen, M. F., Kobayashi, R. M., Muth, E. A., & Jacobowitz, D. M. Biochemical mapping of noradrenergic projections of axons in the dorsal noradrenergic bundle. *Brain Reserach*, 1976, **104**, 384–389.

Ross, R. A., Joh, T. H., & Reis, D. J. Reversible changes in the accumulation and activities of tyrosine hydroxylase and dopamine-β-hydroxylase in neurons of nucleus locus coeruleus during the retrograde reaction. *Brain Research*, 1975, **92**, 57–72.

Ross, R. A., & Reis, D. J. Effects of lesions of locus coeruleus on regional distribution of dopamine-β-hydroxylase activity in rat brain. *Brain Research*, 1974, **73**, 161–166.

Segal, D. S., & Kuczinski, R. Tyrosine hydroxylase activity: Regional and subcellular distribution in brain. *Brain Research*, 1974, **68**, 261–266.

Sims, K. L., & Bloom, F. E. Rat brain 3,4-dihydroxyphenylalanine and L-5-hydroxytryptophan decarboxylase activities: Differential effect of 6-hydroxydopamine. *Brain Research*, 1973, **49**, 165–175.

Sokal, R. R., & Rohlf, F. J. *Biometry*. San Francisco, California: Freeman, 1969.

Sorimachi, M. Susceptibility of catecholaminergic cell bodies to 6-hydroxydopamine: Enzymic evidence. *Brain Research*, 1975, **88**, 572–575.

Tassin, J. P., Cheramy, A., Blanc, G., Thierry, A. M., & Glowinski, J. Topographical distribution of dopaminergic innervation and of dopaminergic receptors in the rat striatum. I. Microestimation of ³H-dopamine uptake and dopamine content in microdiscs. *Brain Research*, 1976, **107**, 291–301.

Thierry, A. M., Blanc, G., Sobel, A., Stinus, L., & Glowinski, J. Dopaminergic terminals in the rat cortex. *Science*, 1973, **182**, 499–501.

Ungerstedt, U. Histochemical studies on the effect of intracerebral and intraventricular injections of 6-hydroxydopamine on monoamine neurons in the rat brain. In T. Malmfors & H. Thoenen (Eds.), *6-hydroxydopamine and catecholamine neurons*. New York: American Elsevier, 1971. Pp. 101–127. (a)

Ungerstedt, U. Stereotaxic mapping of the monoamine pathway in the rat brain. *Acta Physiologica Scandinavica, 1971 Supplementum,* **367,** 1–48. (b)

Walter, D. S., & Eccleston, D. Increase of noradrenaline metabolism following electrical stimulation of the locus coeruleus in the rat. *Journal of Neurochemistry*, 1973, **21**, 281–289.

Zigmond, M. J., Chalmers, J. P., Simpson, J. R., & Wurtman, R. J. Effect of lateral hypothalamic lesions on uptake of norepinephrine by brain homogenates. *Journal of Pharmacology and Experimental Therapeutics*, 1971, **179**, 20–28.

Zigmond, R. E., & Ben-Ari, Y. A simple method for the serial sectioning of fresh brain and the removal of identifiable nuclei from stained sections for biochemical analysis. *Journal of Neurochemistry*, 1976, **26**, 1285–1287.

Chapter 12

Analysis of Retrograde Degeneration in Cell Soma Following Axon Transection

Richard J. Ravizza

Psychology Department
The Pennsylvania State University
University Park, Pennsylvania

I. Introduction: Questions For Which Analysis of Retrograde Degeneration Is Appropriate

Until the very recent introduction of the protein transport techniques (see Chapter 9 by Henrickson, Chapter 10 by Lynch *et al.*, and Chapter 13 by LaVail in this volume), the available methods for studying pathways within the nervous system were based on the observation that if an axon is severed or a cell body destroyed, dramatic changes take place throughout the neuron. The distal portion of a severed axon, its terminal ramifications, synaptic endings, and myelin sheath all undergo structural changes called anterograde degeneration (Fig. 1). Anatomical techniques based on anterograde degeneration include the Marchi method (which impregnates degenerating myelin sheaths) and the Nauta and Fink–Heimer methods (which reveal degenerating axoplasm and synaptic endings; see Chapter 8 by Giolli and Karamanlidis in this volume). Neuroanatomists have also made extensive use of structural changes that occur proximal to the site of the severed axon (Fig. 1). These changes proximal to the injury are generally labeled retrograde degeneration. Although some effort has been made to trace changes in the proximal portion of the severed axon particularly in young animals (Grant, 1970), neuroanatomists have relied much more heavily on structural changes in the cell body itself in their study of anatomical pathways within the central nervous system. Consequently, this latter form of retrograde change, as revealed by microscopic analysis of Nissl stained material, provides the primary focus of this chapter.

Quite simply, in a typical retrograde degeneration experiment a lesion is placed in some portion of the nervous system, and cell bodies in other portions of the nervous system are examined for evidence of retrograde degenerative changes. If changes are found in a given group of neurons then the axons of these neurons are said to be connected to or pass through the damaged area.

Though simple in conception, in actual practice the analysis of both retrograde and anterograde changes as techniques for tracing anatomical pathways show methodological limitations that have encouraged many anatomists to abandon

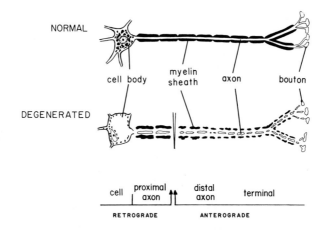

DEGENERATION

Fig, 1.   Schematic representation of typical degenerative changes which take place in neurons following axon transection. Changes distal to the section are called anterograde (or orthograde) degeneration. Proximal changes are called retrograde degeneration. Within the cell body note the swelling, breakdown of Nissl substance, and movement of the nucleus toward the cell periphery.

these techniques in favor of the more recently developed protein transport procedures. While these newer methods have advantages over the analysis of retrograde degeneration for tracing anatomical pathways, they are still only poorly understood and may well have their own idiosyncrasies. For example, not all neuronal systems may be labeled by retrograde transport of peroxidase (Nauta, Pritz, & Lasek, 1974). Analysis of retrograde degeneration, then, appears to remain a useful tool to the neuroanatomist, particularly as a complement to protein transport techniques. In addition, there are at least two other important uses for the analysis of retrograde degeneration. First, the study of retrograde changes can play an important part in the study of regenerative phenomenon following damage to the nervous system. Second, the analysis of retrograde degeneration is of crucial importance to the neuroscientist interested in describing lesions in various parts of the nervous system following long-term physiological or behavioral observations (see Chapter 15 by Meyer and Bresnehan in this volume). For example, a researcher interested in the role of primate neocortex in vision might surgically remove visual cortex from several animals and then behaviorally evaluate the remaining visual capacity for several weeks or even months. Because retrograde changes in the cell body are evident much longer than the accompanying anterograde degeneration, which persists only 2 weeks or so, the analysis of long-term retrograde changes is an essential tool for accurately determining the locus and extent of surgical damage. In a similar way, because these retrograde degenerative changes persist for years, their analysis provides a powerful technique for studying anatomical connections in man with natural lesions resulting from various neurological diseases and strokes.

The purposes of the present chapter are threefold. First, the phenomenon of retrograde degeneration will be briefly described, focusing on those changes which are visible under the light microscope. Next, detailed step-by-step recommendations will be made for conducting analysis of Nissl-stained material for evidence of retrograde degeneration. Finally, the interpretation of retrograde changes and some of the limitations will be discussed.

II.   Phenomena of Retrograde Degeneration

The first detailed description of the changes that appear in the cell body after direct injury to its axon was made by Nissl toward the end of the nineteenth century. Nissl undertook a systematic study of the structure of normal neurons and neuronal changes that accompany both various diseases and direct injury to the nervous system. His work included a careful description of structural changes that occurred in the cell bodies of the facial nucleus following transection of the rabbit's facial nerve. Briefly, Nissl (1892) observed a deterioration of chromatin substance (later named Nissl substance in his honor) within each cell body. This breakdown, called chromatolysis, began centrally within 24 hr and in 2–5 days extended throughout the cytoplasm, leaving only a rim of delicate chromatin granules along the periphery of the cell body. The nucleus was displaced to the periphery, opposite the axon hillock; it frequently appeared flattened but did not appear to suffer much damage. Finally, the entire cell body gradually appeared swollen. These alterations of the cell body were fully developed after a week or 10 days, and rendered the affected neurons strikingly different from normal as illustrated in Fig. 1. Nissl called this stage of retrograde degeneration "Primare Reizung" (primary irritation).

Since Nissl's early observations, researchers have extended the description of retrograde changes in a variety of directions. For example, although Nissl's original analysis dealt with neurons of the facial nucleus of the rabbit, it was believed for years that retrograde reactions were identical for all neurons. One hope was that comparative studies involving a variety of nuclei in different species might lead to a model of retrograde degeneration which fit all neurons. As it turned out, patterns of retrograde degeneration vary considerably. For example, following section of their axons the neurons of some regions of the brain shrink while in other regions they swell. Nuclei are not always peripherally displaced. Some neurons appear to degenerate rapidly without ever passing through a phase of chromatolysis, while the soma of other neurons show virtually no response to interruption of their axons. Furthermore, following comparable lesions, younger animals typically exhibit much more severe neuronal retrograde changes than do adults. The severity of degeneration and its time course are also dependent on the species involved, the distance from the cell body to the site of injury, and possibly the existence of uninjured axon collateral branches proximal to the site of injury, which are somehow capable of preserving the integrity of the cell soma. Taken together, factors such as these make it impossible to present a "typical" picture of the course of retrograde degeneration that describes the sequences of changes in all neurons, both central and peripheral. (For further discussion of these factors, see Bodian & Mellas, 1945; Brodal, 1940; Chow & Dewson, 1966; Cole, 1968; Fry & Cowan, 1972; Geist, 1933; Morton, 1970.)

As anatomists became increasingly aware of the number of variables that influence retrograde degeneration, they also noted that if degeneration is allowed to continue for several weeks or even months, the ensuing structural changes appeared to stabilize in any one of three general states: (1) completely deteriorated neurons, (2) neurons in some state of atrophy, and (3) neurons recovered to an essentially normal state. As illustrated in Fig. 2, if the retrograde process is allowed to continue, some neurons may continue to swell and eventually disintegrate with their remains presumably removed by glia. Other neurons may actually show signs of recovery. For example, Nissl substance reappears in the cell body, first in the central portions and then more peripherally. The cell body becomes less swollen and slowly regains its normal contour with the nucleus

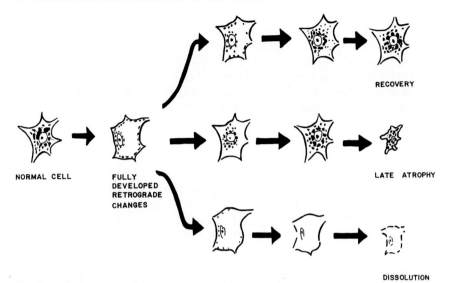

RECOVERY

NORMAL CELL   FULLY   LATE ATROPHY
               DEVELOPED
               RETROGRADE
               CHANGES

DISSOLUTION

FIG. 2.   Diagram showing structural changes which may take place after the initial fully developed retrograde changes. Note that the ensuing structural changes can stabilize in recovery, late atrophy, or dissolution. After Brodal (1969).

resuming its central position. After several weeks, or in some cases months, these partially reintegrated neurons frequently undergo a second series of degenerative changes. These changes are called late atrophy and are characterized by a slow atrophy with considerable shrinkage and more intense staining of the cell body (Brodal, 1969). In some neurons this recovery process may not be interrupted by late atrophy. For example, available evidence indicates that certain damaged *peripheral* motor neurons may completely recover if their regenerating axon is able to establish functional contact with an appropriate muscle group (Bodian, 1947). This discussion emphasizes the need to pay particular attention to time parameters when studying degenerating material; the postoperative survival period is a major factor in the appearance of degenerating neurons. This has a disadvantage, as different cases with different survival times are not always directly comparable. On the other hand, it also has the distinct advantage that one can follow the time course of the degenerative changes. This allows more conclusive interpretations to be drawn from degenerating material, particularly in distinguishing retrograde degeneration from, e.g., anterograde transneuronal degeneration. It should also be kept in mind that these degenerative changes occur more rapidly and are also more severe in infant animals than in the adult (Brodal, 1969), thus perhaps making the infant a more useful subject in experimental investigations.

For the purpose of analyzing retrograde changes within the central nervous system, the affected regions which have undergone either complete deterioration or some stage of retrograde atrophy are quite distinct from adjacent regions containing normal neurons. These long-term retrograde changes are evident in Nissl-stained material and are easily detected with the light microscope. Consequently, Section III focuses on the analysis of late retrograde changes as a tool to determine the locus and extent of neurons whose axons have been severed by some distant lesion.

III. Analysis of Retrograde Degeneration

In a typical experiment involving analysis of long-term retrograde changes a surgical ablation is made, and the animal is permitted to survive 6 weeks or

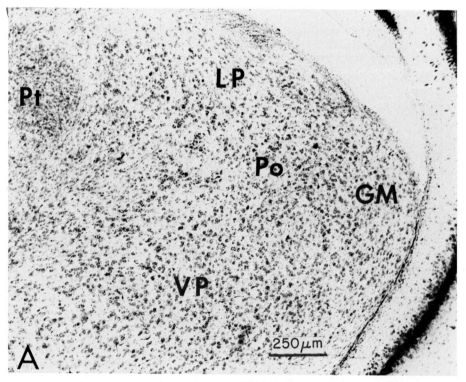

FIG. 3. Low-power photomicrographs of dorsal thalamus (A) in normal hedgehog and (B) in hedgehog with a large ablation of auditory cortex 6 weeks prior to sacrifice. Note in (B) that the medial geniculate and adjacent regions of Po in the animal without auditory cortex lack the deeply staining

longer. The animal is then sacrificed and its brain prepared for histological examination. Because microscopic analysis of retrograde degeneration involves detailed examination of individual cell bodies, the quality of tissue preparation is extremely important. Typically, researchers embed the experimental brains in celloidin and section at 30–50 μm or embed in paraffin and section at 10–20 μm. The sections are then stained with a routine Nissl stain. Step-by-step procedures for celloidin and paraffin embedding and a simple Nissl stain are available in the Chapter 2 by Clark in this volume and in Clark (1973).

In a very real sense the analysis of retrograde degeneration is a two-part process. The first step involves locating those neurons that have undergone retrograde changes. The second, and in many cases the more difficult, is identifying the exact regions of the brain in which the degenerated neurons are located. Thus, meaningful analysis of retrograde degeneration is completely dependent on a thorough knowledge of the *normal* appearance of the region under study. This

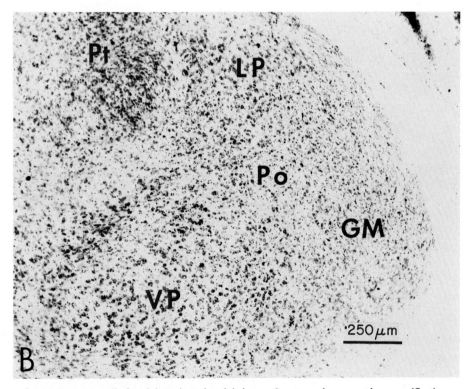

large neurons normally found throughout dorsal thalamus. Consequently, even at low magnification these degenerated areas appear paler (GM, medial geniculate nucleus; Po, posterior nuclear group; VP, ventral posterior nucleus; LP, lateral posterior nucleus; Pt, pretectal nucleus).

is especially true in nuclear groups with poorly defined boundaries. The normal appearance of neurons within these areas—their size, shape, depth of stain, overall packing density relative to adjacent areas, and concentration of associated glia cells—are useful indices of the normal cytoarchitectonics of a given region. A few hours, days, or even weeks spent becoming familiar with the normal cytoarchitectonic appearance of an area will greatly facilitate its subsequent analysis for evidence of retrograde degeneration.

Once the normal appearance of an area is known, any abnormalities can be detected. Perhaps the most striking characteristic of retrograde degeneration is cellular loss: the neurons involved simply deteriorate and eventually their remains are removed by glia cells. The dorsal lateral geniculate nucleus of many mammals is a striking example of an area in which virtually no neurons remain after removal of its cortical projection area. With extensive cell loss it is not uncommon to see a nucleus shrink to less than one-half of its normal volume. In

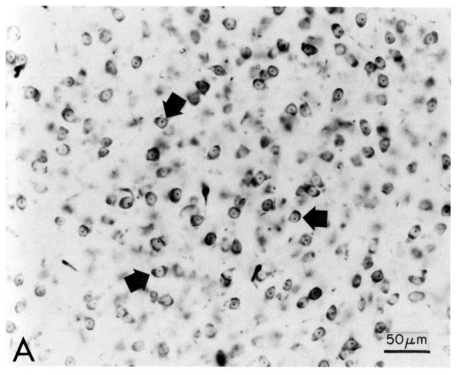

FIG. 4.   Intermediate-power photomicrograph of medial geniculate neurons in cases shown in Fig. 3. Because the tissue is considerably thicker than the plane of focus, many neurons appear out of focus. Consequently, typical normal neurons in (A) and examples of degenerated neurons in (B) are

other regions, the degree of cell loss may vary all the way from a complete loss to instances in which only a small percentage of neurons is missing. A second characteristic of retrograde degeneration concerns the alterations that take place in the remaining neurons (i.e., the degree of retrograde atrophy). In some cases surviving neurons are severely altered. Changes in these neurons may include tremendous shrinkage of the cell body (in some cases to the size of glia cells), general distortion of normal cellular contour, and loss of Nissl substance resulting in a pale washed-out appearance. Such severely degenerated neurons are often called "ghosts." In other regions of the brain the alterations may be considerably less and result in cells that show only slight shrinkage, or perhaps a slight decrease in the intensity with which they are stained. In such cases involving only a slight degree of retrograde change, direct comparison of the affected regions with their normal appearance in control brains or preferably in the un- lesioned hemisphere of the same brain can be very helpful. Obviously, an entire

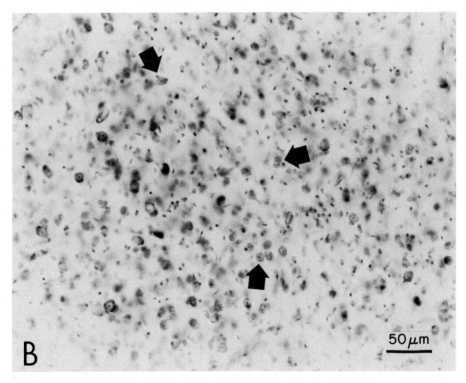

B

50 μm

indicated by arrows. Note that few, if any, normal neurons remain in the degenerated medial geniculate and that the degenerated neurons are smaller and paler than normal.

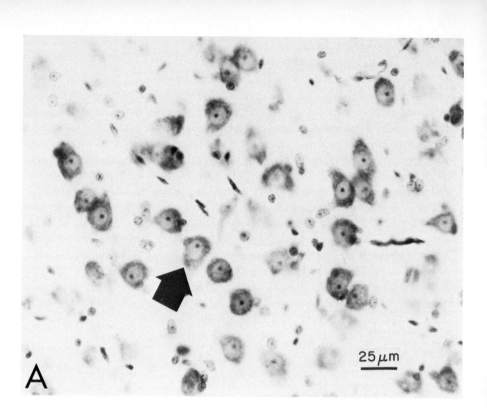

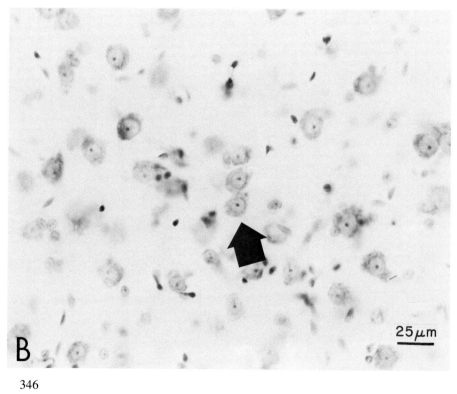

continuum of the severity of changes in remaining neurons is possible from just noticeable changes to a ghostlike appearance. Photomicrographs of normal cells and cell bodies representing various stages of retrograde change are shown in Figs. 3, 4, and 5.

Another aspect of the retrograde reaction involves the reaction of glial elements. In many cases there is a large increase in the number of glia in a degenerated area (a phenomenon called gliosis). Further, the glia cells tend to cluster or group around the affected neurons. Sometimes there is an increased number of glia in an area containing apparently normal neurons. Close observation may reveal that the glia are not grouped around neurons, rather they seem more closely related to fibers whose cell bodies are undergoing retrograde change and

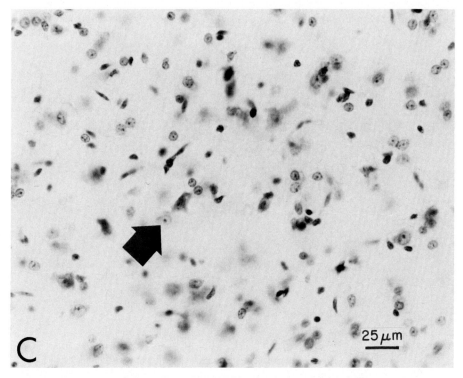

FIG. 5. High-power photomicrographs of dorsal lateral geniculate (DGL) in hedgehog. (A) Normal DGL. (B) DGL following partial removal of visual cortex 6 weeks earlier. Note that although there is little or no cell loss the remaining neurons are considerably paler than their normal counterparts shown in (A). (C) DGL following complete removal of its cortical projection area. Note almost total cell loss. Examples of a severely degenerated neurons sometimes called "ghosts" are indicated by arrow. (Slightly modified from Hall and Diamond, 1968. Printed here by the courtesy of the authors and permission of the S. Karger Publishing Company.)

merely pass through the area. In such a case, the increased number of glia should
not be taken to mean that the neurons in this area are undergoing retrograde
change.

Before concluding that a given population of distorted neurons has undergone
retrograde degeneration, it is frequently useful to check regions that are expected
to contain normal neurons. If these areas also contain distorted neurons, then
some nonspecific factor such as artifacts due to tissue processing or poor health
of the animal may be involved. If lesions are placed unilaterally, and the projection system under study is not a bilateral one, the opposite unlesioned hemisphere
provides an excellent control for nonspecific factors.

Having defined some of the characteristics of neurons that have undergone
retrograde degeneration it is now possible to analyze every section throughout the
degenerated region. One procedure for doing this involves first drawing an
enlarged outline of the area under study with the aid of a microprojector. To

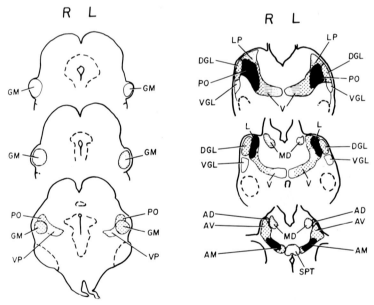

FIG. 6. Pattern of thalamic retrograde degeneration in opossum resulting from removal of neocortex. Black, no neurons remain; cross-hatched, some neurons remain; stripe, many neurons remain;
stipple, no cell loss yet some degenerative changes are present. AD, n. anterior dorsal; AM. n.
anterior medial; AV, n. anterior ventral; DGL, n. dorsal lateral geniculate; GM, n. medial geniculate;
L, n. lateral; LP, n. lateral posterior; PO, posterior nuclear group; VGL, n. ventral lateral geniculate;
VP, n. ventral posterior, V, n. ventral. Except for AD, AV, and portions of V thalamic nuclei are
degenerated to the same degree that follows complete removal of neocortex. Note various degrees of
degeneration in thalamic nuclei ranging from complete cell loss in L and PO to complete presentation
of caudal GM.

facilitate the task of identifying the regions containing affected neurons, the nuclear boundaries of the area under study and the surrounding areas are indicated whenever possible. Next, the same section is examined under high power with the light microscope and any unclear regional boundaries completed. Once these boundaries have been located, the location of any neurons which have undergone retrograde changes is noted. Finally, the same section is returned to the microprojector; the nuclear boundaries are completed, and those areas which have undergone retrograde change are indicated. Obviously, some sections may require several microscopic examinations before the complete pattern of retrograde changes and nuclear boundaries are accurately indicated in the drawings.

If a more refined representation of retrograde change is required, the various degrees of change can be indicated by various symbols. For example, in his now classic analysis of retrograde degeneration in the macaque monkey, Walker (1938) describes four degrees of retrograde change: (a) absence of cells and marked gliosis, (b) absence of cells and slight gliosis, (c) decrease in a number of cells and marked gliosis, and (d) pale cells without gliosis. Figure 6 shows the results of an analysis of retrograde degeneration following removal of all neocortex in the opossum in which four slightly different degrees of severity of retrograde degeneration are indicated. Once each section through the affected zone is drawn it is a simple matter to select five or six representative sections that accurately indicate the overall pattern of retrograde degeneration for each case.

IV. Interpretation of Retrograde Analysis

Before attempting to interpret the total pattern of retrograde degeneration for any given case, several factors should be kept in mind. One such factor is illustrated in Fig. 6. In this case virtually all neocortex was removed bilaterally from an opossum, yet even after a survival period of more than 6 months, not all regions of dorsal thalamus underwent the same degree of retrograde degeneration. Indeed, in this single case the range of changes extend from a total or extensive loss of neurons (indicated by black or cross-hatching) to virtually no change in the appearance of other thalamic neurons. Bodian (1942) and Diamond and Utley (1963) report a similar spectrum of changes for the decorticate opossum. As mentioned above, this spectrum of differences could conceivably be due to any of several factors. For example, the lesion may have severed the various axons at different distances from their respective cell bodies, or perhaps each type of neuron is differentially susceptible to retrograde changes. Another interpretation might be that differences in the extent of degeneration reflect differences in collateral connections en route to neocortex, as an uninterrupted axon collateral may be able to somehow sustain the injured neuron (for a detailed discussion of the possible role of sustaining collaterals, see Fry & Cowan, 1972;

Rose & Woolsey, 1949). The correct interpretation of these different degrees of retrograde changes may be any one or any combination of these and even other factors. Yet, the fact remains that this is the *maximum* extent of degeneration we can expect with a complete removal of all neocortex in this species. This means that any subtotal cortical lesion which results in this maximal extent of retrograde changes in a given nucleus involves that region's entire cortical projection area even though neurons exhibiting various degrees of retrograde change may remain. For example, in the case of the opossum, a considerably smaller lesion involving only auditory cortex will cause degeneration in the medial geniculate identical to that seen following complete removal of neocortex while leaving all other neurons in dorsal thalamus completely intact.

A second factor to keep in mind when interpreting retrograde analysis is that any lesion indiscriminantely destroys neurons, axon terminals, *and* axons from other parts of the brain which may simply pass through the area of the lesion. This problem is particularly acute in interpreting the effects of subcortical lesions but is also relevant for cortical ablations. For example, if the intended cortical lesion is too deep and involves white matter, numerous thalamocortical and corticothalamic axons associated with adjacent cortical regions and cortical–cortical axons to and from other parts of cortex will also be destroyed. This unintentional damage to cortical white matter is referred to as undercutting. In some species where large portions of the internal capsule enter the white matter at a restricted point, a small but relatively deep cortical lesion can disconnect most of the cortex from dorsal thalamus and thus constitute a much larger disruption than intended. One great value of the retrograde technique over a simple cortical reconstruction, photography, or even photomicrograph of a lesion, is that it reflects any inadvertant damage to axons in the underlying white matter that project beyond the intended area to other parts of cortex.

Finally, before attempting to interpret the total pattern of retrograde degeneration for any given projection system, the possibility of confusing retrograde degenerative changes with anterograde transneuronal changes should be kept in mind. Anterograde transneuronal degeneration refers to structural changes that take place in undamaged neurons when a large portion of their afferent input has been removed. For example, it has been known for years that following section of the optic nerve the neurons of the lateral geniculate nucleus in many primates undergo structural changes that, like retrograde degeneration, can result in atrophy of the cell. Fortunately, as it turns out, several known differences between retrograde and the known instances of transneuronal degeneration combine to reduce the likelihood of confusing these two causes of neuronal degeneration. First, relatively few instances of cellular degeneration severe enough to be discernable in Nissl material under the light microscope have been described resulting solely from deafferentation, particularly in mature animals (Cowan, 1970). Second, the limited evidence available on the time course of transneuronal degeneration indicates that it is typically a much slower process than retrograde degeneration. For example, Cook, Walker, and Barr (1951) report that changes in the cat lateral geniculate nucleus did not even begin to take place until 2 months after section of the optic tract. Although such differences in severity of cellular changes and its time course provide a framework on which decisions might be made, the effect of deafferentation remains to be carefully studied in many portions of the nervous system. Consequently the neuroscientist interested in retrograde degenerative changes should keep in mind that whenever a large number of afferents to a nucleus has been removed in the process of severing its axonal projections, it is possible that degenerative changes may be, at least in part, transneuronal in character. The use of varied survival times in experimental material is a distinct aid in distinguishing these two processes.

V. Summary

In recent years neuroanatomists interested in tracing anatomical pathways have increasingly abandoned analysis of anterograde and retrograde degeneration in favor of the various newly developed protein transport labeling procedures. The analysis of retrograde degeneration remains useful to complement or perhaps even to supplement facts uncovered by retrograde transport techniques. To the neuroscientists interested in describing brain damage, the analysis of retrograde degeneration remains virtually the only means of precisely defining the locus and extent of damage to the central nervous system, particularly after long survival periods. This is particularly true in human neuroanatomy where the analysis of retrograde degeneration following natural lesions could have very wide application in the study of human anatomical connections.

In the preceding pages the phenomenon of retrograde degeneration was discussed, and those procedures involved in its analysis were described. Briefly, the analysis of Nissl material for evidence of retrograde degeneration is dependent on two factors: (1) the ability to recognize changes in cell bodies, and (2) the ability to identify the region of the brain in which such changes are located. Practical issues such as its usefulness in revealing unintended interruption of axons of passage and the interpretation of various degrees of degeneration were discussed. Finally, the concept of maximal degeneration was stressed as a means of defining the projection area of a given nuclear complex.

Acknowledgments

Special thanks are extended to Dr. W. C. Hall and Dr. I. T. Diamond, to the S. Karger Publishing Company for permission to reprint Figs. 5A, B, and C, and to Dr. Jeannette P. Ward for suggestions on portions of the manuscript. I am also indebted to Anna Mary Madden and Claudia Keith for typing the manuscript and to Anita Monyok for the artwork. This paper was supported in part by NINCDS Research Grant No. 11554.

References

Beresford, W. A. A discussion on retrograde changes in nerve fibers. *Progress in Brain Research*, 1965, **14**, 33–56.

Bodian, D. Studies on the diencephalon of the Virginia Opossum. III. The thalamocortical projection. *Journal of Comparative Neurology*, 1942, **77**, 525–575.

Bodian, D. Nucleic acid in nerve cell regeneration. *Symposia of the Society for Experimental Biology*, 1947, **1**, 163–178.

Bodian, D., & Mellas, R. C. The regenerative cycle of motoneurons with special reference to phosphatase activity. *Journal of Experimental Medicine*, 1945, **81**, 469–487.

Brodal, A. Modification of Gudden method for study of cerebral localization. *Archives of Neurology and Psychiatry*, 1940, **43**, 46–58.

Brodal, A. *Neurological anatomy: In relation to clinical medicine.* (2nd ed.) London & New York: Oxford University Press, 1969.

Chow, K. L., & Dewson, J. H., III. Neuronal estimates of neurons and glia in lateral geniculate body during retrograde degeneration. *Journal of Comparative Neurology*, 1966, **128**, 63–74.

Clark, G. *Staining procedures.* Baltimore, Maryland: Williams & Wilkins, 1973.

Cole, M. Retrograde degeneration of axon and soma in the nervous system. In G. H. Bourne (Ed.), *The structure and function of nervous tissue.* Vol. I. New York: Academic Press, 1968. Pp. 269–300.

Cook, W. H., Walker, J. H., & Barr, M. L. A cytological study of transneuronal atrophy in the cat and rabbit. *Journal of Comparative Neurology*, 1951, **94**, 267–292.

Cowan, W. M. Anterograde and retrograde transneuronal degeneration in the central and peripheral nervous system. In W. J. H. Nauta & S. O. E. Ebbesson (Eds.), *Contemporary research methods in neuroanatomy.* Berlin & New York: Springer-Verlag, 1970. Pp. 217–251.

Diamond, I. T., & Utley, J. D. Thalamic retrograde degeneration of sensory cortex in opossum. *Journal of Comparative Neurology*, 1963, **120**, 129–160.

Fry, F. J., & Cowan, W. M. A study of retrograde cell degeneration in the lateral mammillary nucleus of the cat, with special reference to the role of axonal branching in the preservation of the cell. *Journal of Comparative Neurology*, 1972, **144**, 1–24.

Geist, F. D. Chromatolysis of efferent neurons. *Archives of Neurology and Psychiatry*, 1933, **29**, 88–103.

Grant, G. Neuronal changes central to the site of axon transection. A method for the identification of retrograde changes in perikarya, dendrites and axons by silver-impregnation. In W. J. H. Nauta & S. O. E. Ebbesson (Eds.), *Contemporary research methods in neuroanatomy.* Berlin & New York: Springer-Verlag, 1970. Pp. 173–186.

Hall, W. C., & Diamond, I. T. Organization of the visual cortex in Hedgehog. I. Cortical cytoarchitectonics and thalamic retrograde degeneration. *Brain, Behavior and Evolution*, 1968, **1**, 181–214.

Morton, A. The time course of retrograde neuron loss in the hypothalamus magnocellular nuclei of man. *Brain*, 1970, **93**, 329–336.

Nauta, H. J. W., Pritz, M., & Lasek, R. J. Afferents to the rat caudoputamen studied with horseradish peroxidase. An evaluation of a retrograde neuroanatomical research method. *Brain Research*, 1974, **67**, 219–238.

Nissl, F. Über die Veränderungen der Ganglienzellen am Facialiskeen des Kaninchens nach Ausseissung der Nerven. *Allgemeine Zeitschrift fuer Psychiatrie und Psychisch-Gerichtliche Medicin*, 1892, **48**, 197–198.

Rose, J. E., & Woolsey, C. N. Cortical connections and functional organization of the thalamic auditory system of the cat. In H. Harlow & C. Woolsey (Eds.), *Biological and biochemical bases of behavior*. Madison: University of Wisconsin Press, 1958. Pp. 127–150.

Walker, A. E. *The primate thalamus*. Chicago, Illinois: University of Chicago Press, 1938.

Chapter 13

A Review of the Retrograde Transport Technique

Jennifer H. LaVail

Department of Anatomy
University of California
San Francisco, California

I.   Introduction: Applications of the Retrograde Transport Technique

Horseradish peroxidase (HRP), an enzymatic marker, originally was adopted by neurobiologists to define the external environment of the nervous system (Becker, Hirano, & Zimmerman, 1968; Brightman, 1965; Turner & Harris, 1974; Zacks & Saito, 1969). More recently it has also been used to determine the cytoarchitecture of neurons (Adams, 1977; Muller & McMahon, 1975; Snow, Rose, & Brown, 1976) or to test the patency of specialized junctions between electrically coupled cells (Reese, Bennett, & Feder, 1971). Further developments in the use of this marker evolved with the discovery by Kristensson and Olsson (1971) that HRP accumulates in cell bodies of axons of the peripheral nervous system after introduction of the enzyme into the vicinity of axon terminals, in a manner analogous to the accumulation of virus particles in infected neurons. Their observations resulted in subsequent demonstrations that (1) axons within the CNS are also capable of retrograde intraaxonal transport (LaVail & LaVail, 1972), (2) the transport and accumulation of marker is not limited to those axons directly injured during the application of the marker, (LaVail & LaVail, 1972, 1974), and (3) the accumulation of HRP within the perikarya might serve to locate those particular neurons that project axons to the site of injection of the enzyme marker (Glatt & Honegger, 1973; Kuypers, Kievit, & Groen-Klevant, 1974; LaVail, Winston, & Tish, 1973; Nauta, Pritz, & Lasek, 1974).

The horseradish peroxidase or retrograde transport method, a more suitable term since other proteins than HRP are presently being used, is uniquely valuable among other experimental neuroanatomical tools in enabling the modern neuroanatomist to scan multiple sources and to identify new sources of afferent fibers to a particular site. For instance, cell bodies in as many as 14 distinct locations within the brain accumulate HRP after it is injected into the inferior parietal lobule of the rhesus monkey (Divac, LaVail, Rakic, & Winston, 1977). The method has a further advantage over previously available anatomical experimental techniques of pinpointing even a single neuron in some cases, whereas the cell degeneration methods, the classic means of identifying the source of axonal projections, require a loss of many cells within a nucleus before the degeneration can be recognized.

Moreover, the transport method is also useful in studies of neuron populations that are difficult to study with cell degeneration methods, either because they are refractory to axotomy and do not become chromatolytic or die, or because the cells do become chromatolytic after lesions but rather as a result of a transneuronal change (LaVail, 1975). For example, some of the cells of the reticular nucleus of the thalamus shrink after cortical lesions. However, after cortical injections of peroxidase, the cells do not accumulate the label, although the thalamic cells that project to the cortex do appear labeled (Jones, 1975a). In this case, it is likely that the reticular cells are affected by retrograde transneuronal changes secondary to the degeneration of the thalamic cells.

From the number of publications based on the retrograde transport technique that appear regularly in neuroscience journals, it is clear that neuroanatomists are already well aware of many of its advantages. Despite this evidence of widespread familiarity, it would be valuable to consider further its real strength, its use in combination with other morphological techniques. For example, the same neurons that contain HRP can also be impregnated with silver salts (Adams, 1977). Thus using standard neurofibrillar impregnating techniques, both the target and the dendritic morphology of a neuron can be discovered.

The presence of collateral branches of axons to different targets has also been defined using the retrograde transport method in combination with lesions resulting in cell degeneration. Jones (1975b) found that somatosensory neurons in the thalamus that shrink in response to lesions in the somatosensory I cortex, became labeled with HRP following injection of the marker into somatosensory II cortex; therefore, the thalamic neurons probably innervate both cortical sites.

By combining immunofluorescence with peroxidase histochemistry, it is possible to define the projections of a population of neurons and its characteristic neurotransmitter. Ljungdahl, Hökfelt, Goldstein, and Park (1975) have followed this approach to identify dopamine neurons that project to the caudoputamen of the rat. Sections from brains of animals injected with HRP in the caudoputamen were first incubated with fluorescent-labeled antibodies prepared to tyrosine hydroxylase, an enzyme involved in the synthesis of catecholamines. The fluorescent cells were examined and photographed with fluorescent light. The same sections were then reacted for HRP activity and rephotographed with the light microscope. A comparison of the fluorescent and light micrographs allowed Ljungdahl and co-workers to distinguish those dopamine neurons that innervated the caudoputamen.

The anterograde transport of radioactively labeled proteins combined with the retrograde transport of peroxidase is another example. By injecting a cocktail of ^{3}H-labeled amino acid and HRP into a single site, several groups of investigators have been able to trace both the afferent and efferent connections of that site in the brain (Colwell, 1975; Graybiel & Devor, 1974; Jacobson & Trojanowski, 1975).

Another use of the retrograde transport technique and autoradiography offers exciting potential to the neuroembryologist. One can define the time of embryonic origin of neurons and also their characteristic axonal projections by injecting fetal animals with ^{3}H-thymidine and later, after the animals have matured, injecting a site in the nervous system with peroxidase. Both the tritium and HRP can be reliably identified and distinguished in the same cell (Nowakowski, LaVail, & Rakic, 1975).

From these few examples the valuable possibilities of combining the retrograde transport technique with other experimental methods can be seen for strengthening the confidence with which one interprets results with the retrograde transport method alone. For instance, the presence of nonprimary afferents to the dorsal column nuclei was originally suggested by Rustioni (1974) on the basis of experiments involving sequential lesions and fiber degeneration. The same year he and Dekker (1974) reported that using the retrograde transport method, they had identified a source of at least some of the afferent fibers in laminae III and IV of the cervical and lumbar spinal cord. Subsequently, Land, Barrett, and Whit-lock (1975) have obtained evidence to support the existence of this projection by injecting radioactive amino acids into the lumbar spinal cord of the cat and tracing the label in an anterograde direction to the medulla.

Furthermore, in some cases, information derived from two techniques may be integrated into larger, functional concepts. The identification of the connectivity of a neuron and its characteristic neurotransmitter is one example. In this case the results may provide important insight into the functional relationship of neurons.

The aim of this chapter is twofold. First, the protocol used by the majority of current workers in the field will be described and some comments made about particular modifications devised by investigators studying particular species or systems. From this general description neuroscientists who wish to employ the method to determine the origin of fibers to a brain locus should be able to begin their studies with some hope of success. However, needless to say, the defini-tive, optimal protocol for demonstrating retrograde-labeled cell cells has not yet been devised, and researchers just beginning to use the technique should not hesitate to adapt and modify the procedure for their own particular needs. Sec-ond, the investigators should be aware of the limitations of the technique as well as its applications, and this chapter may be useful for their critical interpretation of the experimental results. Further discussions of the limitations and advantages can also be found in several of the listed references.

II. Protocol

A. *Choice of Marker*

Horseradish peroxidase is the marker most frequently selected for retrograde transport studies by most neuroanatomists, since it can be identified histochemi-cally fairly easily at both the light and electron microscopic levels by using standard procedures and relatively short incubations. Therefore, the discussion of protocol will be concerned principally with its use. However, several other protein markers have also been proposed and may, in the future, be preferred. Kristensson (1970) has studied the uptake and transport of a fluorescent marker, Evans blue-labeled albumen which, with the proper filter combination, fluoresces red in ultraviolet light. Another fluorescent marker that has been tested is Lucifer yellow VS which reacts with proteins readily (J. H. LaVail & W. Stewart, unpublished observations). This probe holds promise owing to its very intense fluorescence (Fig. 1) and to its ability to fluoresce in sections embedded in glycol methacrylate medium (W. Stewart and N. Feder, unpublished observations).

The combination of either a radioactive or fluorescent tag on carrier molecules has also been proposed and tested. Schwab, Agid, Glowinski, and Thoenen (1977) suggest that ^{125}I-labeled tetanus toxin might offer special advantages such as the restricted diffusion, low concentrations needed for injection and the uptake and transport of the radioactively tagged toxin by a variety of classes of neurons. Nerve growth factor has been investigated as well and provides fascinating clues about its specificity of uptake and transport and about its biological significance (Stöckel & Thoenen, 1975). However, the most significant limitation in using nerve growth factor as a general tracer molecule lies in its apparently restricted

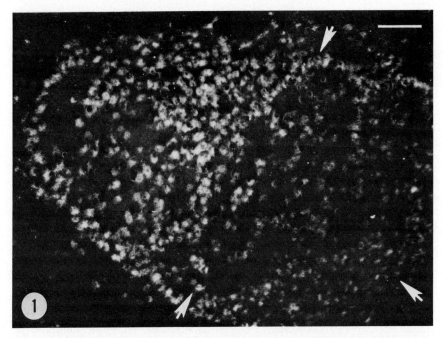

FIG. 1.   This dark-field photomicrograph demonstrates the pattern of fluorescent labeling seen in cells of the chick isthmo-optic nucleus. The contralateral vitreous was injected with Lucifer yellow VS dye, and a portion of the retina was damaged 17 hr before the animal was fixed. Each neuron cell body appears as a tight cluster of tiny white dots. The arrows delimit the region of the nucleus that corresponds to the retinal lesion. See text for further details. Calibration bar: 35μ m. (J. H. LaVail & W. Stewart, unpublished observations).

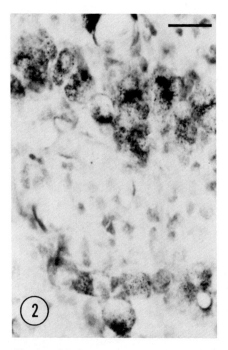

FIG. 2.   A higher magnification of the isthmo-optic nucleus seen in bright-field microscopy. In this experiment HRP was injected into the contralateral vitreous, and 48 hr later the chick was fixed. The peroxidase reaction product appears as small, dark granules in the cell cytoplasm. DAB substrate. Cresyl violet counterstain. Calibration bar: 25 μm.

transport by only sympathetic motor and sensory classes of neurons (Stöckel, Schwab, & Thoenen, 1975).

   In addition to individual alternative markers for transport, several groups of researchers have spent considerable effort looking for two or more markers that might be used in combination in the same experiment and which are distinguishable in single cells after retrograde transport. The hope is to identify collateral branches of the same axon using two markers. By conjugating either fluorescein or tetramethylrhodamine isothiocyanate with HRP, Hanker, Norden, and Diamond (1976) have opened the way to double-marker studies. They have found that the two fluorochrome-tagged HRP species can be distinguished in cells of the lumbosacral spinal ganglia after injection of the markers into either the foot pad or thigh of the hind limb of a young mouse. Alternatively, using another approach, by adding a radioactive tag to HRP and using this marker in low concentrations with more concentrated solutions of nontagged HRP, Geisert

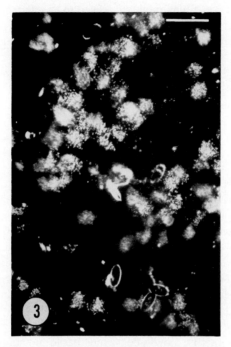

FIG. 3.   A dark-field micrograph of the same experiment as in Fig. 2. In this case the labeled cells contain refractive HRP product and appear as clusters of white dots. The outlines of a few red blood cells that contain endogenous peroxidase can be seen in the bottom of the figure. Calibration bar: 30 μm.

(1976) reports being able to distinguish the histochemically identifiable probe from the autoradiographically identifiable probe in cells of the lateral geniculate nucleus after injections into visual cortex.

The most widely used marker, HRP, is available from several supply houses in different grades of purity as determined by specific activity assays.[1] Bunt, Haschke, Lund, and Calkins (1976) have examined the efficiency of uptake and transport of type VI from Sigma and of the various isoenzymes extractable from horseradish root. They found that over 80% of the enzyme activity of type VI was due to isoenzyme C, one of the several isoenzymes present in the type VI commercial preparation. Furthermore, they proposed that the specificity of up-

[1]The composition of different lots of HRP from a single source may vary. Thus, it may be important to test empirically each enzyme lot to be sure of its usefulness for studies of connectivity.

take and transport they found might be related to the characteristic carbohydrate composition or net positive charge of the molecule. For this reason, it is assumed that the more purified preparations are likely to result in more labeled cells.

B. Injection

The HRP is generally measured out in small quantities and diluted with phosphate buffered saline (pH 7.4) to a concentrated solution varying from 30 to 50%; we routinely use a 30% solution, i.e., 3 mg dissolved in 10 μl of saline delivered into the vial with a microcapillary pipette. The solution can be used for several injections and stored at least overnight at 4°C.

A small amount of the enzyme is drawn up into a 1-μl volume syringe and is injected either stereotaxically or under visual control into the brain region. For larger animals such as monkeys or cats, multiple injections may be needed to fill the site. In our own experience a single injection of 0.1 to 0.5 μl is sufficient to label an area several millimeters in diameter in chick brainstem. Since the marker may spread varying distances depending at least in part on characteristics of the site of injection (LaVail & LaVail, 1974; Walberg, Brodal, & Hoddevik, 1976), the choice of volume should be made after several trial injections have been carried out.

Beitz and King (1976) devised a method to reduce the spread of HRP back along the pipette tract by slowly delivering sterile mineral oil from the pipette after HRP has been injected and while the pipette is withdrawn from the brain. They compared results from animals injected with HRP or with HRP followed by mineral oil and found that the spread of marker from the site of injection was approximately the same, but in those cases in which mineral oil was also passed from the pipette, there was no "spur" of stained tissue superficial to the deepest point of injection and along the needle tract.

The optimal postinjection interval before fixation may be estimated if the distance for transport is considered. The rate of retrograde transport may vary for different fiber systems, but, in general, it is thought to be about 80–100 mm/day (Kristensson, 1975; LaVail & LaVail, 1974; Walberg et al., 1976).

C. Fixation and Buffer Wash

After a suitable time the animals are fixed by vascular perfusion with a mixture of 1% paraformaldehyde and 1.25% glutaraldehyde in a phosphate buffer solution (pH 7.4). The brains are stored in the same fixative for several hours or overnight at 4°C, and the next day the tissue is transferred to a solution of 0.1 M phosphate buffer with 5% sucrose. It is stored in this solution for at least 24 hr at 4°C. The larger the block of tissue the longer the buffer wash, since the presence of fixative in the tissue at the time of incubation may interfere with the enzymatic reaction (Bunt, Lund, & Lund, 1974; LaVail, 1975; Llamas & Martínez-Moreno, 1974).

Several investigators have commented on the importance of controlling the concentration of aldehydes in the fixative for maintaining the enzymatic activity of the peroxidase (Llamas & Martínez-Moreno, 1974; Straus, 1964). Jones and Leavitt (1974) noted that concentrations of formaldehyde greater than 0.4% resulted in inadequately labeled cells in squirrel monkey brain after retrograde transport and after a 24–48 hr wash. Kim and Strick (1976) also directed their attention to the effect of fixative concentration on enzymatic activity. They found from many experiments that the use of formaldehyde in concentrations of 4% inhibited the reaction of the enzyme with substrate. It is unlikely that the concentration of paraformaldehyde alone is responsible for the loss in activity, however, since other investigators have obtained satisfactory results with fixatives composed in part of 4% formaldehyde (Graham & Karnovsky, 1966; Ralston & Sharp, 1973; Wong-Riley, 1974).

The possibility that the ratio of glutaraldehyde to formaldehyde might be critical for preservation of enzymatic activity was suggested recently by Hedreen, McGrath, and Warner (1976). They found that either glutaraldehyde or formaldehyde alone in concentrations ranging from 1–5% or 2–8%, respectively, fixed the tissue and maintained sufficient enzyme activity to be useful for light microscopic studies. However, glutaraldehyde combined with formaldehyde in moderate concentrations, e.g., 3.5% glutaraldehyde with 2% formaldehyde, strongly inhibited the activity. They recommend using glutaraldehyde alone, both to prevent postmortum diffusion of the enzyme and to maintain peroxidase activity. Adams (1977) has also recommended the exclusive use of glutaraldehyde. The combination of both aldehydes has generally been preferred, however, to preserve the fine structure satisfactorily (Karnovsky, 1965).

Clearly the last word on the ideal fixative and optimum concentrations for studies using retrograde transport of HRP remains to be heard. It would appear that additional factors other than a particular concentration of formaldehyde or wash time are important, and each investigator may have to try several fixative combinations or concentrations before the most suitable one for that particular project is found.

The block is frozen and cut at 20–80 μm on a freezing microtome, and the sections are collected in a bath of the same sucrose buffer solution. The sections should then be reacted as quickly as is convenient, since the activity seems to decrease with storage time.

D. Incubation

A saturated solution of 0.5% diaminobenzidine (DAB), e.g., 10 mg DAB/20 ml of phosphate buffer (pH 5.5) is prepared. This is a modification of the original protocol of Graham and Karnovsky (1966) in that phosphate buffer is substituted for the tris·HCl buffer and the reaction is carried out at a lower pH than that suggested by Graham and Karnovsky since the pH optimum of the enzyme is nearer 5.5 for this particular substrate (Clarke & Cowan, 1976; Straus, 1964; Weir, Pretlow, Pitts, & Williams, 1974). To this solution 0.01% hydrogen peroxide is added, and tissue sections are incubated in this substrate at room temperature for 20 min. It is important to prepare the substrate solution fresh every 45 min to 1 hr, since if allowed to stand at room temperature for too long, background staining of endogenous enzymes in the tissue increases.

After incubation the sections are rinsed in buffer and distilled water and then mounted on slides in a solution of 1.5% aqueous gelatin and 80% alcohol (Albrecht, 1954), air dried, and subsequently dehydrated and coverslipped. They may also be counterstained with 0.5% cresyl violet before dehydration, but it is important to counterstain only faintly, since the stain may obscure very lightly stained, small granules.

TABLE I

Suggested Substrates for HRP and Their Characteristics

Substrate	Color	Relevant technique[a]	Ref.
Benzidine[b] (nitroferricyanide as stabilizer)	Blue/brown	LM, EM	Mesulam, 1976; Straus, 1964
Diaminobenzidine[c] [3,3',4,4'-tetraaminobiphenyl; bis(orthophenylene diamine)]	Light brown	LM, EM	Graham & Karnovsky, 1966
Pretreatment with cobalt salt	Black	LM, EM	Adams, 1977
o-Dianisidine[c] (3,3'-dimethoxybenzidine dihydrochloride)	Blue/green	LM, EM	Colman et al., 1976; J. de Olmos, 1977; Graham & Karnovsky, 1966
α-Naphthol (followed by pyronine staining)	Reddish pink	LM	Nakane, 1968
4-Cl-1-Naphthol	Grayish blue	LM	Nakane, 1968
3-Amino-9-ethylcarbazole	Red	LM	Graham et al., 1965
Homovanillic acid	Brown/black	LM, EM, FM	Papadimitrou et al., 1976
p-Phenylenediamine and pyrocatechol	Blue/violet	LM, EM	Hanker et al., 1977
3,3',5,5' Tetramethylbenzidine	Blue	LM	Holland et al., 1974

[a] LM, Light microscopy: EM, electron microscopy; FM, fluorescence microscopy.

[b] Carcinogenic; see *Federal Register* **39**; 3755 (1974) for regulations concerning use of benzidine.

[c] Possibly carcinogenic; substrate should be handled with care.

E. *Alternative Substrates*

For studies using both light and electron microscopy, DAB is generally the first choice of substrate. However, when only bright- or dark-field microscopy is planned, several additional histochemical substrates are available and may be preferred (Table I). For example, the recent modification of Straus's (1964) protocol as described by Mesulam (1976) allows the observer to recognize sites of HRP transport very easily with benzidine as the substrate. The blue reaction product (Fig. 4) stands out against neutral red counterstained neuronal cytoplasm, and it is more easily distinguished from lipofuscin or neuromelanin than is the brown reaction product obtained with DAB as the substrate (Figs. 2 and 5). The principal handicap of this substrate lies in the carcinogenicity of benzidine. However, several new substrates have been suggested as substitutes for ben-

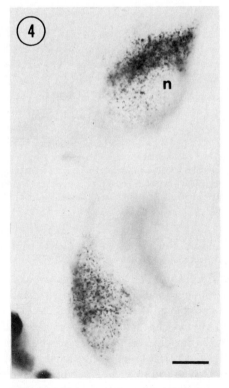

FIG. 4.   Two neurons of the red nucleus of a rhesus monkey labeled after HRP injection into the spinal cord. The granules appear dark blue against neutral red counterstain. n, Nucleus. Mesulam protocol. Calibration bar: 15 μm.

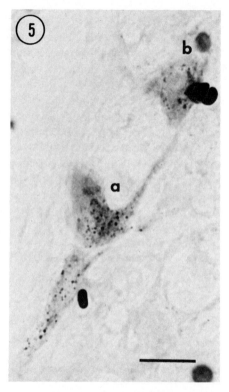

FIG. 5.   Two cells in Clarke's column of the mouse labeled as a result of injection of HRP in the cerebellum. One cell (a) contains both brown granules of enzyme product against a diffuse dispersion of the brown marker throughout the cell body and into two processes. The other cell (b) contains only the brown granules in the cytoplasm which has been counterstained with cresyl violet. DAB substrate. Calibration bar: 15 μm.

zidine. Holland, Saunders, Rose, and Walpole (1974) have proposed using 3,3', 5,5'-tetramethyl benzidine [see Hardy and Heimer (1977) for its application to nervous tissue]. Hanker, Yates, Metz, Carson, Rosen, and Rustioni (1977) have provided another substrate combination, p-phenylenediamine and pyrocatechol, which is easily recognizable at both the light and electron microscope levels.

In addition to DAB Graham and Karnovsky (1966) also described using o-dianisidine, another benzidine derivative as substrate in their ultrastructural study of the uptake of peroxidase by mouse kidney cells. Since then several workers have attempted to improve the visibility of the product by varying parameters such as the oxidizing agents, incubation times, temperatures, and buffers. Two modifications that seem particularly promising are those of Col-

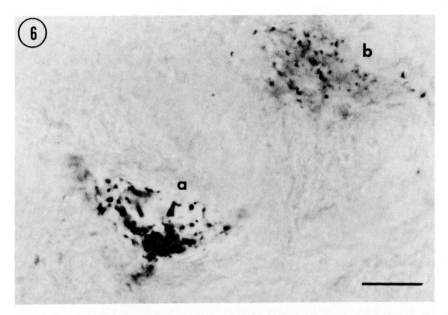

FIG. 6. Two neurons from Clarke's column of a macaque monkey labeled after an HRP injection in the anterior lobe of the cerebellum. One cell (a) contains more obvious green granules than does the other cell (b). de Olmos modification. *o*-dianisidine substrate. Calibration bar: 15 μm. (Tissue courtesy of G. Grant and B. Robertson.)

man, Scalia, and Cabrales (1976) and of de Olmos (1977). The dark green product formed from the incubation of *o*-dianisidine and H_2O_2 in the presence of HRP followed by treatment with nitroferrocyanide stands out sharply against the background tissue (Fig. 6). In some cases the sensitivity of these methods appears greater than that obtained with DAB. For example, several days after injection the area of the intense reaction and greatest concentration of product located near the injection site as seen with the light microscope is significantly larger with de Olmos' modification (Fig. 8) than the area as visualized in adjacent sections reacted with DAB according to the Graham and Karnovsky protocol (Fig. 7). The intense color also permits fairly lightly labeled cells to be noticed more readily. Since the ultrastructural evidence suggests that within 6–24 hr the HRP diffuses well beyond the several millimeters of very heavily stained brain parenchyma of the injection site (LaVail & LaVail, 1974), it remains to be determined whether the limits of the enlarged intensely stained area can be correlated with the actual area of uptake of the marker by axon endings. However, the practical advantage of identifying labeled cells more easily cannot be overlooked.

An additional advantage suggested by de Olmos is that cut sections can be

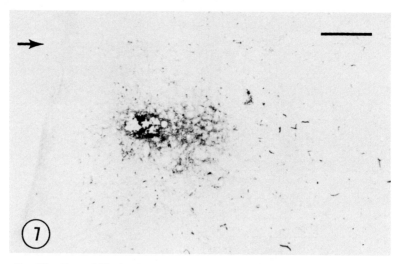

FIGS. 7 and 8. Two neighboring frontal sections taken near an iontophoretic injection of HRP into the brainstem of a cat. The arrows mark the midlines of the sections which are taken from a series of serial section. The distance between sections is within about 200 μm.

Fig. 7. An illustration of the apparent spread of HRP after treatment with DAB.

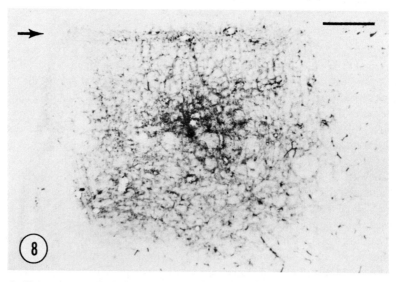

Fig. 8. The apparent spread after treatment with de Olmos' modification using *o*-dianisidine. Additional sections from the same series located more rostrally and more caudally illustrated the same differences in apparent spread, depending on the histochemical substrates used. Calibration bars: 0.5 mm. (A. I. Basbaum & J. H. LaVail, unpublished observations.)

stored at low temperature for several weeks. Peroxidase activity may be lost from the tissue with extended storage in buffer solution (Bunt *et al.*, 1974), and the use of ethylene glycol, methanol, and dimethyl sulfoxide as cryoprotective measures to extend the storage time is certain to be appreciated by those neuroscientists working with larger brains such as monkey or cat, which often involve large numbers of sections.

Two other substrates, α-naphthol and 4-chloro-1-naphthol are also of interest because of the color of product produced, i.e., pink and blue, respectively. Neither of these produces a product as intense as that produced with *o*-dianisidine or DAB, and, unfortunately, they are also soluble in organic solvents. On the other hand, they do have the advantage of environmental safety.

F. Electron Microscopy

Tissues prepared for examination with the electron microscope can be selected from frozen sections, from sections cut on a Vibratome, or from small pieces of tissue 0.5 to 1.0 mm in thickness (Westergaard & Brightman, 1973).

If frozen or vibratome sections are to be used, the sections are incubated for 10 to 20 min in the DAB without the addition of H_2O_2 before a second incubation that is carried out in the complete substrate including both DAB and H_2O_2 for another 10–20 min. The sections are then washed briefly in buffer solution and postfixed in 2% osmium tetroxide in 0.1 M buffer with 5% sucrose for 1 hr. The remaining steps are similar to standard plastic embedding procedures (see Chapter 7 by Bernstein and Bernstein in this volume). Smaller areas of interest can then be cut out of the plastic block with a jeweler's saw.

If small blocks of tissue are to be reacted, larger pieces of tissue may be first fixed and washed in buffer, then cut into smaller blocks about ½ to 1 mm in thickness. They are then incubated in the complete substrate at a temperature of 5°C for 3 hr while being agitated. The blocks may then be treated with 2% osmium tetroxide in phosphate buffer for 1 hr. The remaining steps are identical to the standard plastic embedding procedures.

Plastic sections cut 1–2 μm and stained with toluidine blue may serve to orient the tissue for thin sectioning. After staining with lead citrate or with uranyl acetate and lead citrate, the thin sections can be examined. It is also important to examine some unstained sections, since heavy metal staining can cause dense bodies and some lysosomes to appear as if they also contain reaction product.

Following this procedure, sites with the greatest concentration of HRP such as injection sites often may fail to react evenly after incubation in the substrate. Instead there may be a 15–20 μm border around the edge of the frozen section or small block that is reacted, but the center of the block appears to be free of enzyme, although presumably it is present there. However, in sites to which the HRP is transported, where the concentration of HRP is much less, there should be no difficulty in identifying labeled cells even in the center of a 1-mm-thick slab of tissue (LaVail & LaVail, 1974).

III. Appearance of the Reaction Product after Transport

A. Light Microscopy

If DAB has been the substrate, HRP-positive perikarya can be identified by the fine, dustlike sprinkling of light brown granules in the cytoplasm (Fig. 2). The outline of the nucleus is often suggested by the absence of granules in that region of the cell (Fig. 5, cell b; see also Fig. 4). Generally, the product in vesicles does not extend into the dendrites of the cells, although in some larger cells such as hippocampal pyramidal cells they may be seen in the primary dendrites. With other substrates such as that of de Olmos (Figs. 6 and 9) the product may be packed in large granules up to 3 μm, and they may extend farther into the dendritic processes. Since the organelles carrying the HRP (LaVail & LaVail, 1974) may become integrated into the lysosomal system of the cell, it is not surprising to find many of the labeled granules located near the Golgi region of the cell body.

In plastic-embedded material and after osmication, the DAB positive material appears as dark brown to black granules in the cytoplasm of cells (Fig. 10).

Dark-field microscopy often allows an investigator to scan large areas of tissue somewhat faster than can be done easily in bright-field microscopy. In this case, the cluster of dense granules in the cytoplasm that marks the presence of HRP often appears as pastel colored, closely packed dots (Fig. 3). Care must be taken to check the identity of the cells containing the dots by comparing the bright- and dark-field images, in order to avoid confusing labeled neurons with pericytes or glial cells that may also incorporate the HRP. Neurons labeled with fluorescent markers also appear to contain clusters of densely packed dots (Fig. 1). The background fluorescence, particularly that resulting from lipofuscin, must also be considered carefully.

The intensity of retrogradely transported label found in a neuron cell body may be a result of several physiological factors: the spread of the terminal field of the neuron and the state and activity of the axon incorporating the marker. Jones (1975a) has focused his attention on the first factor. He has compared the size and density of HRP grains in the intralaminar nucleus after separate HRP injections into two sites receiving projections from the intralaminar nucleus pre-

FIG. 9.   A neuron from Clarke's column from the same experiment described in Fig. 6 to illustrate the dispersion of granules into the dendrites. Calibration bar: 12 μm. (Tissue courtesy of G. Grant and B. Robertson.)

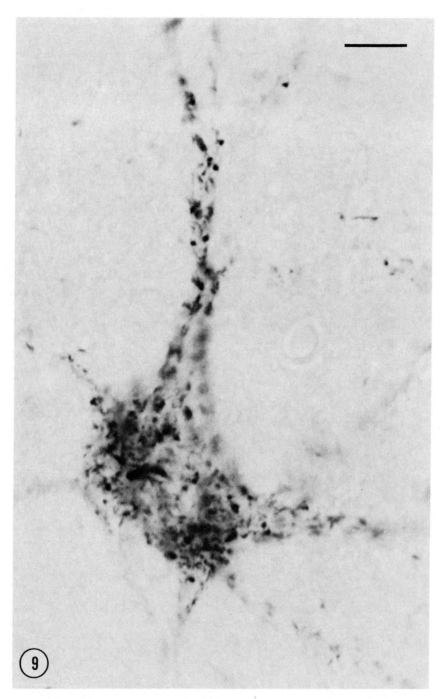

9

viously injected with tritiated amino acids. His findings suggest that the number and size of axon terminals and possibly the number of preterminal branches present may be important determinants of a cell's capacity to accumulate HRP.

The functional activity of the terminals may also affect the intensity of labeling of a nucleus. If a cat is unilaterally enucleated 1 day before HRP is injected into both visual cortices, the number of labeled cells in two active laminae of the geniculates, i.e., A ipsilateral to and A_1 contralateral to the enucleation, is greater than that in the two inactive laminae (Singer, Hollander, & Vanegas, 1977). This finding suggests that neurons of the geniculate that are driven by the intact eye incorporate and/or transport more (or inactivate less) HRP than do neurons in the lateral geniculate nucleus that are no longer driven physiologically.

Direct trauma to axons of passage and exposure to HRP may also result in labeling of the cell bodies of origin of the transected axons (DeVito, Clausing, & Smith, 1974; Kristensson & Olsson, 1974). Whether or not the cell bodies appear labeled, however, depends on, among other factors, the delay between injection and fixation. In one case Halperin and LaVail (1975) found uninjured neurons were definitely more intensely labeled between 3.5 and 6 hr after injection and injury than injured neurons exposed presumably to the same concentration of peroxidase. Surprisingly, however, they found that the traumatized cells contained more marker between 6.75 and 18 hr after injury. Thereafter, the cells whose axons were cut accumulated less peroxidase than nearby cells that had not been cut. By inference one might suspect that an axon interrupted during the course of injection may take up more marker or perhaps inactivate the marker less rapidly in the cell body than would intact axons passing through or terminating in the injection site. Depending on the time at which the animal is killed, one may find that neurons with axons only passing through the injection site may or may not appear labeled. This may complicate the interpretation of results (see below).

An example of the differential labeling seen after injury is seen in Fig. 1. The isthmo-optic nucleus of the chick contains neurons whose axons project to the contralateral retina (McGill, Powell, & Cowan, 1966). At the time of introduction of the fluorescent marker intravitreally, one area of the retina was damaged. Seventeen hours later the chick was fixed, and the brain was processed to demonstrate the fluorescent marker that had accumulated in the cell bodies of the isthmo-optic nucleus. The cells appear as dots composed of many smaller dots in the dark-field photomicrograph. The arrows highlight the region of the nucleus in this section that contains less intensely fluorescent clusters of dots or labeled neurons. This region corresponds topographically to the area of the nucleus that projects to the area of injury in the retina. By hr 17 the neurons whose axons were damaged are only faintly labeled, and in some cases it is difficult to make out the granules in the cell.

The granular pattern of labeling in the damaged cells represents only one manifestation of labeling after axotomy. The HRP may disperse diffusely throughout the axoplasm both distal and proximal to the site of transection and extend as far proximally as the perikaryon and dendrites and as far distally as the axon terminals (Adams & Warr, 1976). In this case the Golgi-like appearance of the cell body may be helpful in determining the finer cytological details of the neuron (Adams, 1977).

Using DAB as substrate, uninjured axons and axon terminals distant from the injection site usually cannot be seen with the light microscope. This is presumably due to the fact that the organelles that carry the marker are too small (see below) to be resolved with the light microscope. However, results obtained using alternative substrates and protocols (Colman et al., 1976; de Olmos, 1977; Lynch, Gall, Mensah, & Cotman, 1974) indicate that much more peroxidase may appear in the axons after entering the neuron at the level of the cell body than was previously suspected. Using these procedures, several authors have been able to label axons and axon terminals. Whether this is due to a greater sensitivity with these modifications, greater amplification of product, or less restricted containment of the enzyme or product in organelles remains to be clarified. The obvious benefits for tracing afferent and efferent connections with a single injection of peroxidase are clear, however. For further details and evaluation of this procedure, see Chapter 10 by Lynch et al. in this volume.

B. Electron Microscopy

The peroxidase is pinocytosed by terminals and presumably preterminal portions of axons as well as by other cells and processes in the area infused with HRP. The axonal pinocytosis involves two different classes of membrane-bound vesicles. The incorporation of tracer into small 43–50 nm (synaptic) vesicles in terminals is similar to that described by other investigators of neuromuscular and neurosecretory terminals (see LaVail & LaVail, 1974, for review). Larger 100–125 nm coated vesicles make up the second class of vesicles incorporating tracer near the injection site.

In addition to neurons, glial cells, endothelial cells, and pericytes are also readily labeled with HRP after injection. These cells may preferentially accumulate the exogenous protein and act to clear the extracellular spaces of HRP. Certainly by 24 hr the extracellular space is free of obvious marker (LaVail & LaVail, 1974; Turner & Harris, 1974), although the brown stain may appear homogeneous at the level of the light microscope.

Despite a rapid rate of extracellular HRP diffusion, the HRP in axons moves even faster by retrograde transport, about 80–100 mm/day. Several classes of organelles serve to carry it back to the cell body: multivesicular bodies, cup-shaped organelles, round or ovoid vesicles about 100–150 nm in diameter, and cisternae of the endoplasmic reticulum. However, HRP does not appear to move within axons in vesicles that resemble synaptic vesicles.

With the electron microscope the HRP can be seen in the cell body in irregularly shaped organelles about 65 nm to 0.65 μm or more in diameter (Fig. 11). These labeled structures are dispersed within the perikaryal cytoplasm, frequently near the region of Golgi cisternae, although sometimes dense organelles may also be seen in the proximal portions of dendrites. With time the product disappears from the cells, presumably as a result of fusion of HRP-labeled organelles with primary lysosomes (Sellinger & Petiet, 1973).

Peroxidase marker may also be transported in an anterograde direction to the axon terminal via the same four kinds of organelles (Hansson, 1973; LaVail & LaVail, 1974; Nauta, Kaiserman-Abramof, & Lasek, 1975; Sotelo & Riche, 1974). However, the amount and rate of transport in an anterograde direction may be markedly less than that found in the retrograde direction. In every case,

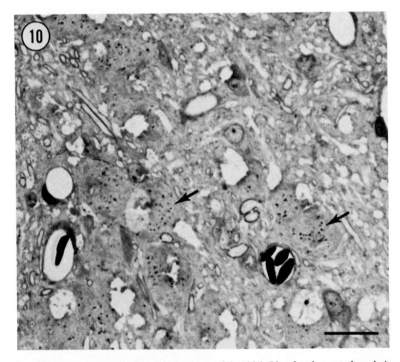

FIG. 10.   Plastic section of the isthmo-optic nucleus of the chick 5 hr after the contralateral vitreous was injected with HRP. This region of the nucleus corresponds to an area of retina not deliberately damaged at the time of injection. The HRP product appears as dark granules (arrows) against the toluidine blue counterstained cytoplasm. Calibration bar: 25 μm.

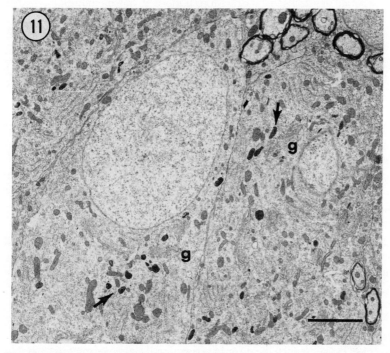

FIG. 11. Electron micrograph of two labeled neurons of the chick isthmo-optic nucleus. The enzyme-positive organelles (arrows) appear clustered near regions of the Golgi apparatus. Calibration bar: 3 μm.

special care must be taken in interpreting dense bodies seen in electron micrographs as HRP-containing bodies, since the normal tissue may contain many osmophilic organelles. In these cases, it is always wise to compare the distribution of dense bodies on the control and experimental sides and to examine sections unstained with heavy metals.

IV. Limitations of the Technique

A. Definition of the Region of Effective Uptake

Since the original proposal to use the retrograde transport of HRP to identify sources of axonal afferents (LaVail & LaVail, 1972), there has been continued concern about the difficulty in defining the limits of spread of HRP after injection. It seemed obvious that axons in the center of the most intensely stained area could incorporate and transport label. More important was the question whether these axons are the sole source of the retrograde labeling, perhaps as a result of injury incurred during the injection process (Bunt *et al.*, 1974; Jones & Leavitt, 1974). This possibility was based at least in part on findings that multiple penetrations and injections of HRP in cerebral cortex resulted in discrete bands of HRP-positive cells in the thalamus, although the appearance of HRP in the injection sites suggested that it had spread fairly homogeneously from one injection site to an adjacent one. Further experience with the technique seems to have resolved the question (Jones, 1975a; LaVail *et al.*, 1973; Walberg *et al.*, 1976). Terminals near by but not actually in the path of the needle also take up and transport enough HRP to label cell bodies.

If the site of uptake is greater than the actual tissue surrounding the needle tract, can it be directly related to the tissue stained most intensely? As a general rule, one empirically defines the limits of spread using the light microscope to trace out the most intensely stained area. This region probably represents a zone of high concentration in which glial cells and neurons have not as yet inactivated the marker.

However, the rapid diffusion of HRP and the inactivation of it with time (LaVail & LaVail, 1974; Walberg *et al.*, 1976) seriously limit the precision with which one can demarcate the site of effective uptake. Labeled cell bodies apparently may project to an area beyond the limits of the most intense staining. The use of more sensitive substrates such as those proposed by Mesulam (1976) and de Olmos (1977) may reduce the error in estimating the zone of intense stain, but the delay necessary between injection and fixation permits the HRP to diffuse away from the site (see below) at a rapid rate, and the extreme limits of diffusion well beyond the stained area seen with the light microscope, are only visible at the ultrastructural level. Until a marker is developed that is pinocytosed and transported by all neurons exposed to the HRP but which does not diffuse readily extracellularly, this handicap of the technique will remain a problem.

B. Further Problems of Diffusion

Because neuron cell bodies are capable of at least limited pinocytosis of extracellular markers (Brightman, 1965; Holtzman & Peterson, 1969; LaVail & LaVail, 1974; Turner & Harris, 1974), it is essential for investigators to define carefully the pathway by which substances arrive at the cell body. Several pathways are possible. The most obvious, diffusion, was the earliest examined and coincided with studies on the definition of extracellular space within the CNS and the blood–brain barrier (Brightman, 1965). In one study Brightman and Reese (1969) identified HRP 1 mm away from the ventricular surface 90 min after intraventricular administration of the marker. From our own studies we have found that HRP apparently diffuses at a rate of at least 1 mm/hr (LaVail & LaVail, 1974). Ten hours after injection of about 1 mg of the protein marker into the chick tectum, HRP was identified extracellularly in the optic nerve, a distance of about 10 mm from the site of injection.

In the context of a technical application of HRP and transport, this diffusion may result in a confusion of neurons that are labeled as a result of diffusion and subsequent pinocytotic uptake by the cell body with neighboring labeled cells that project axons to the injection site. If sufficient time elapses between the time of injection and that of fixation, both populations of neurons may appear equally labeled, one as a result of retrograde transport of HRP to the cell body and the other as a result of diffusion of HRP to the vicinity of the cell body. It may be essential to choose shorter survival periods in order to distinguish the two processes. However, if this difficulty is overcome, the transport technique may provide positive evidence of the identity of cells whose axons innervate the injection site as compared to neurons whose axons terminate elsewhere or locally within the nucleus (Ralston, 1975).

Transfer from the blood supply is a second pathway by which HRP may reach the cell body. In most cases, the amount of peroxidase is too small to reach a level in the blood for this to introduce artifactual labeling. However, in sites in which the blood–brain barrier is less effective, e.g., the medial eminence or area postrema, nearby neurons may be labeled via extracerebral and cerebral blood (Broadwell & Brightman, 1976). As a control for this extraneous labeling of cells, HRP may be injected intravenously and this pattern of labeling compared to the pattern found after an injection into CNS parenchyma. Further details of the labeling of motor, sensory, autonomic, and neurosecretory neurons after significant quantities of HRP (90–150 mg) are introduced intravenously may be found in the report of Broadwell and Brightman (1976).

C. Problem of Fibers of Passage

Several early experiments using the retrograde transport technique appeared to indicate that neither intact fibers nor transected fibers passing through but not terminating in the site of injection were capable of uptake and retrograde transport of peroxidase (Jones & Leavitt, 1974; LaVail et al., 1973; Nauta et al., 1974). Soon thereafter, however, reports appeared indicating that transected axons would accumulate label and result in additional labeled cells (Bunt, Hendrickson, Lund, & Fuchs, 1975; DeVito et al., 1974; Kristensson & Olsson, 1974). The apparently capricious reproducibility of labeling in damaged neurons may reflect the differential labeling seen in comparisons of injured cells and neurons not directly damaged (see Section III,A). Whatever the cause, the complications introduced as a result of transport and accumulation of marker by fibers of passage represents a major limitation of the retrograde transport technique.

D. False Negatives and False Positives

It is tempting to conclude that the axon of an unlabeled neuron does not project to the HRP injection site. However, several factors may affect the amount of peroxidase accumulated by a neuron. For example, the axon of the cell may have failed to incorporate sufficient marker to result in effective label. Nauta *et al.* (1974) observed an important example of this false negative result in their study of retrograde transport of HRP from the rat caudoputamen. Although many other labeled populations of cells could be identified, only a very few cortical cells accumulated the marker, despite a well-documented corticocaudate projection. With modifications in the fixative and amount of HRP injected, J. C. Hedreen (1977) subsequently has successfully found HRP-positive cortical cells that project to the caudoputamen. The point remains, nevertheless, that certain neurons may fail to label with HRP due among other factors to the ill-defined process of inactivation of the marker or simply as a result of technical limitations. Until an investigator can be confident of the labeling capability of a particular class of neurons or until a result is verified with other experimental techniques, the unlabeled neuron cannot be interpreted unequivocally.

One should be aware of the presence of significant amounts of endogenous pigment, including neuromelanin or lipofuscin (Barden, 1969) or of biogenic amines which are themselves possible pigment precursors and which may react with oxidizable substances such as DAB (A. B. Novikoff, Biempica, Beard, & Dominitz, 1971; Pearse, 1972). Furthermore, endogenous heme enzymes, such as catalase or cytochrome oxidases, can also oxidize the substances used to demonstrate HRP, and these should also be considered carefully (Bunt *et al.*, 1974; Wong-Riley, 1976). The identification of peroxidase product solely on the basis of light brown granules in the cytoplasm after reaction with DAB or on the basis of the absence of granules in unreacted sections cannot be considered definitive, conclusive evidence. Cytochemical reaction of sections from an uninjected, control animal may clarify whether the product is neuromelanin, although it may not discriminate neurons containing HRP from those with endogenous heme enzymes or pigment precursors. The use of very young animals without discernible lipofuscin may provide a solution to this particular artifact. In some cases treatment with inhibitors of catalase, such as 2,6-dichlorophenolindophenol (P. M. Novikoff & Novikoff, 1972), may be necessary to avoid confusing the product of catalase and peroxisomes with HRP in vesicles. From this brief discussion it is obvious that the use of HRP as a marker entails careful consideration of the possible presence of endogenous enzymes. For this reason, alternative tracers with radioactive or immunohistochemical labels may be preferred in the future.

E. Environmental Safety of Substrates

The carcinogenicity of benzidine is well known and has limited its usefulness as a substrate for HRP in retrograde tracing studies. Conflicting reports of the dangers involved with derivatives of benzidine, such as DAB and o-dianisidine have also appeared (Gorrod, Carter, & Roe, 1968; Hadidian, Fredrickson, Weisburger, Weisburger, Glass, & Mantel, 1968; Miller, Sandin, Miller, & Rusch, 1956). Until a definitive study is made it would seem prudent to treat these substrates with the same caution as recommended for benzidine.

Several precautions are relatively simple and inexpensive to carry out. Use gloves, lab coat, and mask whenever handling the benzidine or its derivatives. Weigh out multiple aliquots of substrate at one time in a safe hood. We use 10-mg aliquots deposited into 20-ml volume vials that are sealed and stored with desiccant in a freezer. The aliquots can then be dissolved in the buffer solution in the same vial and thus, one passage of the powdered substrate is eliminated. Dispose of the used and unused substrate in containers in accordance with Environmental Safety standards.

V. Summary

Two limitations of the retrograde transport method are particularly critical for understanding and interpretating experimental results. First, a clear definition of the zone of uptake of the marker for effective labeling of cell bodies cannot be precisely determined, since many parameters such as rapid diffusion, clearance from the extracellular space, and minimum concentration recognizable in neuron perikarya are not as yet clearly understood. Second, the uptake of HRP by fibers of passage is an equally significant handicap of the method. Neurons whose axons pass through the injection site but terminate elsewhere may also appear labeled and be confused with neurons that project to and terminate in the region of the injection. Despite these reservations, the combination of the retrograde transport method and other experimental, neuroanatomical tools offers exciting and powerful means of analyzing the development and connectivity of neurons.

Acknowledgments

This review was based on information available in late 1976. It was written with the support of U. S. Public Health Service Grant No. NS-13533 and of a fellowship from the Alfred P. Sloan Foundation.

It is a pleasure to thank Drs. J. C. Adams, A. Basbaum, J. de Olmos, M.-M. Mesulam, W. Stewart, I. T. Diamond, M. Biber, L. Kneisley, and G. Grant for permission to refer to their material. I am also grateful to Dave Akers for his photographic assistance and to Terry Perkins for his technical assistance.

References

Adams, J. C. Technical consideration of the use of horseradish peroxidase as a neuronal marker. *Neuroscience*, 1977, **2**, 141–146.

Adams, J. C., & Warr, W. B. Origins of axons in the cat's acoustic striae determined by injection of horseradish peroxidase into severed tracts. *Journal of Comparative Neurology*, 1976, **170**, 107–121.

Albrecht, M. H. Mounting frozen sections with gelatin. *Stain Technology*, 1954, **29**, 89–90.

Barden, H. The histochemical relationship of neuromelanin and lipofuscion. *Journal of Neuropathology and Experimental Neurology*, 1969, **28**, 419–441.

Becker, N. H., Hirano, A., & Zimmerman, H. M. Observations of the distribution of exogenous peroxidase in the rat cerebrum. *Journal of Neuropathology and Experimental Neurology*, 1968, **27**, 439–452.

Beitz, A. J., & King, G. W. An improved technique for the microinjection of horseradish peroxidase. *Brain Research*, 1976, **108**, 175–179.

Brightman, M. W. The distribution within the brain of ferritin injected into cerebrospinal fluid compartments. II. Parenchymal distribution. *American Journal of Anatomy*, 1965, **117**, 193–200.

Brightman, M. W., & Reese, T. S. Junctions between intimately apposed cell membranes in the vertebrate brain. *Journal of Cell Biology*, 1969, **40**, 648–677.

Broadwell, R. D., & Brightman, M. W. Entry of peroxidase into neurons of the central and peripheral nervous systems from extracerebral and cerebral blood. *Journal of Comparative Neurology*, 1976, **166**, 257–284.

Bunt, A. H., Haschke, R. H., Lund, R. D., & Calkins, D. F. Factors affecting retrograde axonal transport of horseradish peroxidase in the visual system. *Brain Research*, 1976, **102**, 152–155.

Bunt, A. H., Hendrickson, A. E., Lund, J. S., Lund, R. D., & Fuchs, A. F. Monkey retinal ganglion cells: Morphometric analysis and tracing of axonal projections, with a consideration of the peroxidase techniques. *Journal of Comparative Neurology*, 1975, **164**, 265–286.

Bunt, A. H., Lund, R. D., & Lund, J. S. Retrograde axonal transport of horseradish peroxidase by ganglion cells of the albino rat retina. *Brain Research*, 1974, **73**, 215–228.

Clarke, P. G. H., & Cowan, W. M. The development of the isthmo-optic tract in the chick, with special reference to the occurrence and corrections of developmental errors in the location and connections of isthmo-optic neurons. *Journal of Comparative Neurology*, 1976, **167**, 143–163.

Colman, D. R., Scalia, F., & Cabrales, E. Light and electron microscopic observations on the anterograde transport of horseradish peroxidase in the optic pathway in the mouse and rat. *Brain Research*, 1976, **102**, 156–163.

Colwell, S. A. Thalamo-cortical-corticothalamic reciprocity: A combined anterograde-retrograde tracer technique. *Brain Research*, 1975, **92**, 443–449.

de Olmos, J. S. An improved HRP method for the study of central nervous connections. *Experimental Brain Research*, 1977, **29**, 541–551.

DeVito, J. L., Clausing, K. W., & Smith, O. A. Uptake and transport of horseradish peroxidase by cut end of the vagus nerve. *Brain Research*, 1974, **82**, 269–271.

Divac, I., LaVail, J. H., Rakic, P., & Winston, K. R. Heterogeneous afferents to the inferior parietal lobule of the rhesus monkey revealed by the retrograde transport method. *Brain Research*, 1977, **123**, 197–207.

Geisert, E. E. The use of tritiated horseradish peroxidase for defining neuronal pathways: A new application. *Brain Research*, 1976, **117**, 130–135.

Glatt, H. R., & Honegger, C. G. Retrograde axonal transport for cartography of neurones. *Experientia*, 1973, **29**, 1515–1517.

Gorrod, J. W., Carter, R. S., & Roe, F. J. C. Induction of hepatomas by 4-aminobiphenyl and three of its hydroxylated derivatives administered to newborn mice. *Journal of the National Cancer Institute*, 1968, **41**, 403–410.

Graham, R. C., Jr., & Karnovsky, M. J. The early stages of absorption of injected horseradish peroxidase in the proximal tubules of mouse kidney: Ultrastructural cytochemistry by a new technique. *Journal of Histochemistry and Cytochemistry*, 1966, **14**, 291–302.

Graham, R. C., Jr., Lundholm, V., & Karnovsky, M. J. Cytochemical demonstration of peroxidase activity with 3-amino-9-ethyl carbazole. *Journal of Histochemistry and Cytochemistry*, 1965, **13**, 150–152.

Graybiel, A. M., & Devor, M. A microelectrophoretic delivery technique for use with horseradish peroxidase. *Brain Research*, 1974, **68**, 167–173.

Hadidian, Z., Fredrickson, T. N., Weisburger, E. K., Weisburger, J. H., Glass, R. M., & Mantel, N. Tests for chemical carcinogens. Report on the activity of derivatives of aromatic amines, nitrosamines, quinolines, nitroalkanes, amides, epoxides, azeridines and purines antimetabolites. *Journal of the National Cancer Institute*, 1968, **41**, 985–1025.

Halperin, J. J., & LaVail, J. H. A study of the dynamics of retrograde transport and accumulation of horseradish peroxidase by injured neurons. *Brain Research*, 1975, **100**, 258–269.

Hanker, J. S., Norden, J. J., & Diamond, I. T. Horseradish peroxidase tracers with fluorescent reporter groups. *Journal of Histochemistry and Cytochemistry*, 1976, **24**, 609.

Hanker, J. S., Yates, P. E., Metz, C. B., & Rustioni, A. A new specific, sensitive and noncarcinogenic reagent for the demonstration of horseradish peroxidase. *Histochem. J.*, 1977, **9**, 789–792.

Hansson, H.-Å. Uptake and intracellular bidirectional transport of horseradish peroxidase in retinal ganglion cells. *Experimental Eye Research*, 1973, **16**, 377–388.

Hardy, H., & Heimer, L. A safer and more sensitive substitute for diaminobenzidine in the light of microscopic demonstration of retrograde and anterograde axonal transport of HRP. *Neuroscience Letters*, 1977, **5**, 235–240.

Hedreen, J. C. Corticostriatal cells identified by the peroxidase method. *Neuroscience Letters*, 1977, **4**, 1–7.

Hedreen, J. C., McGrath, S., & Warner, C. Survival time, aldehyde fixation, and axonal transport of horseradish peroxidase. *Society for Neuroscience Abstracts*, 1976, **2**, 38.

Holland, V. R., Saunders, B. C., Rose, F. L., & Walpole, A. L. A safer substitute for benzidine in the detection of blood. *Tetrahedron*, 1974, **30**, 3299–3302.

Holtzman, E., & Peterson, E. R. Uptake of protein by mammalian neurons. *Journal of Cell Biology*, 1969, **40**, 863–869.

Jacobson, S., & Trojanowski, J. Q. Corticothalamic neurons and thalamocortical terminal fields: An investigation in rat using horseradish peroxidase and autoradiography. *Brain Research*, 1975, **85**, 385–401.

Jones, E. G. Possible determinants of the degree of retrograde neuronal labeling with horseradish peroxidase. *Brain Research*, 1975, **85**, 249–253. (a)

Jones, E. G. Some aspects of the organization of the thalamic reticular complex. *Journal of Comparative Neurology*, 1975, **162**, 285–308. (b)

Jones, E. G., & Leavitt, R. Y. Retrograde axonal transport and the demonstration of non-specific projections to the cerebral cortex and striatum from thalamic intralaminar nuclei in the rat, cat and monkey. *Journal of Comparative Neurology*, 1974, **154**, 349–378.

Karnovsky, M. J. A formaldehyde-glutaraldehyde fixative of high osmolality for use in electron microscopy. *Journal of Cell Biology*, 1965, **27**, 137A.

Kim, C. C., & Strick, P. L. Critical factors involved in the demonstration of horseradish peroxidase retrograde transport. *Brain Research*, 1976, **103**, 356–361.

Kristensson, K. Transport of fluorescent protein tracer in peripheral nerves. *Acta Neuropathologica*, 1970, **16**, 293–300.

Kristensson, K. Retrograde axonal transport of protein tracers. In W. M. Cowan & M. Cuénod (Eds.), *The use of axonal transport for studies of neuronal connectivity*. New York: American Elsevier, 1975: Pp. 70–82.

Kristensson, K., & Olsson, Y. Retrograde axonal transport of protein. *Brain Research*, 1971, **29**, 363–365.

Kristensson, K., & Olsson, Y. Retrograde transport of horseradish peroxidase in transected axons. I. Time relationships between transport and induction of chromatolysis. *Brain Research*, 1974, **79**, 101–110.

Kuypers, H. G. J. M., Kievit, J., & Groen-Klevant, A. C. Retrograde axonal transport of horseradish peroxidase in rat's forebrain. *Brain Research*, 1974, **67**, 211–218.

Land, L. J., Barrett, J., & Whitlock, D. G. Ascending pathways originating in different laminae of the rat spinal gray. *Anatomical Record*, 1975, **181**, 404.

LaVail, J. H. Retrograde cell degeneration and retrograde transport techniques. In W. M. Cowan & M. Cuénod (Eds.), *The use of axonal transport for studies of neuronal connectivity*. New York: American Elsevier, 1975. Pp. 218–248.

LaVail, J. H., & LaVail, M. M. Retrograde axonal transport in the central nervous system. *Science*, 1972, **176**, 1416–1417.

LaVail, J. H., & LaVail, M. M. The retrograde intraaxonal transport of horseradish peroxidase in the chick visual system: A light and electron microscopic study. *Journal of Comparative Neurology*, 1974, **157**, 303–358.

LaVail, J. H., Winston, K. R., & Tish, A. A method based on retrograde intraaxonal transport of protein for identification of cell bodies of origin of axons terminating within the CNS. *Brain Res.*, 1973, **58**, 470–477.

Ljungdahl, A., Hökfelt, T., Goldstein, M., & Park, D. Retrograde peroxidase tracing of neurons combined with transmitter histochemistry. *Brain Research*, 1975, **84**, 303–319.

Llamas, A., & Martínez-Moreno, E. Modificaciones al método del transporte axonal retrograde de la horseradish peroxidasa en el sistema nervioso central en el gato adulto. *Anales de Anatomia*, 1974, **23**, 431–443.

Lynch, G., Gall, C., Mensah, P., & Cotman, C. W. Horseradish peroxidase histochemistry: A new method for tracing efferent projections in the central nervous system. *Brain Research*, 1974, **65**, 373–380.

McGill, J. I., Powell, T. P. S., & Cowan, W. M. The retinal representation upon the optic tectum and isthmo-optic nucleus in the pigeon. *Journal of Anatomy*, 1966, **100**, 5–33.

Mesulam, M.-M. The blue reaction product in horseradish peroxidase neurohistochemistry: Incubation parameters and visibility. *Journal of Histochemistry and Cytochemistry*, 1976, **24**, 1273–1280.

Miller, E. C., Sandin, R. B., Miller, J. A., & Rusch, H. P. The carcinogenicity of compounds related to 2-acetylaminofluorene. III. Aminobiphenyl and benzidine derivatives. *Cancer Research*, 1956, **16**, 525–534.

Muller, K. R., & McMahon, U. J. The arrangement and structure of synapses formed by specific sensory and motor neurons in segmental ganglia of the leech. *Anatomical Record*, 1975, **181**, 432.

Nakane, P. K. Simultaneous localization of multiple tissue antigens using the peroxidase-labeled antibody method: A study on pituitary glands of the rat. *Journal of Histochemistry and Cytochemistry*, 1968, **16**, 557–560.

Nauta, H. J. W., Kaiserman-Abramof, I. R., & Lasek, R. J. Electron microscopic observations of horseradish peroxidase transported from the caudoputamen to the substantia nigra in the rat: Possible involvement of the agranular reticulum. *Brain Research*, 1975, **85**, 373–384.

Nauta, H. J. W., Pritz, M. B., & Lasek, R. J. Afferents to the rat caudoputamen studied with horseradish peroxidase. An evaluation of a retrograde neuroanatomical research method. *Brain Research*, 1974, **67**, 219–238.

Novikoff, A. B., Biempica, L., Beard, M., & Dominitz, R. Visualization by diaminobenzidine of norepinepherine cells, premelanosomes and melanosomes. *Journal of Microscopy (Oxford)*, 1971, **12**, 297–300.

Novikoff, P. M., & Novikoff, A. B. Peroxisomes in absorptive cells of mammalian small intestine. *Journal of Cell Biology*, 1972, **53**, 532–560.

Nowakowski, R. S., LaVail, J. H., & Rakic, P. The correlation of the time of origin of neurons with their axonal projection: The combined use of ^{3}H-thymidine autoradiography and horseradish peroxidase histochemistry. *Brain Research*, 1975, **99**, 343–348.

Papadimitriou, J. M., Van Duijn, P., Brederoo, P., & Streefkerk, J. G. A new method for the cytochemical demonstration of peroxidase for light, fluorescence and electron microscopy. *Journal of Histochemistry and Cytochemistry*, 1976, **24**, 82–90.

Pearse, A. G. E. (Ed.) *Histochemistry: theoretical and applied*. Vol. 2. London: Churchill, 1972 pp. 1076–1082.

Ralston, H. J., III. Methods for the analysis of local neuronal circuits. Anatomical techniques. *Neurosciences Research Program, Bulletin*, 1975, **13**, 320–326.

Ralston, H. J., III, & Sharp, P. V. The identification of thalamocortical relay cells in the adult cat by means of retrograde axonal transport of horseradish peroxidase. *Brain Research*, 1973, **62**, 273–278.

Reese, T. S., Bennett, M. V. L., & Feder, N. Cell-to-cell movement of peroxidases injected into the septate axon of crayfish. *Anatomical Record*, 1971, **169**, 409.

Rustioni, A. Non-primary afferents to the cuneate nucleus in the brachial dorsal funiculus of the cat. *Brain Research*, 1974, **75**, 247–259.

Rustioni, A., & Dekker, J. J. Non-primary afferents to the dorsal column nuclei of cat: Distribution pattern and cells of origin. *Anatomical Record*, 1974, **178**, 454–455.

Schwab, M., Agid, Y., Glowinski, J., & Thoenen, H. Retrograde axonal transport of ^{125}I-tetanus toxin as a tool for tracing fiber connections in the central nervous system: Connections of the rostral part of the rat neostriatum. *Brain Research*, 1977, **126**, 211–224.

Sellinger, O. Z., & Petiet, P. D. Horseradish peroxidase uptake *in vivo* by neuronal and glial lysosomes. *Experimental Neurology*, 1973, **38**, 370–385.

Singer, W., Hollander, H., & Vanegas, H. Decreased peroxidase labelling of lateral geniculate neurons following deafferentation. *Brain Research*, 1977, **120**, 133–137.

Snow, P. J., Rose, P. K., & Brown, A. G. Tracing axons and axon collaterals of spinal neurons using intracellular injection of horseradish peroxidase. *Science*, 1976, **191**, 312–313.

Sotelo, C., & Riche, D. The smooth endoplasmic reticulum and the retrograde and fast orthograde transport of horseradish peroxidase in the nigro-striato-nigral loop. *Anatomy and Embryology*, 1974, **146**, 209–218.

Stöckel, K., Schwab, M., & Thoenen, H. Comparison between the retrograde axonal transport of nerve growth factor and tetanus toxin in motor, sensory, and adrenergic neurons. *Brain Research*, 1975, **99**, 1–16.

Stöckel, K., & Thoenen, H. Retrograde axonal transport of nerve growth factor: Specificity and biological importance. *Brain Research*, 1975, **85**, 337–341.

Straus, W. Factors affecting the cytochemical reaction of peroxidase with benzidine and the stability of the blue reaction product. *Journal of Histochemistry and Cytochemistry*, 1964, **12**, 462–469.

Turner, P. T., & Harris, A. B. Ultrastructure of exogenous peroxidase in cerebral cortex. *Brain Research*, 1974, **74**, 305–326.

Walberg, F., Brodal, A., & Hoddevik, G. H. A note on the method of retrograde transport of horseradish peroxidase as a tool in studies of afferent cerebellar connections, particularly those from the inferior olive; with comments on the orthograde transport in Purkinje cell axons. *Experimental Brain Research*, 1976, **24**, 383–401.

Weir, E. E., Pretlow, T. G., Pitts, A., & Williams, E. E. A more sensitive and specific histochemical peroxidase stain for the localization of cellular antigen by the enzyme-antibody conjugate method. *Journal of Histochemistry and Cytochemistry*, 1974, **22**, 1135–1140.

Westergaard, E., & Brightman, M. W. Transport of proteins across normal cerebral arterioles. *Journal of Comparative Neurology*, 1973, **152**, 17–44.

Wong-Riley, M. T. T. Demonstration of geniculocortical and collosal projection neurons in the squirrel monkey by means of retrograde axonal transport of horseradish peroxidase. *Brain Research*, 1974, **79**, 267–272.

Wong-Riley, M. T. T. Endogenous peroxidatic activity in brain stem neurons as demonstrated by their staining with diaminobenzidine in normal squirrel monkeys. *Brain Research*, 1976, **108**, 257–277.

Zacks, S. I., & Saito, A. Uptake of exogeneous horseradish peroxidase by coated vesicles in mouse neuromuscular junctions. *Journal of Histochemistry and Cytochemistry*, 1969, **17**, 161–170.

Chapter 14

Electrophysiological Mapping Techniques

John Schlag

Department of Anatomy and Brain Research Institute
University of California, Los Angeles

I. Introduction: Electrophysiology as a Neuroanatomical Tool

The terms *electrophysiological* and *anatomical* are usually contrasted as characterizing two approaches with different objectives in neurological research. One is primarily concerned with function, the other with structure. Yet, there are many problems of neuronal connectivity which can be worked out by electrophysiological means. In certain cases, the latter even offers the most direct solution for obtaining anatomical information not readily available by other methods.

As a neuroanatomical tool, electrophysiological mapping is an antique. Fritsch and Hitzig's (1870) elicitation of movements by electrical stimulation of an area of cerebral cortex was probably one of the earliest experiments leading to the discovery of a nervous pathway. At that time, the neuron doctrine was still in limbo. Many of the most significant contributions to the core of classical material presented in neuroanatomy textbooks were made by stimulation and recording. For the present and the future, the electrophysiological method still offers unique advantages, especially for investigating interactions between components of a neuropil and for tracing whole macroscopic pathways. This chapter will be concerned with the latter problem.

II. Scope and Problems

In its most general form, the electrophysiological paradigm of mapping consists of applying a stimulus at a site S and seeking whether a response is elicited at site R. Thus, the first question asked is always: *"Can one get there from here?"* The occurrence of a response implies that a path between S and R exists. Once this conclusion is accepted, a number of secondary questions can be posed: What is stimulated in S? What is responding in R? Is it an axon? A cell body? What type of cells? Do the cells in R receive other inputs? (convergence of inputs). Do the cells in S project elsewhere? (divergence of outputs). Is there some kind of orderly arrangement in the projection between elements of the S region and those of the R region? (e.g., a topological projection such as a point-to-point representation in primary sensory systems). Are synapses interposed in the hypothetical pathway? What is the normal direction of impulses along a fiber path? (Backward—i.e., antidromic—conduction can be elicited by electrical stimulation, thereby providing an easy way of identifying the projection of a neuron.) How fast is the conduction between S and R? (Under appropriate circumstances, the size of nerve fibers can be deduced from conduction time.) Are some of the excited neurons inhibitory? What is the course of the hypothetical pathway? (The approach to this particular problem requires additional stimulation or placement of lesions along the presumed path.)

To deal with this variety of questions exists an even larger variety of experimental designs and tests. They draw on practically all the principles applied in neurophysiology. Since it would be impossible to cover this material in a single chapter, reference should be made to general textbooks for basic neurophysiological notions and to specialized books (e.g., Thompson & Patterson, 1973) for technical details of procedures. We shall mainly consider problems of strategy in selecting methods of stimulation, recording, and experimental design.

III. Methods

A. Natural Stimulation

A pathway is not any arbitrary ensemble of neurons that happen to be interconnected. In the concept of pathway, there is, in addition, the requirement that information of a particular kind be carried from neuron to neuron. Obviously, this information is altered as it progresses, but its nature can remain recognizable at least on some portion of the circuits. Pathways are useful, though artificial, constructs.

The easiest pathways to identify are those that convey sensory signals. For tracing them, natural stimuli (light, sound, pressure, movement, specific chemical, temperature changes, etc.) should be used whenever possible. They are ideally suited since the experiment mimics the normal conditions of functioning of the system under investigation. Natural stimuli should also be physiological, i.e., kept within a reasonable range of intensities, and they should be adequate, i.e., of such modality as to elicit the "best" response. This last requirement is probably the most difficult to fulfill since there is no straightforward strategy to find the adequate stimulus and no safe criterion to determine that the optimum has been achieved. The adequate stimulus is often selected by trial and error, and, in practice, the solutions proposed can differ greatly. Finally, it is desirable that the stimulus be spatially as limited as feasible (i.e., affecting a minimum number of sensory receptors) in order to obtain the greatest resolution in defining the inputs to a neuron or neuronal population. These are necessary requirements for receptive field studies (e.g., see Jacobs, 1969, for a review of visual receptive field explorations).

Electrophysiological mapping techniques have been highly successful in revealing organizational designs in several sensory systems. The important concepts of point-to-point representation, cortical homunculi, cerebral columnar organization, lateral inhibition, and others derive, in fact, from electrophysiological studies. Often, these pioneer works preceded by far the visual demonstration of the postulated structural substrate (e.g., T. A. Woolsey & Van der Loos, 1970).

B. Electrical Stimulation

Natural stimuli, obviously, can no longer be used when the problem is to trace a portion of a circuit that does not include sensory receptors. Artificial stimuli are then substituted to trigger a localized action accurately controlled, specified, and reproducible. In the past, attempts have been made to use strychnine, which acts on synapses or synaptic terminals, and produces either spontaneous or easy-to-evoke "strychnine spikes." This procedure called "strychnine neuronography" has served for tracing pathways (e.g., Dusser de Barenne & McCulloch, 1936) by injecting or topically applying minute amounts of the drug at selected locations. Other chemical agents have been tried; some may become very valuable tools in the future to the extent that their site of action will prove to be limited to a specific neuronal population. But, up to now, in comparison with all other means, electrical stimulation has overwhelming advantages; it is universally adopted to excite nervous structures directly.

Ideally, in mapping studies, one would wish to restrict the electrical stimulation to a limited number of identified neuronal elements. How should parameters of stimulation be selected? Little attention is generally given to this problem except for avoiding noxious or destructive effects. Yet data exist—and Ranck (1975) has recently reviewed them—for the purpose of determining which elements are excited in electrical stimulation of the mammalian central nervous system. All these data concern monofocal stimulation (also called monopolar, i.e., one electrode being away from the generator) rather than bifocal stimulation. Thus, monofocal stimulation seems advisable if comparison with parametric studies is sought.

Bifocal stimulation (with two parallel wires or a wire inserted within a needle, protruding at the tip and forming the so-called concentric electrode) sometimes is preferred for one of two reasons. First, the contamination of records by shock artifacts can be minimized. This is particularly helpful when the sites of stimulation and recording are very close to each other (a few millimeters) and the latency of response is short. Second, the overall spread of stimulus current can be restricted. This applies when relatively large currents are needed to obtain a response, for instance, if the location of the structure to be stimulated is not easily determined or if this structure occupies a large volume of tissue. Such circumstances are not ideal for delineating a pathway. As much as possible, one wishes the stimulating electrodes to be so adequately placed that minimum stimulation would be required. The problem with bifocal stimulation is that the effects produced by a cathode and an anode within the excitable tissue are not readily analyzable.

The principal parameters of stimulation to consider are discussed below.

1. PULSE DURATION

Short pulses (e.g., 0.05 msec) are effective on myelinated axons whereas longer pulses (e.g., 1 msec) stimulate axons, cell bodies, and dendrites. Short pulses should be used if the purpose is to selectively excite fibers in the gray matter of nervous tissue.

2. CURRENT INTENSITY

The strength of stimulation is best expressed in terms of current (not voltage). What really matters is the current density which is a function of the current and the conducting surface of the electrode. Less current is needed when delivered by a microelectrode than by a gross electrode to stimulate a small population of neurons. Microelectrode stimulation is clearly to be preferred since it allows a better identification of the elements stimulated. Currents adequate for extracellular stimulation are in the order of microamperes (10^{-6} A). Thus, the threshold for exciting a cortical pyramidal cell by microelectrode may be as low as 1 μA (Stoney, Thompson, & Asanuma, 1968). By comparison, nanoamperes (10^{-9} A) can be sufficient, when applied intracellularly, to drive single nerve cells (e.g., Granit, Kernell, & Shortess, 1963; Woody & Black-Cleworth, 1973).

Theoretically, one can argue that the effective spread of a stimulating current, i, is proportional to its square root. If one assumes that cell bodies are uniformly distributed, their number in a sphere centered on the tip of the microelectrode is proportional to the cube of the radius of the sphere. Consequently, in a medium of homogeneous conductivity, the number n of cell bodies excited is given by $n = ki^{3/2}$, where k is a constant (Asanuma & Hunsperger, 1975). Thus, Stoney *et al.* (1968) have estimated that an average of 28 cell bodies are stimulated by 10-μA current pulses of 0.2-msec duration in the anterior sigmoid gyrus of cat. However, such figures provide only an order of magnitude, and they concern cell bodies only, not passing axons. There are several sources of variability in experimental data collected under different conditions (see Ranck, 1975), and many factors have to be taken into consideration. Thus, for instance, the actual site of spike initiation (virtual cathode) at the level where the neuronal membrane has the lowest threshold may be some distance away from the stimulating electrode. It should also be noted that too large cathodal currents may block action potentials in axons (Ranck, 1975).

3. POLARITY

The most effective polarity of stimulation can be readily tested during the experiment. In general, one would expect cathodal stimulation (negative electrode) to be best, but there are exceptions. Thus, when applied at the surface of the cerebral cortex, anodal stimulation (positive electrode) is more effective for exciting pyramidal neurons directly. In the cortex, it appears that anodal currents penetrate superficial dendrites, leave at the axon or cell body, and depolarize the latter (axon and cell body have a lower threshold than dendrites). Such findings, analyzed in relation with the known cytoarchitectonics of the structure studied, can provide a cue on the geometry and, thus, on the nature of the elements stimulated.

4. PULSE FREQUENCY

Single action potentials (spikes) can be evoked in fibers by applying single pulses of current, but repetition is often needed to obtain postsynaptic responses. This phenomenon is known as temporal summation; effective transmission depends on the number and rate of stimuli. Here again, practical values depend on the structure.

In general, it is advantageous to routinely try several values of the parameters of stimulation (duration, intensity, frequency) and to reverse the polarity. Such trials help find the optimal conditions of minimum stimulation and, also, may provide useful cues for the interpretation of the results.

C. Electrophysiological Recording

Recording techniques used for tracing of pathways include recording of gross evoked potentials; nerve or tract action potentials; and multiunit, single-unit extracellular, and single-unit intracellular activity (for details of procedures, see Thompson & Patterson, 1973). To this list could also be added the recording of electromyographic (EMG) activity since it provides an objective test of motor response apparently sensitive, for instance, even to stimulation of single motor cortex neurons (Woody & Black-Cleworth, 1973).

In the past, most of the groundwork of outlining primary sensory pathways in mammals was laid down by recording gross evoked potentials (e.g., Adey & Kerr, 1954; Cowey, 1964; C. N. Woolsey, 1958; Mountcastle & Henneman, 1952; C. N. Woolsey & Walzl, 1942). However, the interpretation of this type of mass potentials is hampered by the difficulty of identifying the origin and nature of the wave components in the responses (Schlag, 1973). Under favorable circumstances, the earliest of these components can be given significance in mapping because they are unambiguously due to the arrival of impulses along direct routes. But, as latencies increase, the possibilities of indirect pathways can no longer be excluded and the data lose their usefulness.

Multiunit recording with thin wires (e.g., around 50 μm, Schlag & Balvin, 1963; also Chapters 7, 8, and 9 in Thompson & Patterson, 1973) has a clear advantage. As many firing elements are sampled (probably including some which can be hard to isolate in the form of single spikes with finer microelectrodes) one gets a rapid evaluation of the shortest latencies to be expected in a given population. However, the problem of identifying these firing elements (e.g., presynaptic afferents, passing fibers, cell bodies, dendrites, departing axons) is not much easier than that of interpreting the significance of waves in evoked potentials. Therefore, it appears more advantageous to rely on single-unit recording.

Today, most of the mapping work is done with micro- rather than macroelectrodes. In a cell population, the microelectrode can pick up the so-called local field potential which is, on a smaller scale, the equivalent of a gross evoked potential. In addition, the microelectrode records discrete events due to the activity of individual neurons, named *units*. Unit data are certainly less ambiguous than gross evoked potentials and topographical analyses can be made with a much finer grain than with large electrodes (e.g., for receptive fields: Poggio & Mountcastle, 1963; Whitsel, Dreyer, & Roppolo, 1971). Therefore, microelectrode techniques will be discussed in more detail considering both the intracellular and extracellular modes of derivation.

Microelectrodes are either glass micropipettes filled with a conducting solution (for fabrication, see La Vallee, Schonne, & Hebert, 1969; Thompson & Patterson, 1973) or metal rods with varnish or glass insulation. Various metals are commonly used; they differ by their electrical properties (Geddes, 1972), and other characteristics are more or less important depending on specific objectives (Snodderly, 1973). Thus, platinum (and metals plated with platinum black) has a lower impedance (Kinnard & MacLean, 1967; Wolbarsht, MacNichol, & Wagner, 1960); tungsten is particularly stiff, allowing penetration through the dura (Cool, Crawford, & Soheer, 1970; Hubel, 1957); stainless steel is inexpensive and permits easy marking at the tip site by iron deposit in the tissue (Grundfest, Sengstaken, & Oettinger, 1950). Recent techniques have been developed for tip marking with iron (nickel–cobalt–iron alloy, see Suzuki & Asuma, 1976) or silver-plated tungsten or platinum (Spinelli, 1975). For tip marking with micropipettes, see Lee, Mandl, and Stean (1969), Barrett and Graubard (1970), Jankowska and Lindström (1970), Globus (1973), West, Deadwyler, Cotman, and Lynch (1975), and Tweedle, Chapter 6 in this volume.

1. INTRACELLULAR DERIVATION

Strictly from the viewpoint of information yield, the intracellular derivation appears advantageous for it shows both the input (postsynaptic potentials) and the output (spikes) of a given cell. The detection of postsynaptic potentials provides direct evidence of the arrival of impulses. By extracellular recording of spikes, one gets the answer to a slightly different question: what is the cell responding to? This can be a significant difference since the output of a neuron is not a straightforward image of its input. Hence, the absence of firing is not definite proof that a neuron does not receive afferents from a given source. Practically, the intracellular derivation is a more sensitive method to the extent that it can show subthreshold events. Illustrative examples of postsynaptic potentials and of their time course in relation to the cell action potentials can be found in Shepherd (1974).

Intracellular recordings are quite useful in solving certain specific problems pertinent to almost any nervous structure. Practically all these problems concern detection and characterization of inputs. The most obvious case is the identification of inhibitory inputs. The effect of an inhibitory input passes unnoticed if a neuron, recorded extracellularly, does not fire spontaneously. Even if it does, it may be difficult to decide whether a pause in its activity is due to postsynaptic inhibition, to the temporary depression of a postsynaptic excitation, or to any other cause. The direct observation of fluctuations of the membrane potential can show hyperpolarization, absence of depolarization, or inactivation (prolonged, intense depolarization) and, thereby, give evidence on the nature of the input. The latter can be even better characterized by passing current through the membrane and measuring its changes of conductance.

By recording postsynaptic potentials, one also detects the earliest sign of a response, which can conveniently be used to measure a latency (see Sections IV B and IV C). This is the only way by which the timing of an inhibitory input can be determined, and it is also useful to estimate the latency of an excitatory input whenever the rise of the excitatory postsynaptic potential is slow and spike initiation occurs with a delay.

Knowing that given sets of afferents impinge on a particular cell population, it is important to determine whether these afferents synapse directly on the same or different cells in that population. The clearest demonstration can be provided by recording postsynaptic potentials in the same neuron to the stimulation of the various afferents (for an illustrative example, see Desijaru & Purpura, 1969). The latency and shape of the postsynaptic potentials, their modifications with repetitive stimulation (e.g., Phillips & Porter, 1964), and their interactions when different inputs are excited often can lead to hypotheses on the location of synapses (for actual examples, see Shepherd, 1974).

Finally, intracellular recording permits the injection of dyes or traceable substances which ulteriorly will show the detailed processes of the cell investigated (see Chapter 6 by Tweedle, this volume). However, puncturing a neuron is a delicate technique and, even under the best circumstances, the accumulation of a sufficient amount of data remains a time-consuming process. This, certainly, is the most serious limitation.

2. Extracellular Derivation

The extracellular derivation is undoubtedly much easier to perform with success. The advantages of convenience and simplicity make it extremely attractive, especially for gathering information on as many units as possible in the same preparation. Moreover, extracellular recording can now be performed easily in chronic animals, thereby avoiding problems inherent in the use of anesthetics, curarizing agents, or lesioned preparations.

Electrophysiology is a blind procedure. The experimeter does not "see" what he is doing at the time when he tries to draw inferences from the data collected. The data are in the form of sequences of spikes which, when their shape and amplitude remain constant (all-or-none principle), can be said to be generated by a single "unit." But what exactly is the unit recorded?

First of all, there is an obvious problem of sampling bias (Towe, 1973). Once the recording site is selected, units are picked up one by one along the microelectrode track as they signal their presence either by spontaneous firing, the appearance of a D.C. potential and/or spikes elicited by stimulation. Neurons with large cell bodies are more likely to be picked up and to survive the pressure or penetration by the microelectrode. This source of bias is generally recognized. What is less often mentioned is that the properties of the microelectrode, e.g., its material, its size, particular shape, and the extent of the uninsulated part of the tip (in a metal electrode) also play a role in sampling the neurons and in shaping the potentials recorded (Bishop, Burke, & Davies, 1962). Most experimenters share the belief that certain types of electrodes are better than others to record from certain types of neurons. The optimal characteristics of microelectrodes are determined empirically in each case. Thus, it is probably not exaggerated to state that the results of a unit study depend not only on the type of neurons available for sampling, but also on the type of probes selected for this sampling.

Very often, it is desirable to establish that a recording is made from a cell body rather than from fibers. Fibers can be afferent, efferent, or even passing through the structure explored without functional relation to it. If the record is from a cell body, the number of options as to the nature and type of the neuron can be reduced, and the results become easier to interpret.

Theoretically, chances are greater of recording from soma-proximal dendrites than from axons or terminal dendrites. This is due to the relative size of these different parts of a neuron. However, one may wish for more than a probable inference. Are there other criteria? In intracellular derivation, the occurrence of postsynaptic potentials excludes the possibility that the microelectrode picks up from a neuronal part far away from synapses, such as the distal portion of the axon. In extracellular derivation, unfortunately, the uncertainty cannot be resolved in any simple manner. Criteria were suggested, based on observations in particular cases where the identification of cell bodies and axons could be verified by input–output tests (Bishop et al., 1962; Fussey, Kidd, & Whitman, 1970; Hubel, 1960). According to these studies, an axonal spike would be distinguished from a somatic spike by: (1) its shorter duration (e.g., less than 0.2 msec, Fussey et al., 1970; however actual values may vary with the characteristics of the recording system); (2) its rapid initial stroke (Bishop et al., 1962); and (3) the absence of inflections on its initial stroke (Bishop et al., 1962; Hubel, 1960). By comparison, a somatic spike would have a longer duration (2–5 msec,

Bishop *et al.*, 1962) and initial notches. These are the most frequently cited criteria. However, they are not infallible. Distinctive features may be absent (e.g., initial inflections in a somatic spike) or ambiguous (e.g., spike duration of intermediate value), and their occurrence may depend on the type of microelectrode used (Bishop *et al.*, 1962). Finally, it is not known whether the criteria listed above can apply in general. If they have to be used, it is certainly advisable to test their validity at sites where only axons are present (e.g., in tracts versus nuclei, or in white matter versus gray matter). Some techniques for discriminating axon from cell body spikes have been proposed (e.g., Fries & Zieglgänsberger, 1974) but they are not yet simple enough to become routinely used in the near future. In summary, the identification of units by the characteristics of the spikes in extracellular derivation is still questionable. Therefore, it is preferable to rely on tests which can establish the input and/or output connectivity of the units (see Section IV C). Such tests do not help in determining the exact site of recording but they can provide information on the type of neuron studied.

IV. Limitations of the Technique and Interpretations of Results

A. *The State of the Organism*

One should always keep in mind that the results obtained by electrophysiological mapping do not depend exclusively on the particular wiring under exploration. Thus, results are affected by the state of the organism, its level of arousal or attention, its history, the anesthetics used, the stimulation of other sensory systems, etc. For example, receptive field properties of cells in the visual cortex were found to vary with bodily tilt (Horn & Hill, 1969). Actually, many such extraneous influences can be considered as the contribution of inputs other than those of immediate interest. These inputs interact, and it may be practically impossible to keep such interactions invariant in a subject placed in conditions as natural as possible. In selecting an experimental situation, there is a necessary trade-off between conditions that isolate a particular circuit and those that allow this circuit to operate normally. Different results will occur under different conditions (e.g., certain responses disappear under anesthesia or after placement of lesions away from the pathway studied) and all can yield meaningful results.

B. *Zero, One, or Several Synapses?*

Figure 1 presents a series of possible circuits through which a stimulus at S produces a response at R. The arrows indicate the normal (i.e., orthodromic) direction of impulses as they circulate under natural conditions. Is it possible to infer the general layout of such circuits from electrophysiological data without

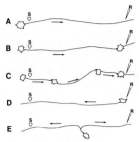

FIG. 1. Possible circuits through which response is obtained at R to stimulation at S. Arrows indicate direction of natural impulses. A. Nonsynaptic connection. B. Monosynaptic connection. C. Polysynaptic path. D. Nonsynaptic connection, neuron invaded antidromically. E. Nonsynaptic connection, antidromic invasion of an axonal branch followed by orthodromic invasion of another branch of the same neuron.

actually seeing any of the neurons involved? The inferences should be based on the response characteristics using any stimulation paradigm that seems appropriate. Two main questions will be faced: Are there synapses on the pathway (as in Figs. 1B and C)? Is the pathway invaded antidromically on, at least, part of its course (as in Figs. 1D and E)?

The strategy proposed takes advantage of differences existing between synaptic transmission and the conduction of impulses along nerve fibers. These differences, despite exceptions, are reliable enough to serve as a basis of diagnosis. In the following summary, it is assumed that synapses are chemically operated, which is by far the most common case in mammalian central nervous systems.

1. Transmission across synapses is unidirectional (Fig. 2A). By contrast, the conduction of impulses artificially generated in fibers is bidirectional.

2. Transmission in synapses involves a delay called postsynaptic reaction time (Fig. 2B). It consists of the time taken for initiating an excitatory postsynaptic potential (EPSP) (called synaptic delay; measured from stimulus to onset of EPSP) plus the time for the EPSP to grow and initiate an action potential (measured from onset of EPSP to onset of first spike).[1] Synaptic delays are generally of the order of 0.2 to 0.5 msec. The postsynaptic reaction time can last from about 0.8 msec to several hundred msec in some cases (transmission across electrical synapses is much faster: 0.1 msec or less).

3. The postsynaptic reaction time tends to fluctuate much more widely than the time interval between the onset of an electrical stimulus and the initiation of an axonal spike (sometimes called reaction time). Therefore, response latencies

[1]In actual measurements, the postsynaptic reaction time also includes a presynaptic delay due to slower conduction along the fine presynaptic terminals. Theoretically, this presynaptic delay should not be included; practically, however, it cannot easily be measured independently.

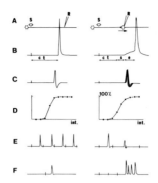

FIG. 2. Differences between characteristics of nonsynaptic (left-hand column) and synaptic (right-hand column) responses. A. Location of stimulation (*S*) and recording (*R*) sites. Arrow indicates unidirectionality of synapse. B. Latencies include presynaptic conduction time (*ct*), plus synaptic delay (*s*), and time for EPSP (*e*) to reach threshold. C. Several responses superimposed for comparison of fluctuations of latency. D. Percentage of response occurrences as a function of stimulus intensity (int.). E. Frequency following; in right-hand column, second response is only a partial spike, subsequent responses are absent. F. Single or multiple responses and effect of stimulus repetition.

vary more often when synaptic transmission is involved in the circuit tested (Fig. 2C). These variations are greatest when the stimulus current is close to threshold.

4. The threshold of stimulation is sharper when no synaptic transmission is involved: a minute increase of stimulus intensity is sufficient to pass from 0 to nearly 100% of response occurrence (Fig. 2D).

5. The refractory period of fibers is relatively short and stable. Hence, responses can be elicited by repetitive stimuli applied at a high rate (one spike per stimulus at frequencies up to several hundred per second). In laboratory jargon, this is called frequency following (Fig. 2E). Synaptic transmission cannot follow at such high frequencies and, in addition, it tends to fatigue: the latency increases and then responses fail to appear at the rate of stimulation.

6. In its normal environment, a nerve fiber responds by a single spike to each single electrical shock applied (Fig. 2F). Often, synaptic transmission is effective only when the stimulus is repeated (summation).

7. Multiple responses very frequently result from stimulation of a pathway involving chemically operated synaptic transmission (Fig. 2F). The number of evoked spikes and their intervals depend on the time course of the EPSP. They are relatively independent of the number and frequency of stimuli in contrast with the situation where no synapse is interposed in the pathway tested.

By relying on these differences, it is often possible to determine whether synaptic transmission is involved in the pathway tested. The general procedure calls for the use of single and repeated stimuli of various strengths, numbers, and frequencies. Accurate measurements are made of latency and its fluctuations, threshold, and frequency following. Practically, it is advisable to perform several of these tests, for none of them is absolutely reliable. Moreover, their results may be ambiguous: for instance, latency measurements are helpful only when latencies happen to be very short; as latencies increase, so does the number of possible interpretations (e.g., slow conduction and monosynaptic or polysynaptic connections).[2] In the past, long latencies tended to be uniformly attributed to the existence of a polysynaptic network but such a conclusion is certainly not warranted by this evidence alone (Orem & Schlag, 1971).

Standard quantitative criteria on time delays, threshold, and stimulus frequency would be useful for drawing conclusions. Unfortunately, these parameters vary depending on the structure investigated. Ideally, the data collected on a given unit population should be analyzed to see whether subgroups can be distinguished in a consistent manner on the basis of several tests. Thus, it may become evident that synaptic transmission is involved in the activation of some units and not in the activation of others. But one should be prepared to find ambiguities for at least a few units.

It is certainly more complicated to determine whether one or more (i.e., successive) synapses are interposed in a circuit. This problem has been recently discussed by Berry and Pentreath (1976) who have suggested some pharmacological tests in addition to electrophysiological tests similar to those described above. This review article should be consulted for further details.

C. Antidromic Response

The term antidromic indicates a direction opposite to the normal direction of impulses along nerve fibers (Fig. 1D). Impulses can go the wrong way only when artificially elicited, and then they are stopped at the first synapse encountered backward. In electrophysiological tracing, the significance of antidromic is equivalent to that of retrograde in techniques of retrograde degeneration and retrograde transport.

Until recently, antidromic conduction was not given equal status to orthodromic conduction as a hypothesis to explain effects caused by electrical stimula-

---

[2]The detection of a delay which can be a postsynaptic reaction time (thereby leading to the conclusion that a synapse is interposed) can be facilitated if the conduction velocity of the stimulated fibers is known. For this purpose, one can measure the conduction time if the point R stimulated is or a tract and if recording from the same tract between R and S can be performed. Then, the measured conduction velocity will be used to calculate the theoretical conduction time on the distance R to S under the assumption that no synapse is interposed. If the actual latency is not greater than the theoretical conduction time by 0.5 msec, there is not sufficient time left for synaptic transmission. If the latency is greater than the theoretical conduction time plus at least 0.5 msec, the result is more ambiguous. The discrepancy may be due to synaptic transmission or else to a reduced conduction velocity along finer axonal branches in the later portion of the path.

tion. However, stimulating at S and recording a response at R does not imply more probably the existence of a path $S{\rightarrow}R$ than $R{\rightarrow}S$ or $S{\leftarrow}P{\rightarrow}R$. The observation that stimulating nucleus ventralis lateralis (VL) of the thalamus evokes muscle contractions can be offered as an illustrative example. The first interpretations to come to mind are that projections descend from VL or are relayed through the cerebral cortex to reach finally the motoneurons. By an ingenious combination of experiments, Asanuma and Hunsperger (1975) have demonstrated that the impulses triggered in VL antidromically invade axonal branches of deep cerebellar nuclear cells, then orthodromically course along other axonal branches of the same neurons (see Fig. 1E) to reach the red nucleus and, indirectly from there, the motoneurons. Electrophysiological techniques now exist to identify antidromic conduction.

In intracellular derivation, the elicitation of a very consistent spike without EPSP (in a cell which shows EPSP's in other conditions) is taken as strong probability of antidromic invasion. In extracellular derivation, the usual tests of nonsynaptic activation (i.e., invariant latency, stability of threshold, and high-frequency following) can be determinant in cases where orthodromic conduction is *a priori* excluded. Such tests (see Section B) are used to identify motoneurons by stimulation of muscle afferents, cortical pyramidal cells by pyramidal tract stimulation, etc. Otherwise, the collision test seems to be the most reliable procedure.

The principle of the collision test rests on the observation that two impulses traveling toward each other along a nerve fiber "collide" and cancel each other. Referring to the schema of Fig. 3, let us assume that a stimulus is applied at S (t_1) shortly after a spontaneous spike is recorded at R (t_0). Leaving R, the spontaneous spike travels toward S, whereas the spike triggered in S travels toward R. It will never arrive, there will be no response to the stimulus, and the record will be as shown in Fig. 4 (C1 to C3). The collision (at t_5) results in a dead time after a spontaneous spike, during which no response is elicitable. This is called the

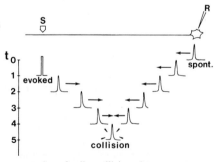

FIG. 3. Diagrammatic representation of spike collision. At t_0, spontaneous spike recorded at R. At t_1, stimulus applied at S. From t_2 to t_4, impulses travel toward each other and collide at t_5.

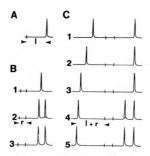

FIG. 4. Measurement of latency in A, refractory period in B, and collision interval in C. B. Interstimulus interval progressively increased from 1 to 3; second response absent in 1, inconsistently occurring in 2, always present in 3; refractory period r corresponds to interval measured in 2. C. Pair of stimuli triggered at increasing intervals from 1 to 5 after a spontaneous spike; second stimulus always evokes a response; first stimulus evokes no response in 1 to 3; shortest interval at which response sometimes appears in 4; this interval should be equal to $l + r$ if blocking is due to collision.

collision interval between spontaneous spike and stimulus which, theoretically, should be equal to the travel time of the spike from R to S (or latency l) plus the refractory period r of the fiber at S: $c = l + r$.

In practice, the collision test is performed by automatically triggering the stimulus at various test intervals after a spontaneous spike. The shortest interval at which a response is obtained is noted [Fig. 4 (C4)]. The values of l and r are experimentally determined in the usual way (Fig. 4A and B) in order to establish whether the prediction $c = l + r$ is satisfied. If so, the existence of a fiber normally conducting from R to S is demonstrated. No collision occurs if the fiber normally conducts from S to R (Fig. 1A).

It should be noted that the collision test allows a quantitative prediction of the outcome if the neuron is antidromically invaded by the evoked spike. This is important because there could be other reasons (e.g., recurrent inhibition) for unresponsiveness after a spike, but only by chance would such a blocking have a duration equal to c.

Recently, the accuracy of the test has been reappraised (Fuller & Schlag, 1976), and it has been shown that the experimental values of c are consistently slightly smaller than the predicted values (by a fraction of millisecond). This discrepancy comes from the way the estimations of l and r are made. Hence, the test can give ambiguous results when l and r are particularly short. In order to uncover the sources of possible errors of measurement, it is advisable to perform the test using three or four different stimulus strengths.

There are several variants for performing the collision test. Thus, one useful control worth adding to the test is to apply not one but a couple of shocks separated by an interval slightly longer than the refractory period, as depicted in Fig. 4C. If the conduction is antidromic, the first evoked spike collides with the

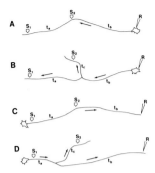

FIG. 5. Possible circuits in cases of collision between impulses evoked at S_1 and S_2, and recorded at R. Arrows indicate direction of natural impulses; $t_{a,b,c}$ represent conduction time along corresponding segments of axons. Differentiation between A, B, and C, based on collision with spontaneous spike in A and B (antidromic conduction of spontaneous spike). Branching of axon (in B and D) detected if latency S_2-R (l_2) is larger than t_b.

spontaneous spike and clears the way for the second and subsequent spikes. This clearing away, when it occurs during the delay when the unit is supposedly blocked after a spontaneous spike [Fig. 4(C1)], provides further evidence of collision. Another advantage of programming repeated stimuli is that a single program can serve to measure all parameters: latency l, refractory period r, and collision interval c.

Strategies based on the principle of collision can be adapted for specific purposes. For instance, consider the circuits of Fig. 5 where it is assumed, in all cases, that a response is elicited by stimulating at either S_1 or S_2. Can these circuits be differentiated? Collision occurs between a spontaneous spike and either the S_1- or S_2-evoked spike in A and B, but not in C. In D, the situation is slightly more complicated: collision with a spike triggered in S_2 after a spontaneous spike arriving at R is to be expected if $t_c > t_b - r$. Collision between both evoked spikes occur in all four cases. The interstimulus collision interval is given by the conduction time between S_1 and S_2 plus the refractory period (i.e., $t_a + r$ in A and B; $t_a + t_c + r$ in C and D), and it is the same whether the sequence of stimuli is S_1S_2 or S_2S_1. Assuming that the conduction velocity is uniform, it is possible to recognize the existence of axonal branching (cases B and D) by measuring the latencies, refractory period, and interstimulus collision interval (see Tolbert, Bantli, & Bloedel, 1976).

In A and C

$$l_1 = t_a + t_b$$
$$l_2 = t_b$$
$$c_{1,3} = c_{2,1} = t_a + r$$

In B and D

$$l_1 = t_a + t_b$$
$$l_2 = t_c + t_b$$
$$c_{1,2} = c_{2,1} = t_a + t_c + r$$

In both situations, the conduction time t_b can be calculated as

$$t_b = \tfrac{1}{2}(l_1 + l_2 - c_{1,2} + r)$$

If separate branches of the cell were excited one should expect to have $l_2 > t_b$.

The possibility to fire a neuron backward from its axon is probably the most reliable electrophysiological method of identification available. The collision principle is widely utilized for this purpose.

V. Conclusions

It is appropriate to distinguish two different anatomical purposes for which electrophysiological techniques are commonly used: (1) the identification of neurons by some known features of their connectivity and (2) the discovery of neuroanatomical connections.

The objective of identification arises practically in all experiments with unit recording—numerous nowadays. By stimulating presumed afferents and/or efferents, the experimenter can characterize the units immediately before or after studying their behavior. For this purpose, the electrophysiological technique cannot be superseded by other neuroanatomical techniques for the latter necessitate that the tissue be removed shortly after the experiment. Marking procedures (by depositing an identifiable trace of substance at the tip of a microelectrode) are certainly indispensable, but they have limitations for only a limited number of marks can be made, otherwise they can be confused with each other.

As neuroanatomical research tools, electrophysiological techniques offer unique advantages for revealing certain features such as: whole pathways, topological projections, convergence of inputs, divergence of outputs (i.e., branching of axons), neurons with very long axons (as the pyramidal tract cell), etc. More generally, they also provide evidence on the functional efficiency of connections seen under the microscope. Electrophysiological and traditional neuroanatomical techniques are complementary tools; they are most productive when used in combination by the same observer.

Acknowledgments

I am deeply grateful to Dr. E. Decima for comments and suggestions on the manuscript. Part of this work was supported by U.S. Public Health Service Grant No. NS-04955.

References

Adey, W. R., & Kerr, D. I. B. The cerebral representation of deep somatic sensibility in the marsupial phalanger and the rabbit: An evoked potential and histological study. *Journal of Comparative Neurology*, 1954, **100**, 597–626.

Asanuma, H., & Hunsperger, R. W. Functional significance of projection from the cerebellar nuclei to the motor cortex in the cat. *Brain Research*, 1975, **98**, 73–92.

Barret, J. N., & Graubard, K. Fluorescent staining of cat motoneurons *in vivo* with beveled micropipettes. *Brain Research*, 1970, **18**, 565–568.

Berry, M. S., & Pentreath, V. W. Criteria for distinguishing between monosynaptic and polysynaptic transmission. *Brain Research*, 1976, **105**, 1–20.

Bishop, P. O., Burke, W., & Davies, R. Single unit recording from anti-dromically activated optic radiation neurones. *Journal of Physiology (London)*, 1962, **162**, 432–450.

Cool, S. J., Crawford, M. L. J., & Soheer, I. L. Tungsten microelectrode preparations for CNS recording. *Review of Scientific Instruments*, 1970, **41**, 1506–1507.

Cowey, A. Projection of the retina on the striate and prestriate cortex in the squirrel monkey, *Saimiri sciureus*. *Journal of Neurophysiology*, 1964, **27**, 366–393.

Desijaru, T., & Purpura, D. P. Synaptic convergence of cerebellar and lenticular projection to thalamus. *Brain Research*, 1969, **15**, 544–547.

Dusser de Barenne, J. G., & McCulloch, W. S. Functional boundaries in the sensory-motor cortex of the monkey. *Proceedings of the Society for Experimental Biology and Medicine*, 1936, **35**, 329–331.

Fries, W., & Zieglgänsberger, W. A method to discriminate axonal from cell body activity and to analyse "silent cells." *Experimental Brain Research*, 1974, **21**, 441–445.

Fritsch, G., & Hitzig, E. Über die elektrische Erregbarkeit des Grosshirns. In G. von Bonin (Ed.), *The cerebral cortex*. Springfield, Illinois: Thomas, 1870. Pp. 73–96.

Fuller, J. H., & Schlag, J. D. Determination of antidromic excitation by the collision test: Problems of interpretation. *Brain Research*, 1976, **112**, 283–298.

Fussey, I. K., Kidd, C., & Whitman, J. G. The differentiation of axonal and soma-dendritic spike activity. *Pfluegers Archives*, 1970, **321**, 283–292.

Geddes, L. A. *Electrodes and the measurement of bioelectric events*. New York: Wiley (Interscience), 1972.

Globus, A. Iontophoretic injection techniques. In R. F. Thompson & M. M. Patterson (Ed.), *Bioelectric recording techniques*. Part A. *Cellular Processes and brain potentials*. New York: Academic Press, 1973. Pp. 24–38.

Granit, R., Kernell, D., & Shortess, G. K. Quantitative aspects of repetitive firing of mammalian motoneurones, caused by injected currents. *Journal of Physiology (London)*, 1963, **168**, 911–931.

Grundfest, H., Sengstaken, R. W., & Oettinger, W. H. Stainless steel micro-needle electrodes made by electrolytic pointing. *Review of Scientific Instruments*, 1950, **21**, 360–361.

Horn, G., & Hill, R. M. Modifications of receptive fields of cells in the visual cortex occurring spontaneously and associated with bodily tilt. *Nature (London)*, 1969, **221**, 186–188.

Hubel, D. H. Tungsten microelectrodes for recording from single units. *Science*, 1957, **125**, 549–550.

Hubel, D. H. Single unit activity in lateral geniculate body and optic tract of unrestrained cats. *Journal of Physiology (London)*, 1960, **150**, 91–104.

Jacobs, G. H. Receptive fields in visual systems. *Brain Research*, 1969, **14**, 553–573.

Jankowska, E., & Lindström, S. Morphological identification of physiologically defined neurones in the cat spinal cord. *Brain Research*, 1970, **20**, 323–326.

Kinnard, M. A., & MacLean, P. D. A platinum microelectrode for intracerebral exploration with a chronically fixed stereotaxic device. *Electroencephalography and Clinical Neurophysiology*, 1967, **22**, 183-186.

La Vallee, M., Schonne, O. F., & Hebert, N. C. (Eds.) *Glass microelectrodes*. New York: Wiley, 1969. Pp. 1-446.

Lee, B. B., Mandl, G., & Stean, J. P. B. Micro-electrode tip position marking in nervous tissue: A new dye method. *Electroencephalography and Clinical Neurophysiology*, 1969, **27**, 610-613.

Mountcastle, V. B., & Henneman, E. The representation of tactile sensibility in the thalamus of the monkey. *Journal of Comparative Neurology*, 1952, **97**, 409-440.

Orem, J., & Schlag, J. Direct projections from cat frontal eye field to internal medullary lamina of the thalamus. *Experimental Neurology*, 1971, **33**, 509-517.

Phillips, C. G., & Porter, R. The pyramidal projections to motoneurones of some muscle groups of the baboon's forelimb. *Progress in Brain Research*, 1964, **12**, 222-242.

Poggio, G. F., & Mountcastle, V. B. The functional properties of ventrobasal thalamic neurons studied in unanesthetized monkeys. *Journal of Neurophysiology*, 1963, **26**, 775-806.

Ranck, J. B., Jr. Which elements are excited in electrical stimulation of mammalian central nervous system: A review. *Brain Research*, 1975, **98**, 417-440.

Schlag, J. Generation of brain evoked potentials. In R. F. Thompson & M. M. Patterson (Eds.), *Bioelectric recording techniques*. Part A. *Cellular processes and brain potentials*. New York: Academic Press, 1973. Pp. 273-316.

Schlag, J., & Balvin, R. Background activity in the cerebral cortex and reticular formation in relation with the electroencephalogram. *Experimental Neurology*, 1963, **8**, 203-219.

Shepherd, G. M. *The synaptic organization of the brain. An introduction*. London & New York: Oxford Univ. Press, 1974.

Snodderly, D. M., Jr. Extracellular single unit recording. In R. F. Thompson & M. M. Patterson (Eds.), *Bioelectric recording techniques*. Part A. *Cellular processes and brain potentials*. New York: Academic Press, 1973. Pp. 137-163.

Spinelli, D. N. Silver tipped metal microelectrodes: A new method for recording and staining single neurones. *Brain Research*, 1975, **91**, 271-275.

Stoney, S. D., Jr., Thompson, W. D., & Asanuma, H. Excitation of pyramidal tract cells by intracortical microstimulation: Effective extent of stimulating current. *Journal of Neurophysiology*, 1968, **31**, 659-669.

Suzuki, H., & Azuma, M. A glass-insulated "Elgiloy" microelectrode for recording unit activity in chronic monkey experiments. *Electroencephalography and Clinical Neurophysiology*, 1976, **41**, 93-95.

Thompson, R. F., & Patterson, M. M. (Eds.) *Bioelectric recording techniques*. Part A. *Cellular processes and brain potentials*. New York: Academic Press, 1973.

Tolbert, D. L., Bantli, H., & Bloedel, J. R. Anatomical and physiological evidence for a cerebellar nucleo-cortical projection in the cat. *Neuroscience*, 1976, **1**, 205-217.

Towe, A. L. Sampling single neuron activity. In R. F. Thompson & M. M. Patterson (Eds.), *Bioelectric recording techniques*. Part A. *Cellular processes and brain potentials*. New York: Academic Press, 1973. Pp. 79-93.

West, J. R., Deadwyler, S. A., Cotman, C. W., & Lynch, G. A dual marking technique for microelectrode tracks and localization of recording sites. *Electroencephalography and Clinical Neurophysiology*, 1975, **39**, 407-410.

Whitsel, B. L., Dreyer, D. A., & Roppolo, R. F. Determinants of body representation in postcentral gyrus of macaques. *Journal of Neurophysiology*, 1971, **34**, 1018-1034.

Wolbarsht, M. L., MacNichol, E. F., Jr., & Wagner, H. G. Glass insulated platinum microelectrode. *Science*, 1960, **132**, 1309-1310.

Woody, C. D., & Black-Cleworth, P. Differences in excitability of cortical neurons as a function of motor projection in conditioned cats. *Journal of Neurophysiology*, 1973, **36**, 1104–1116.

Woolsey, C. N. Organization of somatic sensory and motor areas of the cerebral cortex. In H. F. Harlow & C. N. Woolsey (Eds.), *Biological and biochemical bases of behavior*. Madison: Univ. of Wisconsin Press, 1958. Pp. 63–81.

Woolsey, C. N., & Walzl, E. M. Topical projection of nerve fibers from local regions of the cochlea to the cerebral cortex of the cat. *Bulletin of the Johns Hopkins Hospital*, 1942, **71**, 315–344.

Woolsey, T. A., & Van der Loos, H. The structural organization of layer IV in the somatosensory regions (SI) of mouse cerebral cortex. The description of a cortical field composed of discrete cytoarchitectonic units. *Brain Research*, 1970, **17**, 205–242.

Perspective

Chapter 15

Histological Techniques in Neuropsychology: The Past and Some Trends for the Future

Patricia Morgan Meyer and Jacqueline C. Bresnahan

*Department of Psychology
and Department of Anatomy
The Ohio State University,
Columbus, Ohio*

I. Introduction

Traditionally, neuropsychologists studying brain–behavior relationships have directed their interests toward understanding the functional role of neural systems previously characterized with regard to their anatomy. Histological techniques such as cell body and fiber stains were used primarily for purposes of reconstruction to identify specific neural regions which had been lesioned, stimulated, or

407

recorded from. Since most neuropsychological studies were aimed at elucidating the neural bases of psychological constructs, the primary contributions of these experiments have been for the interpretations of the behavioral data. The anatomical results were necessary for designating the neural structure under investigation, but were ancillary to the psychological outcomes. On the other hand, experimental anatomical and physiological studies have often employed techniques comparable to those used by behaviorists, e.g., stimulating, recording, and lesioning, but the thrust of these studies has been the elucidation of the neural connectivities and physiological properties of the nervous system. Rarely was the behavior of the organism of primary concern. Classically, researchers in these disciplines have defined, on anatomical and physiological grounds, the neural systems which psychologists later manipulate in attempts to understand their behavioral relevance. Behaviorists and anatomists share a common goal in attempting to understand central nervous system (CNS) functions but their approaches diverge.

The consequences of this divergence unfortunately have led to some emotional biases which are reflected in casual statements such as "Behaviorists do lousy anatomy," and conversely, "Anatomists do lousy behavior." We are, of course, overstating the case, but to the extent that such remarks are valid, these prejudices interfere with the integration of information from the two camps. Fortunately, with the advent of interest in neuroscience, a more comprehensive approach has emerged which includes the use of techniques and constructs from more than one discipline to aid in the development of general principles of CNS functions. The authors believe that neuropsychologists with proper training in anatomical techniques and neuroanatomists with proper training in behavioral procedures will provide insights into the neural substrates of behavior which might well be lost if each discipline were to adhere strictly to its own methodological approaches.

II. Some Examples of Experiments with Simultaneous Anatomical and Behavioral Manipulations

To illustrate the appropriateness of an integrated approach to studying CNS function, a small group of experiments will be described whose results yield far broader implications than if only the behavioral or only the anatomical procedures had been employed. Studies have been intentionally selected in which a variety of histological techniques were employed in conjunction with the assessment of several different levels of behavioral complexity. Thus, it will be seen how the Nauta stain and its modifications, the Nissl and the Golgi, as well as the horseradish peroxidase (HRP) and the autoradiographic techniques have been used to assess the relationship between neural structure and function.

A. *Anatomical Substrates for Some Aggressive Behaviors*

Since the early demonstration by Hess (1957) that complex behaviors could be elicited by electrical stimulation of structures deep within the brain, this method has been widely utilized by physiological psychologists in attempts to delineate those regions of the CNS responsible for the integration of various behaviors. Mapping studies have been conducted to determine the location of regions which contribute to complex species-typical behaviors such as eating, drinking, and aggressive behaviors. Such experiments have confirmed Hess' observation that the hypothalamus is an important region for the integration and modulation of many types of complex species-typical behaviors. However, some controversies have developed which we think are particularly relevant to our thesis that concurrent behavioral and anatomical studies are extremely valuable tools for the understanding of brain–behavior relationships.

One such controversy, which in a general way extends to nearly all studies of brain–behavior systems, is the question of functional specificity. For example, electrical and chemical stimulation of various regions within the hypothalamus can produce a wide variety of species-typical behaviors. In fact, stimulation at a single locus may elicit more than a single behavior or even a class of behaviors. This phenomenon has led some investigators (e.g., Valenstein, Cox, & Kakolewski, 1970) to conclude that behaviors produced by electrical stimulation of the hypothalamus may result merely from general activation or arousal. Thus, depending on the past experience of the animal and/or the environmental conditions during stimulation, a particular behavior or class of behaviors is more or less likely to occur. Others have argued that the production of multiple behaviors from a single electrode site simply reflects the fact that the neural substrates for those behaviors overlap anatomically (Berntson & Beattie, 1975; Roberts, 1969; Wise, 1968). The fact that in many cases only a single behavior can be obtained from a specific locus argues for the latter position. And yet it is true that the results of stimulation-mapping studies alone do not provide definitive evidence with regard to the specificity and extent of the neural systems underlying particular behaviors. Indeed, such maps have been shown to overestimate the extent of overlap of such systems in many cases (e.g., Berntson & Beattie, 1975).

An elegant study by Chi and Flynn (1971) combined electrical stimulation and neuroanatomical fiber tracing techniques in a demonstration of the specificity of fiber systems within the hypothalamus mediating two varieties of aggressive behavior in the cat—affective (defensive) attack and quiet biting (predatory) attack on a rat. In this experiment, Chi and Flynn located stimulation points which produced only one of the behaviors in question, and then through the same electrode made electrolytic lesions which were sufficient to abolish the elicitability of the behavior. One or two weeks later, the animals were sacrificed and their brains were stained by a modified Nauta technique to trace the resulting degeneration. They observed that lesions at sites producing quiet biting attack yielded degeneration confined mainly to the medial forebrain bundle, which interconnects the hypothalamus rostrally with the basal olfactory and limbic structures, and caudally with the midbrain tegmentum. Degeneration also was present in the midline nuclei of the thalamus, an area which when stimulated has been shown to produce quiet biting attack (MacDonnell & Flynn, 1968). Lesions at points producing only affective attack resulted in degeneration which was confined primarily to the periventricular system. The authors note that this projection from the affective attack points in the hypothalamus to the central gray is particularly interesting in light of the fact that affective attack can be elicited from stimulation of the central gray (Flynn, Vanegas, Foote, & Edwards, 1970), and lesions in this area can block affective attack elicited by hypothalamic stimulation (e.g., Berntson, 1972; Skultety, 1963).

This study demonstrated that even aggressive behaviors which ostensibly have similar components, have anatomically distinct fiber systems subserving them. Not only are these behaviors more likely to be elicited from one hypothalamic region than another (e.g., Berntson & Beattie, 1975), but their projections, when stimulated, yield behaviors that are clearly differentiable. Indeed, the projection targets from affective and predatory attack points were found to correspond closely to widely differing regions where each behavior also can be elicited by electrical stimulation or blocked by lesions. The Chi and Flynn study is a prime example of how neuroanatomical procedures have helped to define brain–behavior relationships by demonstrating neural networks which underly specific behaviors.

B. A Suggested Anatomical Basis for Recovery of Function following Spinal Cord Lesions

One of the earliest studies which suggested that a behavioral alteration resulting from a lesion in the CNS could be correlated with an anatomical reorganization was the demonstration of collateral sprouting in the spinal cord of the cat by Liu, Chambers, McCouch, and their colleagues. McCouch, Austin, Liu, and Liu (1958) noted that the development of hyperflexia following spinal cord transection occurred concomitantly with electrophysiologically recorded increases in afferent terminal potentials and internuncial potentials elicited by stimulation of the dorsal roots. These results were interpreted as suggesting an increase in the number of dorsal root axons and terminals. Consequently, they bilaterally sectioned dorsal root fibers in animals which had undergone cord hemisections several months previously, and after appropriate time intervals, the animals were sacrificed. The Nauta technique for degenerating axons was used to process the cord tissue. More degeneration was found on the lesioned side than on the control side, confirming their hypothesis that there was an increased distribution of primary afferents following lesions of the descending systems.

More recently a replication and extension of this work was carried out by Murray and Goldberger (1974). Using the cat, these investigators analyzed the pattern of reflex alterations and recovery of locomotor behavior following partial unilateral hemisection, unilateral deafferentation of the hind limb, or both. In addition, they attempted to correlate anatomical alterations with the behavioral recoveries that were observed after surgery.

In the first experiment, Murray and Goldberger (1974) performed partial low thoracic cord hemisections, excluding the dorsal columns. After an initial period of reflex depression, the intrinsic hindlimb reflexes, which are dependent on input from the ipsilateral dorsal roots, were measured both behaviorally and electromyographically. They were observed to recover and in some cases to become hyperresponsive. Furthermore, the reinstatement of the reflexes occurred concurrently with improvement in the use of the limb. No such increase was observed for descending or crossed reflexes.

Murray and Goldberger next mapped the intraspinal distribution of L5 or L6 dorsal root afferents using several modifications of the Nauta technique for anterograde fiber degeneration and autoradiography. In the first part of the experiment, as illustrated in Fig. 1A, the dorsal roots were cut bilaterally in cats previously prepared with spinal cord hemisections. In the second part of the

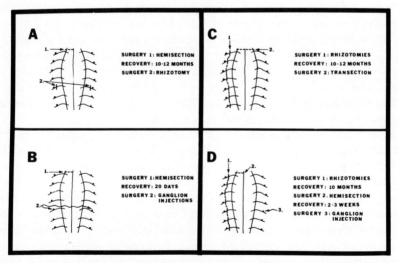

FIG. 1.   Schematic diagram illustrating the location and sequential order of the spinal cord surgeries. Straight arrows indicate transections and wavy arrows indicate tritiated amino acid injections. Detailed explanation is in the text.

experiment, as shown in Fig. 1B, the dorsal root ganglia were injected bilaterally with tritiated proline at a time just after the development of the hyperreflexia. Both techniques revealed an asymmetrical distribution of primary afferents. On the hemisected side, the overall distribution of fibers was greater than on the normal side, and their pattern was altered such that there was more spread to the medial portion of the base of the dorsal horn and a lateral expansion in the zona intermedia.

Murray and Goldberger interpreted these anatomical changes as signs of collateral axonal sprouting of dorsal root afferents in response to the denervation resulting from degeneration of the ipsilateral descending systems. Further, they noted that regions of increased fiber density corresponded to the areas where the interneurons in the path of cutaneous reflexes and interneurons which facilitate stretch reflexes have been identified electrophysiologically. Thus, they suggested that the dorsal root sprouting observed after cord hemisection might be causally related to the hyperreflexia.

In a second study, Goldberger and Murrry (1974) examined the course of recovery of reflexes and movements following spinal deafferentation. Transections of the dorsal roots from approximately T12 to S1 were made unilaterally. Subsequently, all reflex activity was depressed, but with time, hyperactivity of some descending reflexes was observed. Subsequently, locomotor behavior also returned. Goldberger and Murray next subjected one group of cats to total transection of the spinal cord and sacrificed them shortly thereafter to determine the pattern of termination of the descending systems in the deafferented zone (Fig. 1C). They employed several modifications of the Nauta technique for histological analyses. Following cord transection, the degeneration on the deafferented side was greater than on the control side in nucleus proprius, the base of the dorsal horn, and in the zona intermedia. The degeneration in the base of the dorsal horn spread more medially than on the normal side, and the degeneration in the ventral horn spread more laterally. There was also degeneration in Clarke's nucleus, which was not present on the normal side.

As seen in Fig. 1D, a second group of animals was prepared with cord transections ipsilateral to the prior deafferentation. The cats were maintained for longer time intervals than those with total transections so that recovery of spinal reflexes and locomotion following the second lesion could be observed. Under these experimental conditions, exaggerated crossed reflexes could be elicited. Subsequently, the dorsal root ganglion at L5 on the control side was injected with tritiated proline to determine if the contralateral dorsal root input was altered as a consequence of the two sequential surgeries. Histological analyses revealed that those animals which showed hyperactive crossed reflexes exhibited differences in the distribution of contralateral dorsal root afferents when compared to unoperated control cats. The distribution of afferents was increased in the medial base of the dorsal horn, in the basal commissural nucleus of Von Lehossek, and in the ventral commissural nucleus of Cajal. Additionally, a projection to the contralateral ventral horn was observed which was not present in the normal control spinal cord.

The results of the second study suggested that there was a correlation between anatomical changes and reflex and locomotor reinstatements following spinal cord damage. Specifically, they found exaggerated descending reflexes and increased descending input to the cord after deafferentation. The authors inferred that recovery of useful movement is preceded, and perhaps based on, the appearance of the exaggerated reflexes, and that the hyperactivity may be related to collateral sprouting to the denervated zones. In the case of deafferentation followed by an ipsilateral hemisection, hyperactive crossed reflexes were observed along with increased input to the commissural nuclei from the contralateral dorsal root.

These studies are important for a number of reasons. First, they suggest that recovery after spinal cord injury occurs in an orderly fashion. The reflexes return, and then become hyperactive, and subsequently, locomotor behavior improves. Second, collateral sprouting was demonstrated using two different histological techniques, an anterograde degeneration procedure and autoradiography. The latter technique made it possible to establish the temporal correlation between the anatomical reorganization and the behavioral recovery. Since the hyperreflexia occurred only a few weeks following the first surgery, anterograde degeneration techniques could not be used. Long interoperative intervals are necessary in dual lesion studies so that the degeneration produced by the first lesion will be reabsorbed and will not be confused with that from the second lesion. Autoradiographic procedures are dependent on the normal process of anterograde axonal transport and therefore could be used at any time following the first surgery. Thus, Goldberger and Murray used the autoradiographic technique to pinpoint the temporal correlation between anatomical alterations and behavioral reinstatements. Having confirmed this temporal correlation in both of their studies, Murray and Goldberger concluded that there was a distinct possibility that collateral sprouting might mediate the recovery of the spinal cord functions.

C. A Possible Anatomical Alteration Resulting from Rearing in Deprived versus Enriched Environments

Golgi stains have rarely been used in conjunction with psychological experiments, but a series of studies by Greenough and his colleagues at the University of Illinois illustrates the appropriateness of this technique as a tool to supplement behavioral experiments (for review, see Greenough, 1975). In one of these studies, Greenough and Volkmar (1973) investigated whether changes in visual cortical neurons could be detected as a consequence of rearing rats in three environments that differed with respect to the amount of extraneous stimulation encountered by the animals. Prior behavioral research with animals raised under such conditions has indicated that rats living in the most enriched environments often learn mazes and discrimination problems more quickly than those reared in isolated situations (Rosenzweig, Bennett, & Diamond, 1971). Increases in brain weight and increases in the thickness of the neocortex in rats exposed to complex environments have also been reported (Rosenzweig *et al.*, 1971). Greenough and Volkmar attempted to determine if any specific morphological alterations could be detected as a consequence of different rearing conditions.

Greenough and Volkmar raised littermates in enriched (EC), social (SC), or isolated (IC) environments from postweaning to 55 days of age. The EC group lived with several other rats in large cages containing toys and metal and wooden objects which could be explored and manipulated. These items were changed daily. The SC and IC groups were housed, respectively, in pairs or alone in standard laboratory rodent cages. The animals were subsequently sacrificed and their brains were stained with a rapid Golgi technique to analyze patterns of dendritic branching in stellate and pyramidal cells in the visual cortex. Although, in a given area, the Golgi procedure stains a maximum of 10% of the neurons, and usually far fewer than this, the cells are stained relatively completely, and the morphology of cell bodies and their processes can be analyzed with some accuracy.

Four types of cells were analyzed: (1) layer 2 pyramidal cells; (2) layer 4 stellate cells; (3) layer 4 pyramidal cells; (4) layer 5 pyramidal cells. Measurements of dendritic volume, branching patterns, and length of branches were assessed. Total dendritic volume was obtained by counting the number of intersections of dendrites of a particular cell with successive concentric rings arranged at 20 μm intervals from the cell body. Branching patterns were determined by counting the number of branches at each successive bifurcation away from the cell body. Thus, a dendritic branch from the soma was considered first order, the two branches past the first bifurcation were considered the second order, etc., up to the fifth order. Finally, all of the data were analyzed so that the effects of genetic factors, i.e., differences between litters, could be estimated.

For all cell types, the number of branches intersecting with the rings was significantly greater for the EC group than for the SC and IC groups. Smaller but significant differences were observed between the SC and IC groups (SC > IC) in layer 4 stellate and layer 5 pyramidal cells. The increased branching was relatively evenly distributed throughout the dendritic tree as opposed to being confined, for example, to the proximal portion of the dendritic tree.

Order of dendritic branching was significantly greater in the EC group than in the SC and IC groups. Stellate cells in layer 4 exhibited more branching at higher orders in the EC group than in the other two groups. In the pyramidal cells, the bulk of the differences were observed in the basal dendrites at the third, fourth, and fifth orders (EC > IC or SC). Apical dendritic branching was slightly but inconsistently greater in the EC group as compared to the SC and IC groups. Interestingly, further statistical analyses suggested that the environmental factor was more important in determining the degree of higher order branching, while the genetic factor (i.e., litter) was more crucial in determining the degree of lower order branching. The length of the branches at each order away from the cell body was approximately the same for all of the groups.

Greenough's data suggest that the pattern of dendritic elaboration in some of the cortical neurons in the posterior neocortex may be correlated with the degree of environmental complexity during postweaning development. A causal relationship, of course, cannot be inferred, but it is interesting to speculate about the extent to which environmental experiences influenced the morphological changes. If these anatomical alterations are indeed brought about by the complexity of the environments, then we have important information about the interaction between genetic and environmental factors in the determination of CNS morphology. These changes may reflect a unique susceptibility of the distal dendrites to environmental stimulation during this late stage of development, corresponding to something like a "critical period" (Greenough, 1975).

One of the obvious implications of these results is that there may be more postsynaptic space available for synaptic contacts in those subjects reared in the enriched environment, thereby providing more potential substrates for information processing. If so, the contacts would most likely be from other cortical neurons as the increased branching was observed most consistently in the basal dendrites which receive their main input from other cortical neurons (Globus & Scheibel, 1967; Scheibel & Scheibel, 1971). The authors are cautious, however, about drawing such conclusions with certainty. Enriched environments are complex situations indeed: The animal is not only stimulated visually, but aurally; tactile stimulation is increased with objects and other animals; activity differences are present; stress is varied; and a host of other motor and sensory stimulation is available. Thus, it would be difficult to isolate the condition which produced the morphological changes. Nevertheless, the data raise some very provocative issues dealing with the influence of the environment on the genetically predetermined organization of the brain, and ultimately on the determination of behavior.

D. An Anatomical and Behavioral Analysis of the Function of Certain Components of the Visual System in Discrimination Learning

A dissertation recently completed by Hughes at The Ohio State University (1976) illustrates how the results from three different histological techniques have yielded new information about the thalamocortical organization of the rat posterior cortex. These results in conjunction with data from behavioral testing have modified our concepts concerning the functions of the striate and extrastriate neocortex in black–white and pattern discrimination learning.

Hughes was primarily concerned with the role of striate neocortex in pattern discrimination learning. Early research (Lashley, 1934) had indicated that striatectomies prevented pattern learning, but the ablations could have extended into extrastriate cortex as well. Consequently, Hughes was interested in investigating both striate and extrastriate functions. However, ablation of extrastriate cortex posed some technical problems. In the process of selectively removing areas 18 and 18a, the blood supply or the optic radiations to area 17 could accidently be interrupted, thereby producing a striate–extrastriate preparation. To avoid this problem, Hughes attempted to deafferent areas 18 and 18a by placing electrolytic lesions in nucleus lateralis posterior (NLP). He confined his neocortical ablations to area 17 as defined cytoarchitectonically by Krieg (1946a, 1946b; see Fig. 2). Following removals of area 17, however, he consistently found spared normal cells in the dorsolateral portion of the dorsal lateral geniculate nucleus (LGNd), which previously has been reported for the cat by Doty (1971) and by Niimi and Sprague (1970).

To determine the origin of this localized sparing, Hughes conducted a series of anatomical experiments. Using autoradiography for anterograde analyses and HRP for retrograde analyses, he demonstrated that fibers from LGNd, which

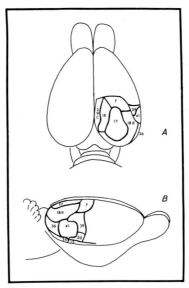

FIG. 2.   Cytoarchitectonic fields of the posterior neocortex in the rat. (Adapted from Krieg, 1946   .)

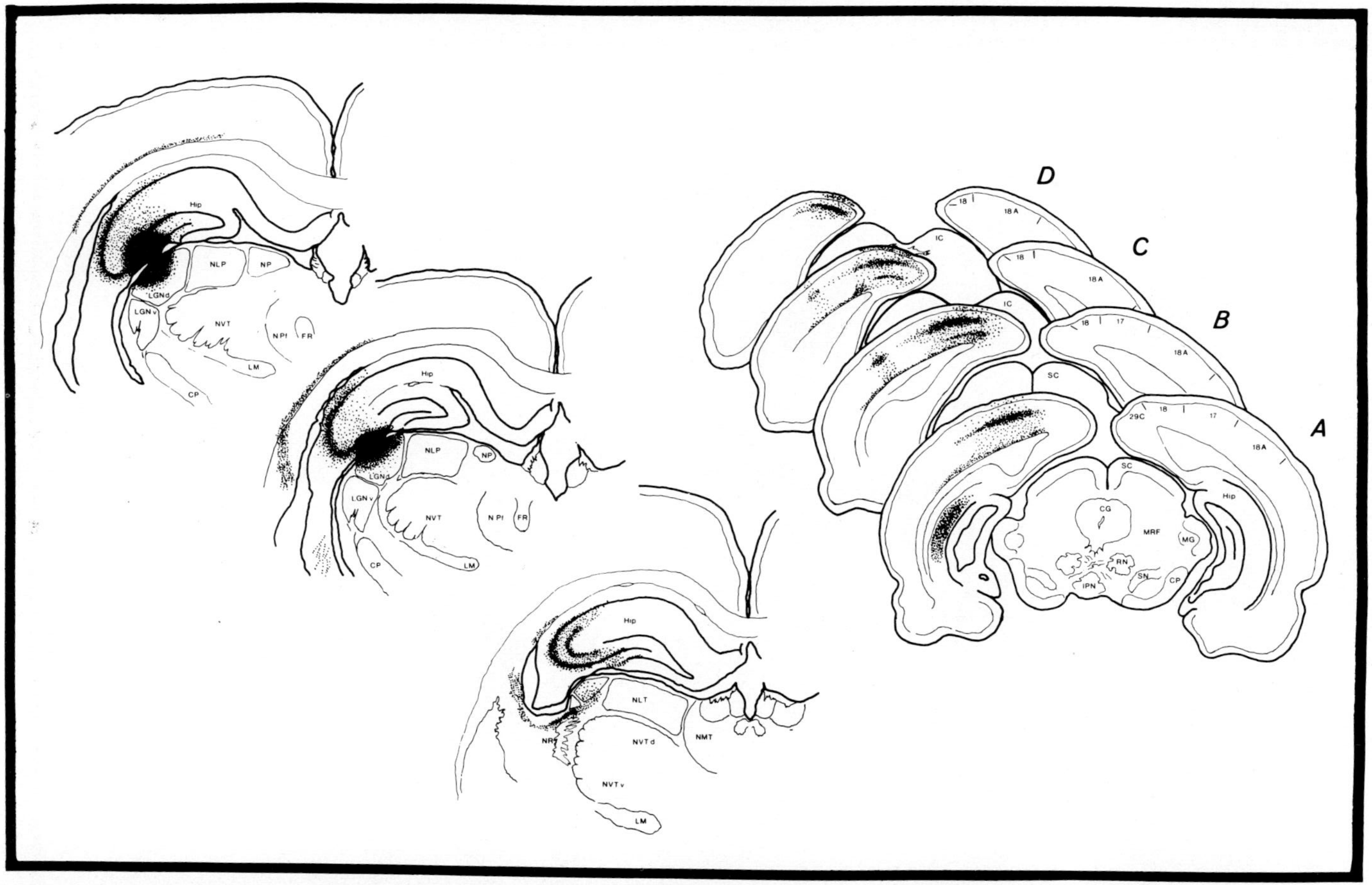

FIG. 3. Neocortical distribution of silver grains after injections of tritiated leucine into the dorsolateral portion of LGNd. (Adapted from Hughes, 1976.)

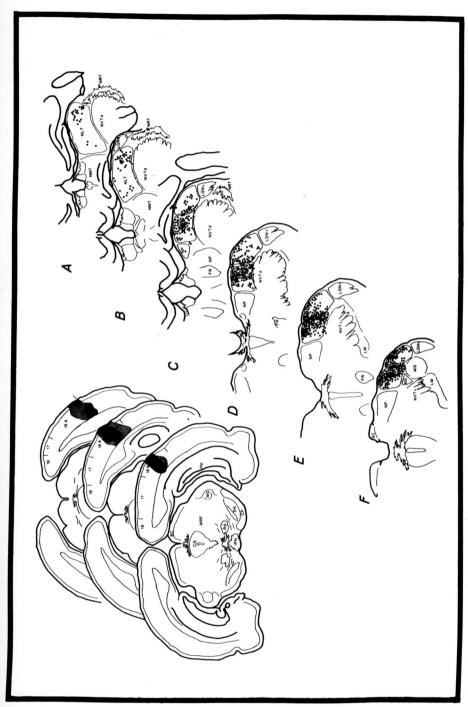

FIG. 4. Retrograde incorporation of HRP in the thalamus after injection into area 18a. Light labeling is indicated by open circles and heavy labeling is indicated by filled circles. (Adapted from Hughes, 1976.)

since the time of Lashley (1939) have been thought to terminate exclusively in area 17, also project to area 18a. By selectively injecting tritiated leucine into the dorsolateral portion of LGNd, the area where sparing was observed following striatectomies, he found afferents from LGNd to areas 17 and 18a (Fig. 3). Hughes next substantiated the extrastriate projection of the LGNd by placing HRP into area 18a (Fig. 4). Interestingly, labeled cells were found in that area of the LGNd which is spared after striatectomies, as well as in NLP. The reaction product was more dense, however, in NLP than in LGNd. These results constitute the first demonstration of geniculate projections to extrastriate cortex in the rat.

In addition, Hughes provided evidence for extrageniculate projections to the striate cortex. HRP was injected into area 17 (Fig. 5) and a column of labeled cells located in an oblique line through the rostrocaudal extent of the LGNd was observed as predicted from an early study by Lashley (1934). Labeled cells were also present in the central portion of NLP, and there was a sparse labeling of the lateral thalamic nucleus (NLT).

Hughes next examined the details of the organization of the extrageniculo-striate pathways by injecting tritiated leucine into NLP. As shown in Fig. 6, he

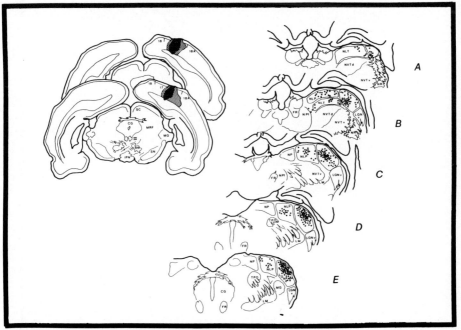

FIG. 5.   Retrograde incorporation of HRP in the thalamus after injection into area 17. (Adapted from Hughes, 1976.)

Fig. 6. Neocortical distribution of silver grains after injection of tritiated leucine into NLP. (Adapted from Hughes, 1976.)

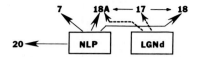

FIG. 7. Some thalamocortical projections to several cortical fields.

detected axons from NLP that followed an initial trajectory similar to the projections of LGNd and which terminated in areas 18, 18a, 7, and 20, with perhaps a weak projection to striate cortex. Placements of HRP into area 18 resulted in labeling of cells in NLP and NLT, but no reaction product was found in LGNd.

Hughes' anatomical findings are summarized diagrammatically in Fig. 7, and show that there are multiple thalamocortical projections to several different cortical fields of the rats 'visual system. Their functional significance remained to be resolved. In the past, visual cortical lesions usually have included removal of not only area 17, but of nearly the whole posterior half of the animal's neocortex. Rats with these radical ablations relearn a black–white discrimination problem in approximately the same number of trials postoperatively as it takes them to learn preoperatively (Horel, Bettinger, Royce, & Meyer, 1966; Lashley, 1935). Furthermore, they are not capable of learning a pattern discrimination following such radical ablations (Horel et al., 1966; Lashley, 1934). Hughes was particularly interested in whether the geniculostriate pathway was singularly important for performances of brightness and pattern habits.

Thus, Hughes trained normal rats on a black–white discrimination problem, and after they had reached a learning criterion, they received ablations of area 17, NLP, area 17 + NLP, or no lesion at all. Following recovery, the animals were first retrained on the black–white problem, and then were given training on a pattern discrimination problem. The discriminanda for the pattern problem were panels of alternating black and white stripes which were oriented 45° left and right from horizontal. These stimuli were employed to eliminate the possibility that the animals would solve the problem by using differences in overall flux, overall contour, local flux, local contour, or from differences in quantities of retinal signals that might arise as a consequence of horizontal scanning of the patterns (see Ritchie, Meyer, & Meyer, 1976).

Figure 8 shows the preoperative learning and postoperative relearning results for the black–white habit. Rats with either NLP lesions or with area 17 ablations relearned at normal rates. These data suggest that the inclusion of area 17 in radical posterior removals is not of any singular importance in producing the black–white retention deficits. Nor are the projections of NLP to the striate and peristriate cortex singularly important for this behavior. However, as is shown in Fig. 8, NLP + striate lesions in combination produce a modest but significant deficit in retention.

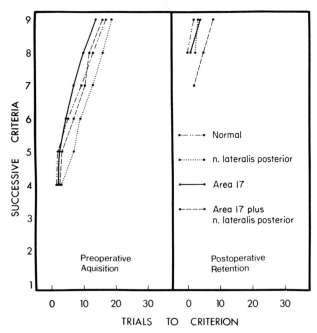

FIG. 8.   Mean trials to successive criteria for preoperative learning and postoperative relearning on the black–white habit. (Adapted from Hughes, 1976.)

The results of the pattern discrimination test are presented in Fig. 9. Unlike the radical posterior preparations, rats with striatectomies can learn the problem, albeit more slowly than normal or NLP lesioned animals. This observation is consistent with results for destriate animals of other species (Killackey & Diamond, 1971; Murphy & Stewart, 1974; Pasik & Pasik, 1971; Sprague, Levy, DiBerardino, & Berlucchi, in press; Sprague, Levy, DiBerardino, & Conomy, 1973; Ware, Diamond, & Casagrande, 1974). Although NLP lesions by themselves were found by Hughes to have no consequential effect upon learning the pattern problem, subjects with ablations of area 17 + NLP were significantly slower to learn the habit than were subjects with only area 17 removals.

These anatomical and behavioral data have led us to modify our concepts about the functions of the components of the visual system in black–white and pattern discrimination learning. The statement that total degeneration of LGNd is an indication of complete ablation of striate cortex in the rat (Lashley, 1934) is not supported. An extrastriate projection from the dorsolateral portion of LGNd to area 18a was found, thus accounting for the spared normal cells in that region following removals only of area 17. Second, striate cortex is not the crucial cortical area involved in black–white and pattern habits. Animals with area 17

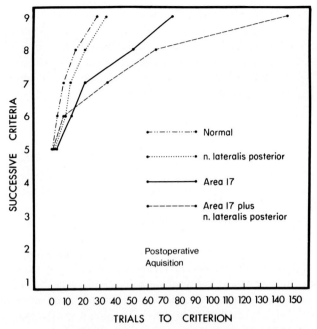

FIG. 9. Mean trials to successive criteria for preoperative learning and postoperative relearning on the pattern habit. (Adapted from Hughes, 1976.)

ablations can master those problems, and when NLP lesions are combined with the striatectomies, the deficits become even more severe than either lesion alone. Nevertheless, the animals can still solve the discriminations, and it remains to be seen what further contributions areas 7 and 20, which receive projections from NLP, make to these tasks. A recent study by Scheff, Wright, Morgen, and Bowers (in press) spoke partially to that issue. Their data suggested that removal of a cortical region in the vicinity of area 20 after sequential unilateral ablations of striate and extrastriate cortex does indeed produce further deficits in black–white habit relearning.

One of the main contributions of the Hughes study was his finding of multiple thalamocortical projections to several different cortical fields in the rat visual system, and their possible behavioral significance. As seen in Fig. 7, the geniculostriate–extrastriate fibers course from LGNd to areas 17 and 18a, and efferent connections from 17 also terminate in areas 18 and 18a (Montero, Bravo, & Fernandez, 1973; Nauta & Bucher, 1954). Furthermore, there is a second connection between NLP and areas 18 and 18a. Thus, there are at least two parallel pathways to peristriate cortex, one from LGNd and one from NLP. As mentioned before, damage to the NLP trajectory alone has an inconsequential effect upon the visual habits tested. Destruction of area 17 also has minor consequences upon the brightness problem, but its intactness is essential for normal pattern discrimination learning. However, when both of these parallel pathways are lesioned, the deficits in both visual tasks are potentiated, which suggests that the normal interaction between these two parallel pathways is more significant than the function of either one of them alone.

The interdisciplanary approach used by Hughes has broadened our understanding of both the structure and function of the rat visual system. We believe that studying the function of connectivities of the parts of neural systems and the interactions between them, will be helpful in elucidating the mechanisms of any system under study.

III.   Concluding Remarks

Although we purposely chose to review experiments dealing with a wide range of behavioral methods and anatomical techniques, a theme emerges that is common to them all. The behavioral results are correlated with defined anatomical networks or alterations of them, and the data contribute significant information to both disciplines. Each study was chosen to illustrate a unique utilization of an anatomical procedure to substantiate or further explicate the neuropsychological results. Examples of a direct and indirect approach to the study of brain–behavior relationships were selected. In the indirect approach, the behavioral data suggested some underlying anatomical correlates, which were subsequently analyzed independently (Greenough & Volkmar, 1973; Hughes, 1976). Not only did the anatomical data supplement the behavioral results, but provided a new dimension to them that broadened our understanding of the problem in question. In the direct approach, the same animals were used in both the behavioral observations and the anatomical analyses (Chi & Flynn, 1971; Goldberger & Murray, 1974; Murray & Goldberger, 1974). Thus, a specific correlation between the behavioral and anatomical results could be suggested.

Not until recently has this interdisciplinary approach been so active or so profitable, for histological techniques just were not available to answer many of the questions. However, with the recent technical revolution in neuroanatomical procedures, many morphological and hodological patterns that escaped us in the past can now be detected and quantified with some degree of accuracy. Appropriate choice of both behavioral and anatomical methodologies must, of course, be made with care. We believe that concurrent behavioral and anatomical investigations of neural systems and their subunits will be helpful in elucidating the properties of any system under study. Furthermore, it is our contention that the ability to utilize sophisticated anatomical techniques to answer structural questions raised by behavioral data, and vice versa, will add an important degree of freedom for the neuroscientist interested in brain–behavior relationships.

Acknowledgments

This chapter was prepared while the first author (P.M.M.) was a U.S. Public Health Service Research Scientist Investigator (5-K2-MH-12, 747) of the National Institute of Mental Health. The second author (J.C.B.) was supported by National Institutes of Health Grant No. NS-10165-05. The experiment reviewed in Section II,D of this chapter was supported in part by National Institute of Mental Health Grant No. MH-06211. We thank H. C. Hughes for permission to present some of his unpublished data. The authors wish to thank their husbands, Dr. Donald R. Meyer and Dr. Michael S. Beattie, for their critical comments and suggestions and for their unusual patience during the preparation of this chapter.

References

Berntson, G. G. Blockard and release of hypothalamically and naturally elicited aggressive behaviors in cats following midbrain lesions. *Journal of Comparative and Physiological Psychology*, 1972, **81**, 541–554.

Berntson, G. G., & Beattie, M. S. Functional differentiation within hypothalamic behavioral systems in the cat. *Physiological Psychology*, 1975, **3**, 183–188.

Chi, C. C., & Flynn, J. P. Neural pathways associated with hypothalamically elicited attack behavior in cats. *Science*, 1971, **171**, 701–703.

Doty, R. W. Survival of pattern vision after removal of striate cortex in the adult cat. *Journal of Comparative Neurology*, 1971, **143**, 341–370.

Flynn, J. P., Vanegas, H., Foote, W., & Edwards, S. Neural mechanisms involved in a cat's attack on a rat. In R. E. Whalen, R. F. Thompson, M. Verzeano, & N. M. Weinberger (Eds.), *The neural control of behavior*. New York: Academic Press, 1970. Pp. 135–173.

Globus, A., & Scheibel, A. B. Pattern and field in cortical structure: The rabbit. *Journal of Comparative Neurology*, 1967, **131**, 155–172.

Goldberger, M. E., & Murray, M. Restitution of function and collateral sprouting in the cat spinal cord: The deafferented animal. *Journal of Comparative Neurology*, 1974, **158**, 37–54.

Greenough, W. T. Experiential modification of the developing brain. *American Scientist*, 1975, **63**, 37–46.

Greenough, W. T., & Volkmar, F. R. Pattern of dendritic branching in occipital cortex of rats reared in complex environments. *Experimental Neurology*, 1973, **40**, 491–504.

Hess, W. R. *The functional organization of the diencephalon*. New York: Grune & Stratton, 1957.

Horel, J. A., Bettinger, L. A., Royce, G. J., & Meyer, D. R. Role of neocortex in the learning and relearning of two visual habits by the rat. *Journal of Comparative and Physiological Psychology*, 1966, **61**, 66–78.

Hughes, H. C. Anatomical and neurobehavioral investigations concerning the thalamo-cortical organization of the rat's visual system. Unpublished Ph.D. dissertation, Ohio State University, 1976.

Killackey, H., & Diamond, I. T. Visual attention in the tree shrew: An ablation study of the striate and extra striate visual cortex. *Science,* 1971, **171,** 696–699.

Krieg, W. J. S. Connections of the cerebral cortex. I. The albino rat. A. Topography of the cortical areas. *Journal of Comparative Neurology*, 1946, **84**, 221–275. (a)

Krieg, W. J. S. Connections of the cerebral cortex. II. The albino rat. B. Structure of the cortical areas. *Journal of Comparative Neurology*, 1946, **84**, 277–284. (b)

Lashley, K. S. The mechanism of vision. VIII. The projection of the retina upon the cerebral cortex of the rat. *Journal of Comparative Neurology*, 1934, **60**, 57–80.

Lashley, K. S. The mechanism of vision. XII. Nervous structures concerned in habits based on reactions to light. *Comparative Psychology Monographs*, 1935, **11**, 43–79.

Lashley, K. S. The functioning of small remnants of the visual cortex. *Journal of Comparative Neurology*, 1939, **70**, 45–67.

MacDonnell, M. P., & Flynn, J. P. Attack elicited by stimulation of the thalamus and adjacent structures of cats. *Behaviour*, 1968, **31**, 185–202.

McCouch, G. P., Austin, G. M., Liu, C. N., & Liu, C. Y. Sprouting as a cause of spasticity. *Journal of Neurophysiology*, 1958, **21**, 205–216.

Montero, V. M., Bravo, H., & Fernandez, V. Striate-peristriate corticocortical connections in the albino and gray rat. *Brain Research*, 1973, **53**, 207–209.

Murphy, E. H., & Stewart, D. L. Effects of neonatal and adult striate lesions on visual discrimination in the rabbit. *Experimental Neurology*, 1974, **42**, 89–96.

Murray, M., & Goldberger, M. E. Restitution of function and collateral sprouting in the cat spinal cord: The partially hemisected animal. *Journal of Comparative Neurology*, 1974, **158**, 19–36.

Nauta, W. J. H., & Bucher, V. M. Efferent connections of the striate cortex in the albino rat. *Journal of Comparative Neurology*, 1954, **100**, 257–281.

Niimi, K., & Sprague, J. M. Thalamo-cortical organization of the visual system in the cat. *Journal of Comparative Neurology*, 1970, **138**, 219–250.

Pasik, T., & Pasik, P. The visual world of monkeys deprived of striate cortex: Effective stimulus parameters and the importance of the accessory optic system. *Vision Research, Supplement*, 1971, **3**, 419–435.

Ritchie, G. D., Meyer, P. M., & Meyer, D. R. Residual spatial vision of cats with lesions of the visual cortex. *Experimental Neurology*, 1976, **53**, 227–253.

Roberts, W. W. Are hypothalamic motivational mechanisms functionally and anatomically specific? *Brain, Behaviour and Evolution*, 1969, **2**, 317–342.

Rosenzweig, M. R., Bennett, E. L., & Diamond, M. C. Chemical and anatomical plasticity of the brain: Replications and extensions. In J. Gaito (Ed.), *Macromolecules and behavior*. New York: Appleton, 1971. Pp. 205–278.

Scheff, S. W., Wright, D. C., Morgen, W. R., & Bowers, R. P. The differential effects of additional cortical lesions in rats with single or multiple stage lesions of the visual cortex. *Physiological Psychology*, 1977, in press.

Scheibel, M. E., & Scheibel, A. B. Selected structural-functional correlations in postnatal brain. In M. B. Sterman, D. J. McGinty, & A. M. Adinolfi (Eds.), *Brain development and behavior*. New York: Academic Press, 1971. Pp. 1–21.

Skultety, P. M. Stimulation of periaqueductal gray and hypothalamus. *Archives of Neurology (Chicago)*, 1963, **8**, 608–620.

Sprague, J. M., Levy, J., DiBerardino, A., & Berlucchi, G. Visual cortical areas mediating form discrimination in the cat. *Journal of Comparative Neurology*, in press.

Sprague, J. M., Levy, J., DiBerardino, A., & Conomy, J. Effect of striate and extra-striate visual cortical lesions on learning and retention of form discriminations. Abstract of the Society for Neuroscience, 3rd Annual Meeting, San Diego, California, 1973.

Valenstein, E. S., Cox, V. C., & Kakolewski, J. W. Reexamination of the role of the hypothalamus in motivation. *Psychological Review*, 1970, **77**, 16–31.

Ware, C. B., Diamond, I. T., & Casagrande, V. A. Effects of ablating the striate cortex on a successive pattern discrimination: Further study of the visual system in the tree shrew (Tupaia glis). *Brain, Behaviour and Evolution*, 1974, **9**, 264–279.

Wise, R. A. Hypothalamic motivational systems: Fixed or plastic neural circuits. *Science*, 1968, **162**, 377–379.

Author Index

Subject Index